Solid State Electronic Devices

PRENTICE HALL SERIES
IN SOLID STATE PHYSICAL ELECTRONICS
Nick Holonyak Jr., Editor

THIRD EDITION

Solid State
Electronic Devices

BEN G. STREETMAN
Microelectronics Research Center
Department of Electrical and Computer Engineering
The University of Texas at Austin

PRENTICE HALL
Englewood Cliffs, New Jersey 07632

Library of Congress Cataloging-in-Publication Data

Streetman, Ben G.
  Solid state electronic devices / Ben G. Streetman. — 3rd ed.
    p.   cm.
  Includes bibliographical references.
  ISBN 0-13-822941-4
  1. Semiconductors. I. Title.
TK7871.85.S77   1990
621.381'52 — dc20

89-23175
CIP

Editorial/production: **Denise Gannon**
Interior design: **Lee Cohen**
Cover design: **Lee Cohen**
Cover illustration courtesy of Motorola, Inc.
Manufacturing buyer: **Donna Douglas**

© 1990, 1980, 1972 by Prentice-Hall, Inc.
A Division of Simon & Schuster
Englewood Cliffs, New Jersey 07632

Printed in the United States of America
10 9 8 7 6 5 4 3

ISBN 0-13-822941-4

Prentice-Hall International (UK) Limited, *London*
Prentice-Hall of Australia Pty. Limited, *Sydney*
Prentice-Hall Canada Inc., *Toronto*
Prentice-Hall Hispanoamericana, S.A., *Mexico*
Prentice-Hall of India Private Limited, *New Delhi*
Prentice-Hall of Japan, Inc., *Tokyo*
Simon & Schuster Asia Pte. Ltd., *Singapore*
Editora Prentice-Hall do Brasil, Ltda., *Rio de Janeiro*

CONTENTS

PREFACE

This book is an introduction to solid state electronic devices for undergraduate electrical engineers, other interested students, and practicing engineers and scientists whose understanding of modern electronics needs updating. The book is organized to bring students with a background in sophomore physics to a level of understanding which will allow them to read much of the current literature on new devices and applications.

GOALS

In my opinion, an undergraduate course in electronic devices has two basic purposes: (1) to provide students with a sound understanding of existing devices, so that their studies of electronic circuits and systems will be meaningful; and (2) to develop the basic tools with which they can later learn about newly developed devices and applications. Perhaps the second of these objectives is the more important in the long run; it is clear that engineers and scientists who deal with electronics will continually be called upon to learn about new devices and processes in the future. For this reason, I have tried to incorporate the basics of semiconductor materials and conduction processes in solids, which arise repeatedly in the literature when new devices are explained. Some of these concepts are often omitted in introductory courses, with the view that they are unnecessary for understanding the fundamentals of junctions and transistors. I believe this view neglects the important goal of equipping students for the task of understanding a new device by reading the current literature. Therefore, in this text most of the commonly used semiconductor terms and concepts are introduced and related to a broad range of devices.

READING LISTS

As a further aid in developing techniques for independent study, I have included in the reading list at the end of each chapter a few articles which students can read comfortably as they study this book. Some of these articles have been selected from periodicals such as *Scientific American* and *Physics Today,* which specialize in introductory presentations. Other articles chosen from books and the professional literature provide a more quantitative treatment of the material. I do not expect that students will read all articles recommended in the reading lists; nevertheless, some exposure to periodicals is useful in laying the foundation for a career of constant updating and self-education.

PROBLEMS

One of the keys to success in understanding this material is to work problems that exercise the concepts. The problems at the end of each chapter are designed to facilitate learning the material. Problems marked with an asterisk are intended for computer solution. In these problems, the ability to vary parameters over wide ranges illustrates important properties of the materials and devices under study.

UNITS   In keeping with the goals described above, examples and problems are stated in terms of units commonly used in the semiconductor literature. The basic system of units is rationalized MKS, although cm is often used as a convenient unit of length. Similarly, electron volts (eV) are often used rather than joules (J) to measure the energy of electrons.

PRESENTATION   In presenting this material at the undergraduate level, one must anticipate a few instances which call for a phrase such as "It can be shown . . ." This is always disappointing; on the other hand, the alternative is to delay study of solid state devices until the graduate level, where statistical mechanics, quantum theory, and other advanced background can be freely invoked. Such a delay would result in a more elegant treatment of certain subjects, but it would prevent undergraduate students from enjoying the study of some very exciting devices.

The third edition includes an expanded treatment of III–V compound semiconductors, to reflect their continuing growth in importance for optoelectronic and high-speed device applications. New topics such as heterojunctions, lattice matching using ternary and quaternary alloys, variation of band gap with alloy composition, and resonant tunneling through quantum wells add to the breadth of the discussion. Not to be outdone by the compounds, silicon-based devices have continued their dramatic record of advancement. I have added new material on FET structures and Si integrated circuits to reflect these advancements. My objective is not to cover all the latest devices, which can only be done in the journal and conference literature. Instead, I have chosen devices to discuss which are broadly illustrative of important principles.

The first four chapters of the book provide background on the nature of semiconductors and conduction processes in solids. Included is a brief introduction to quantum concepts (Chapter 2) for those students who do not already have this background from other courses. Chapters 5 and 6 describe the p-n junction and some of its applications. Chapters 7 and 8 deal with the principles of transistor operation, and Chapter 9 applies these principles to integrated circuits. Chapters 10 through 12 apply the theory of junctions and conduction processes to lasers, switching devices, and microwave devices. All of the devices covered are important in today's electronics; furthermore, learning about these devices should be an enjoyable and rewarding experience. I hope this book provides that kind of experience for its readers.

ACKNOWLEDG-
MENTS   The third edition benefits greatly from comments and suggestions provided by students and teachers of the first two editions. The book's readers have generously provided comments which have been invaluable in developing the present version. I remain indebted to those persons mentioned in the Preface of the first two editions, who contributed so much to the development of the book. In particular, Nick Holonyak has been a source of continuing information and inspiration for all three editions. Additional thanks go to my

colleagues at UT-Austin who have provided special assistance, particularly Sanjay Banerjee, Joe Campbell, Russ Dupuis, Dim-Lee Kwong, Jack Lee, Christine Maziar, Dean Neikirk, and Al Tasch. I am also indebted to my graduate students during this period who helped me collect material for this edition, particularly Tom Block, Andy Campbell, Gentry Crook, Dan Dodabalapur, Craig Farley, Susan Foxworth, and Kayvan Sadra. Many other UT-Austin students provided useful assistance, particularly Garret Okamoto, Vijay Kesan, Mark Somerville, and Oky Setiono. Finally, I thank the many companies and organizations cited in the figure captions for generously providing photographs and illustrations of devices and fabrication processes.

Ben G. Streetman

ABOUT THE AUTHOR

Ben G. Streetman is Professor of Electrical and Computer Engineering and Director of the Microelectronics Research Center at The University of Texas at Austin and holds the Dula D. Cockrell Centennial Chair in Engineering. His teaching and research interests include semiconductor materials and devices, radiation damage and ion implantation, molecular beam epitaxy, transient annealing, deep level impurities and defects in semiconductors, and multilayer heterostructures. After receiving the Ph.D. from The University of Texas at Austin (1966) he was on the faculty (1966–1982) of the University of Illinois at Urbana-Champaign. He returned to The University of Texas at Austin in 1982. In 1989 he was chosen to receive the Education Medal of the Institute of Electrical and Electronics Engineers (IEEE). In 1987 he was elected to membership in the National Academy of Engineering and in the same year received the AT&T Foundation Award of the American Society for Engineering Education (ASEE). In 1981 he received the Frederick Emmons Terman Award of the ASEE, and in 1980 he was elected Fellow of the IEEE. He serves on the Administrative Committee of the IEEE Electron Devices Society and has served on the IEEE Device Research Conference Program Committee (1975–1982). He serves on the Executive Committee of the Electronics Division, Electrochemical Society and is a Divisional Editor of the Journal of the Electrochemical Society. He serves on the NAS/NAE/IoM Government-University-Industry Research Roundtable Council, and is chairman of the Working Group on "New Alliances and Partnerships: Enhancing the Utilization of Scientific and Engineering Advances."

Professor Streetman has published more than 180 technical articles and twenty-four students of electrical engineering and physics have received their Ph.D.s under his direction.

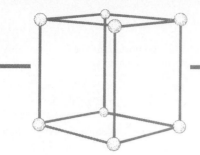

chapter 1

CRYSTAL PROPERTIES AND GROWTH OF SEMICONDUCTORS

In studying solid state electronic devices we are interested primarily in the electrical behavior of solids. However, we shall see in later chapters that the transport of charge through a metal or a semiconductor depends not only on the properties of the electron but also on the arrangement of atoms in the solid. In the first chapter we shall discuss some of the physical properties of semiconductors compared with other solids, the atomic arrangements of various materials, and some methods of growing semiconductor crystals. Topics such as crystal structure and crystal growth technology are often the subjects of books rather than introductory chapters; thus we shall consider only a few of the more important and fundamental ideas that form the basis for understanding electronic properties of semiconductors and device fabrication.

Semiconductors are a group of materials having electrical conductivities intermediate between metals and insulators. It is significant that the conductivity of these materials can be varied over orders of magnitude by changes in temperature, optical excitation, and impurity content. This variability of electrical properties makes the semiconductor materials natural choices for electronic device investigations.

Semiconductor materials are found in column IV and neighboring columns of the periodic table (Table 1-1). The column IV semiconductors, silicon and germanium, are called *elemental* semiconductors because they are composed of single species of atoms. In addition to the elemental materials, compounds

1.1
SEMICONDUCTOR MATERIALS

1

TABLE 1-1. Common semiconductor materials: (a) the portion of the periodic table where semiconductors occur; (b) elemental and compound semiconductors.

(a)	II	III	IV	V	VI
		B	C		
		Al	Si	P	S
	Zn	Ga	Ge	As	Se
	Cd	In		Sb	Te

(b) Elemental	IV compounds	Binary III–V compounds	Binary II–VI compounds
Si	SiC	AlP	ZnS
Ge	SiGe	AlAs	ZnSe
		AlSb	ZnTe
		GaP	CdS
		GaAs	CdSe
		GaSb	CdTe
		InP	
		InAs	
		InSb	

of column III and column V atoms, as well as certain combinations from II and VI, make up the *intermetallic,* or *compound,* semiconductors.

As Table 1-1 indicates, there are numerous semiconductor materials. As we shall see, the wide variety of electronic and optical properties of these semiconductors provides the device engineer with great flexibility in the design of electronic and optoelectronic functions. The elemental semiconductor Ge was widely used in the early days of semiconductor development for transistors and diodes. Silicon is now used for the majority of rectifiers, transistors, and integrated circuits. However, the compounds are widely used in high-speed devices and devices requiring the emission or absorption of light. The two-element (*binary*) III–V compounds such as GaAs and GaP are common in light-emitting diodes (LEDs). As discussed in Section 1.2.4, three-element (*ternary*) compounds such as GaAsP and four-element (*quaternary*) compounds such as InGaAsP can be grown to provide added flexibility in choosing materials properties.

Fluorescent materials such as those used in television screens usually are II–VI compound semiconductors such as ZnS. Light detectors are commonly made with InSb, CdSe, or other compounds such as PbTe and HgCdTe. Si and Ge are also widely used as infrared and nuclear radiation detectors. An important microwave device, the Gunn diode, is usually made of GaAs or InP. Semiconductor lasers are made using GaAs, AlGaAs, and other ternary and quaternary compounds.

One of the most important characteristics of a semiconductor, which distinguishes it from metals and insulators, is its *energy band gap*. This property, which we will discuss in detail in Chapter 3, determines among other things the wavelengths of light that can be absorbed or emitted by the semiconductor. For example, the band gap of GaAs is about 1.43 electron volts (eV), which corresponds to light wavelengths in the near infrared. In contrast, GaP has a band gap of about 2.3 eV, corresponding to wavelengths in the green portion of the spectrum.[†] The band gap E_g for various semiconductor materials is listed along with other properties in Appendix III. As a result of the wide variety of semiconductor band gaps, light-emitting diodes and lasers can be constructed with wavelengths over a broad range of the infrared and visible portions of the spectrum.

The electronic and optical properties of semiconductor materials are strongly affected by impurities, which may be added in precisely controlled amounts. Such impurities are used to vary the conductivities of semiconductors over wide ranges and even to alter the nature of the conduction processes from conduction by negative charge carriers to positive charge carriers. For example, an impurity concentration of one part per million can change a sample of Si from a poor conductor to a good conductor of electric current. This process of controlled addition of impurities, called *doping,* will be discussed in detail in subsequent chapters.

To investigate these useful properties of semiconductors, it is necessary to understand the atomic arrangements in the materials. Obviously, if slight alterations in purity of the original material can produce such dramatic changes in electrical properties, then the nature and specific arrangement of atoms in each semiconductor must be of critical importance. Therefore, we begin our study of semiconductors with a brief introduction to crystal structure.

In this section we discuss the arrangements of atoms in various solids. We shall distinguish between single crystals and other forms of materials and then investigate the periodicity of crystal lattices. Certain important crystallographic terms will be defined and illustrated in reference to crystals having a basic cubic structure. These definitions will allow us to refer to certain planes and directions within a lattice of arbitrary structure. Finally, we shall investigate the diamond lattice; this structure, with some variations, is typical of most of the semiconductor materials used in electronic devices.

1.2
CRYSTAL LATTICES

1.2.1   Periodic Structures

A crystalline solid is distinguished by the fact that the atoms making up the crystal are arranged in a periodic fashion. That is, there is some basic arrange-

---

[†]The conversion between the energy E of a photon of light (eV) and its wavelength λ (μm) is $\lambda = 1.24/E$. For GaAs, $\lambda = 1.24/1.43 = 0.87$ μm.

ment of atoms that is repeated throughout the entire solid. Thus the crystal appears exactly the same at one point as it does at a series of other equivalent points, once the basic periodicity is discovered. However, not all solids are crystals (Fig. 1-1); some have no periodic structure at all (*amorphous* solids), and others are composed of many small regions of single-crystal material (*polycrystalline* solids). The semiconductor materials we shall study are single crystals, with very small departures from perfect periodicity due to the addition of doping atoms for controlling electrical properties.

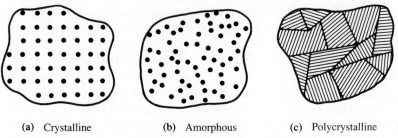

(a) Crystalline	**(b)** Amorphous	**(c)** Polycrystalline

Figure 1-1   Three types of solids, classified according to atomic arrangement: (a) crystalline and (b) amorphous materials are illustrated by microscopic views of the atoms, whereas (c) polycrystalline structure is illustrated by a more macroscopic view of adjacent single-crystalline regions, such as (a).

The periodic arrangement of atoms in a crystal is called the *lattice*. Since there are many different ways of placing atoms in a volume, the distances and orientation between atoms can take many forms. However, in every case the lattice contains a volume, called a *unit cell,* which is representative of the entire lattice and is regularly repeated throughout the crystal. As an example of such a lattice, Fig. 1-2 shows a two-dimensional arrangement of atoms with a unit cell ODEF. This cell has an atom at each corner shared with adjacent cells. Notice that we can define vectors **a** and **b** such that if the unit cell is translated by integral multiples of these vectors, a new unit cell identical to the original is found (e.g., O′D′E′F′). These vectors **a** and **b** (and **c** if the lattice is three dimensional) are called the *basis vectors* for the lattice. Points within the lattice are indistinguishable if the vector between the points is

$$\mathbf{r} = p\mathbf{a} + q\mathbf{b} + s\mathbf{c} \qquad (1-1)$$

where *p, q,* and *s* are integers.

The smallest unit cell that can be repeated to form the lattice is called a *primitive cell*. In many lattices, however, the primitive cell is not the most convenient to work with. The importance of the unit cell lies in the fact that we can analyze the crystal as a whole by investigating a representative volume. For example, from the unit cell we can find the distances between nearest atoms and next nearest atoms for calculation of the forces holding the lattice together; we can look at the fraction of the unit cell volume filled by atoms and relate the density of the solid to the atomic arrangement. But even more important for our interest in electronic devices, the properties of the periodic crystal

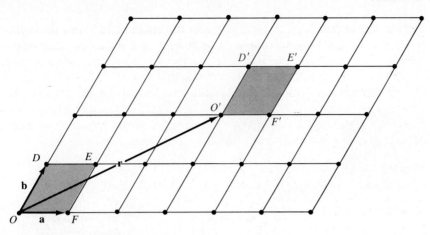

Figure 1-2
A two-dimensional lattice showing translation of a unit cell by $\mathbf{r} = 3\mathbf{a} + 2\mathbf{b}$.

lattice determine the allowed energies of electrons that participate in the conduction process. Thus the lattice determines not only the mechanical properties of the crystal but also its electrical properties.

1.2.2   Cubic Lattices

The simplest three-dimensional lattice is one in which the unit cell is a cubic volume, such as the three cells shown in Fig. 1-3. The *simple cubic* structure (abbreviated *sc*) has an atom located at each corner of the unit cell. The *body-centered cubic* (*bcc*) lattice has an additional atom at the center of the cube, and the *face-centered cubic* (*fcc*) unit cell has atoms at the eight corners and centered on the six faces.

As atoms are packed into the lattice in any of these arrangements, the distances between neighboring atoms will be determined by a balance between the forces that attract them together and other forces that hold them apart. We shall discuss the nature of these forces for particular solids in Section 3.1.1. For now, we can calculate the maximum fraction of the lattice volume that can be filled with atoms by approximating the atoms as hard spheres. For example, Fig. 1-4 illustrates the packing of spheres in a face-centered cubic cell of side a, such that the nearest neighbors touch. The dimension a for a cubic unit cell is called the *lattice constant*. For the fcc lattice the nearest neighbor distance is one-half the diagonal of a face, or $\frac{1}{2}(a\sqrt{2})$. Therefore, for the atom centered on the face to just touch the atoms at each corner of the face, the radius of the sphere must be one-half the nearest neighbor distance, or $\frac{1}{4}(a\sqrt{2})$.

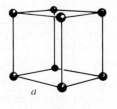

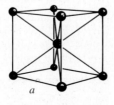

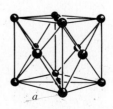

Figure 1-3
Unit cells for three types of cubic lattice structures.

Simple cubic   Body-centered cubic   Face-centered cubic

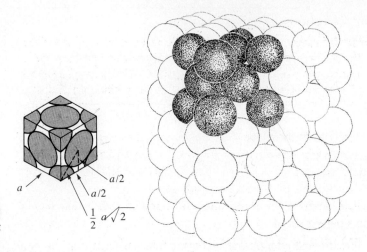

Figure 1-4
Packing of hard spheres in an fcc lattice.

EXAMPLE 1-1:   Find the fraction of the fcc unit cell volume filled with hard spheres as in Fig. 1-4.

SOLUTION   Each corner atom in a cubic unit cell is shared with seven neighboring cells; thus each unit cell contains $\frac{1}{8}$ of a sphere at each of the eight corners for a total of one atom. Similarly, the fcc cell contains half an atom at each of the six faces for a total of three. Thus we have

Atoms per cell = 1 (corners) + 3 (faces) = 4

Nearest neighbor distance = $\frac{1}{2}(a\sqrt{2})$

Radius of each sphere = $\frac{1}{4}(a\sqrt{2})$

Volume of each sphere = $\frac{4}{3}\pi[\frac{1}{4}(a\sqrt{2})]^3 = \dfrac{\pi a^3\sqrt{2}}{24}$

Maximum fraction of cell filled

$$= \frac{\text{no. of spheres} \times \text{vol. of each sphere}}{\text{total vol. of cell}}$$

$$= \frac{4 \times (\pi a^3\sqrt{2})/24}{a^3}$$

$$= \frac{\pi\sqrt{2}}{6} = 74 \text{ percent filled}$$

Therefore, if the atoms in an fcc lattice are packed as densely as possible, with no distance between the outer edges of nearest neighbors, 74 percent of the volume is filled. This is a relatively high percentage compared with some other lattice structures (Prob. 1.3).

1.2.3 Planes and Directions

In discussing crystals it is very helpful to be able to refer to planes and directions within the lattice. The notation system generally adopted uses a set of three integers to describe the position of a plane or the direction of a vector within the lattice. The three integers describing a particular plane are found in the following way:

1. Find the intercepts of the plane with the crystal axes and express these intercepts as integral multiples of the basis vectors (the plane can be moved in and out from the origin, retaining its orientation, until such an integral intercept is discovered on each axis).
2. Take the reciprocals of the three integers found in step 1 and reduce these to the smallest set of integers h, k, and l, which have the same relationship to each other as the three reciprocals.
3. Label the plane (hkl).

The plane illustrated in Fig. 1-5 has intercepts at $2\mathbf{a}$, $4\mathbf{b}$, and $1\mathbf{c}$ along the three crystal axes. Taking the reciprocals of these intercepts, we get $\frac{1}{2}$, $\frac{1}{4}$, and 1. These three fractions have the same relationship to each other as the integers 2, 1, and 4 (obtained by multiplying each fraction by 4). Thus the plane can be referred to as a (214) plane.

EXAMPLE 1-2

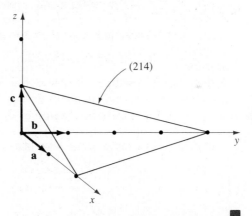

(214)

Figure 1-5
A (214) crystal plane.

    The three integers h, k, and l are called the *Miller indices;* these three numbers define a set of parallel planes in the lattice. One advantage of taking the reciprocals of the intercepts is avoidance of infinities in the notation. One intercept is infinity for a plane parallel to an axis; however, the reciprocal of such an intercept is taken as zero. If a plane contains one of the axes, it is parallel to that axis and has a zero reciprocal intercept. If a plane passes through the origin, it can be translated to a parallel position for calculation of the Miller indices. If an intercept occurs on the negative branch of an axis, the minus sign is placed above the Miller index for convenience, such as $(h\bar{k}l)$.

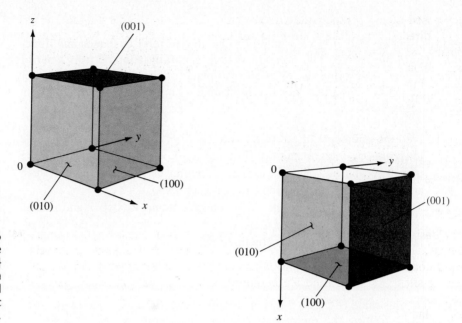

Figure 1-6
Equivalence of the cube faces ({100} planes) by rotation of the unit cell within the cubic lattice.

From a crystallographic point of view, many planes in a lattice are equivalent; that is, a plane with given Miller indices can be shifted about in the lattice simply by choice of the position and orientation of the unit cell. The indices of such equivalent planes are enclosed in braces { } instead of parentheses. For example, in the cubic lattice of Fig. 1-6 all the cube faces are crystallographically equivalent in that the unit cell can be rotated in various directions and still appear the same. The six equivalent faces are collectively designated as {100}.

A direction in a lattice is expressed as a set of three integers with the same relationship as the components of a vector in that direction. The three vector components are expressed in multiples of the basis vectors, and the three integers are reduced to their smallest values while retaining the relationship among them. For example, the body diagonal in the cubic lattice (Fig. 1-7a) is

Figure 1-7
Crystal directions in the cubic lattice.

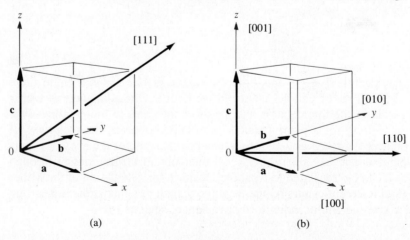

(a)   (b)

composed of the components 1**a**, 1**b**, and 1**c**; therefore, this diagonal is the [111] direction. (Brackets are used for direction indices.) As in the case of planes, many directions in a lattice are equivalent, depending only on the arbitrary choice of orientation for the axes. Such equivalent direction indices are placed in angular brackets ⟨ ⟩. For example, the crystal axes in the cubic lattice [100], [010], and [001] are all equivalent and are called ⟨100⟩ directions (Fig. 1-7b).

Comparing Figs. 1-6 and 1-7, we notice that in cubic lattices a direction [*hkl*] is perpendicular to the plane (*hkl*). This is convenient in analyzing lattices with cubic unit cells, but it should be remembered that it is not necessarily true in noncubic systems.

1.2.4 The Diamond Lattice

The basic lattice structure for many important semiconductors is the *diamond lattice*, which is characteristic of Si and Ge. In many compound semiconductors, atoms are arranged in a basic diamond structure but are different on alternating sites. This is called a *zincblende* lattice and is typical of the III–V compounds. One of the simplest ways of stating the construction of the diamond lattice is the following:

> The diamond lattice can be thought of as an fcc structure with an extra atom placed at **a**/4 + **b**/4 + **c**/4 from each of the fcc atoms.

Figure 1-8a illustrates the construction of a diamond lattice from an fcc unit cell. We notice that when the vectors are drawn with components one-fourth of the cube edge in each direction, only four additional points within the same unit cell are reached. Vectors drawn from any of the other fcc atoms simply determine corresponding points in adjacent unit cells. This method of constructing the diamond lattice implies that the original fcc has associated with it a second interpenetrating fcc displaced by $\frac{1}{4}, \frac{1}{4}, \frac{1}{4}$. The two interpenetrating fcc *sublattices* can be visualized by looking down on the unit cell of Fig. 1-8a from the top (or along any ⟨100⟩ direction). In the top view of

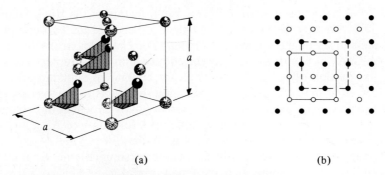

(a)   (b)

Diamond lattice structure: (a) a unit cell of the diamond lattice constructed by placing atoms $\frac{1}{4}, \frac{1}{4}, \frac{1}{4}$ from each atom in an fcc; (b) top view (along any ⟨100⟩ direction) of an extended diamond lattice. The colored circles indicate one fcc sublattice and the black circles indicate the interpenetrating fcc.

Figure 1-8

Fig. 1-8b, atoms belonging to the original fcc are represented by open circles, and the interpenetrating sublattice is shaded. If the atoms are all similar, we call this structure a diamond lattice; if the atoms differ on alternating sites, it is a zincblende structure. For example, if one fcc sublattice is composed of Ga atoms and the interpenetrating sublattice is As, the zincblende structure of GaAs results. Most of the compound semiconductors have this type of lattice, although some of the II–VI compounds are arranged in a slightly different structure called the *wurtzite* lattice. We shall restrict our discussions here to the diamond and zincblende structures, since they are typical of most of the commonly used semiconductors.

EXAMPLE 1-3

Calculate the densities of Si and GaAs from the lattice constants (Appendix III), atomic weights, and Avogadro's number. Compare the results with densities given in Appendix III. The atomic weights of Si, Ga, and As are 28.1, 69.7, and 74.9, respectively.

SOLUTION

For Si: $a = 5.43 \times 10^{-8}$ cm, 8 atoms/cell,

$$\frac{8}{a^3} = \frac{8}{(5.43 \times 10^{-8})^3} = 5 \times 10^{22} \text{ atoms/cm}^3$$

$$\text{density} = \frac{5 \times 10^{22}(\text{atoms/cm}^3) \times 28.1(\text{g/mole})}{6.02 \times 10^{23}(\text{atoms/mole})} = 2.33 \text{ g/cm}^3$$

For GaAs: $a = 5.65 \times 10^{-8}$ cm, 4 each Ga, As atoms/cell,

$$\frac{4}{a^3} = \frac{4}{(5.65 \times 10^{-8})^3} = 2.22 \times 10^{22} \text{ atoms/cm}^3$$

$$\text{density} = \frac{2.22 \times 10^{22}(69.7 + 74.9)}{6.02 \times 10^{23}} = 5.33 \text{ g/cm}^3$$

■

A particularly interesting and useful feature of the III–V compounds is the ability to vary the mixture of elements on each of the two interpenetrating fcc sublattices of the zincblende crystal. For example, in the ternary compound AlGaAs, it is possible to vary the composition of the ternary alloy by choosing the fraction of Al or Ga atoms on the column III sublattice. It is common to represent the composition by assigning subscripts to the various elements. For example, $Al_xGa_{1-x}As$ refers to a ternary alloy in which the column III sublattice in the zincblende structure contains a fraction x of Al atoms and $1 - x$ of Ga atoms. The composition $Al_{0.3}Ga_{0.7}As$ has 30 percent Al and 70 percent Ga on the column III sites, with the interpenetrating column V sublattice occupied entirely by As atoms. It is extremely useful to be able to grow ternary alloy crystals such as this with a given composition. For the $Al_xGa_{1-x}As$ example we can grow crystals over the entire composition range from $x = 0$ to $x = 1$, thus varying the electronic and optical properties of the material from that of GaAs ($x = 0$) to that of AlAs ($x = 1$). To vary the properties even further, it is

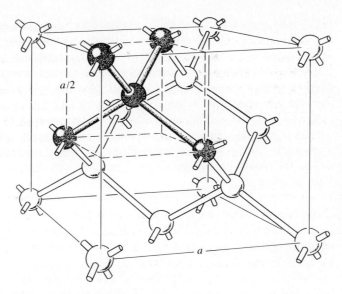

Figure 1-9
Diamond lattice unit cell, showing the four nearest neighbor structure. (From *Electrons and Holes in Semiconductors* by W. Shockley, © 1950 by Litton Educational Publishing Co., Inc.; by permission of Van Nostrand Reinhold Co., Inc.)

possible to grow four-element (quaternary) compounds such as $In_xGa_{1-x}As_yP_{1-y}$ having a very wide range of properties.

It is important from an electronic point of view to notice that each atom in the diamond and zincblende structures is surrounded by four nearest neighbors (Fig. 1-9). The importance of this relationship of each atom to its neighbors will become evident in Section 3.1.1 when we discuss the bonding forces which hold the lattice together.

The fact that atoms in a crystal are arranged in certain planes is important to many of the mechanical, metallurgical, and chemical properties of the material. For example, crystals often can be cleaved along certain atomic planes, resulting in exceptionally planar surfaces. This is a familiar result in cleaved diamonds for jewelry; the facets of a diamond reveal clearly the triangular, hexagonal, and rectangular symmetries of intersecting planes in various crystallographic directions. Semiconductors with diamond and zincblende lattices have similar cleavage planes. As an example, Fig. 1-10 shows a view along a

Figure 1-10
Silicon {111} surface with aluminum alloyed regions, revealing the symmetry of the diamond lattice.

⟨111⟩ body diagonal of a small Si crystal that has been cleaved along intersecting {111} planes. This crystal clearly illustrates the triangular symmetry of the {111} planes when viewed along the body diagonal. As an example of the influence of the lattice upon metallurgical properties, small regions of Al have been alloyed on the face of this crystal.[†] We notice that the alloyed regions reflect the triangular and hexagonal symmetries of the atomic planes. Chemical reactions, such as etching of the crystal, often take place preferentially along certain directions. These properties serve as interesting illustrations of crystal symmetry, but in addition, each plays an important role in fabrication processes for many semiconductor devices. We shall discuss these applications in later chapters.

**1.3
BULK CRYSTAL
GROWTH**

The progress of solid state device technology since the invention of the transistor in 1948 has depended not only on the development of device concepts but also on the improvement of materials. For example, the fact that integrated circuits can be made today is the result of a considerable breakthrough in the growth of pure, single-crystal Si in the early and mid-1950s. The requirements on the growing of device-grade semiconductor crystals are more stringent than those for any other materials. Not only must semiconductors be available in large single crystals, but also the purity must be controlled within extremely close limits. For example, Si crystals now being used in devices are grown with concentrations of most impurities of less than one part in ten billion. Such purities require careful handling and treatment of the material at each step of the manufacturing process.

Elemental Ge and Si are obtained by chemical decomposition of compounds such as GeO_2, $SiCl_4$, and $SiHCl_3$. Once the semiconductor material has been isolated and preliminary purification steps have been performed, it is melted and cast into ingots. Upon cooling from such a casting process, the Si or Ge is polycrystalline (Fig. 1-1c). The atoms are arranged in the diamond lattice over small regions of the ingot, since this is the natural structure for the material. However, unless some control is maintained over the cooling process, the crystalline regions occur with essentially random orientation. For the crystal to grow in a single orientation, it is necessary to maintain careful control over the boundary between the molten material and the solid during cooling.

1.3.1 Growth from the Melt

A common technique for growing single crystals involves selective cooling of the molten material so that solidification occurs along a particular crystal direction. For example, Fig. 1-11a shows a silica (vitreous quartz) crucible containing molten Ge, which may be pulled through a furnace such that solidification begins at one end and slowly proceeds down the length of the bar. Crystal

---

[†]The process of alloying is discussed in Section 5.1.2.

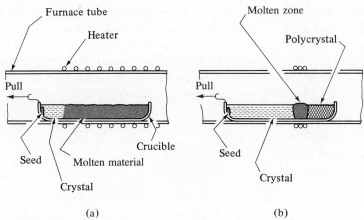

Figure 1-11
Crystal growing from the melt in a crucible: (a) solidification from one end of the melt (horizontal Bridgman method); (b) melting and solidification in a moving zone.

growth can be enhanced if a small "seed" crystal is placed at the end that is cooled first. If the cooling rate is carefully controlled and the position of the interface between solid and melt is moved slowly along the crucible, the Ge atoms will be arranged in a diamond lattice as the crystal cools. The shape of the resulting crystal is determined by the crucible. Germanium, GaAs, and other semiconductor crystals are often grown by this technique, which is commonly called the horizontal *Bridgman* method. In a variation of this procedure, a small region of the polycrystalline material is melted, and the molten zone is moved down the crucible at such a rate that a crystal is formed behind the zone as it moves (Fig. 1-11b).

One disadvantage of growing the crystal in a containing crucible is that the molten material contacts the sides of the container; the resulting interference of the crucible wall introduces stresses during solidification and can cause deviations from the perfect lattice structure. This is a particularly serious problem in Si, which has a high melting point and tends to adhere to crucible materials. An alternative method, which eliminates this problem, involves pulling the crystal from the melt as it grows (Fig. 1-12). In this method a seed crystal is lowered into the molten material and then is raised slowly, allowing the crystal

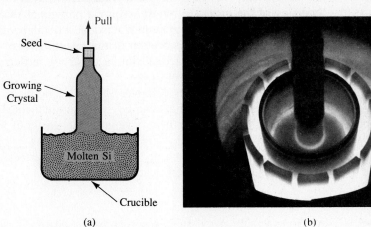

Figure 1-12
Pulling of a Si crystal from the melt (Czochralski method): (a) schematic diagram of the crystal growth process; (b) view through a port in the furnace showing a Si crystal being pulled from the melt. (Photograph courtesy of Texas Instruments, Inc.)

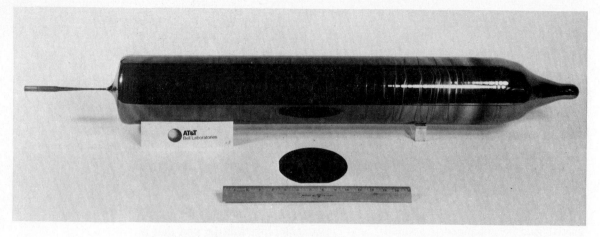

Figure 1-13 Silicon crystal grown by the Czochralski method. This large single crystal, weighing 45 kg, provides 15-cm (6-in.)-diameter wafers such as the one shown here. For size comparison, the ingot and wafer are shown here alongside a ruler. (Photograph courtesy of AT&T Bell Laboratories.)

to grow onto the seed. Generally, the crystal is rotated slowly as it grows to provide a slight stirring of the melt and to average out any temperature variations that would cause inhomogeneous solidification. This technique, commonly called the *Czochralski* method, is widely used in growing Si, Ge, and some of the compound semiconductors.

In pulling compounds such as GaAs from the melt, it is necessary to prevent volatile elements (e.g., As) from vaporizing. In one method a layer of B_2O_3, which is dense and viscous when molten, floats on the surface of the molten GaAs to prevent As evaporation. This growth method is called *liquid-encapsulated Czochralski (LEC)* growth.

In Czochralski crystal growth, the shape of the ingot is determined by a combination of the tendency of the cross section to assume a polygonal shape due to the crystal structure and the influence of surface tension, which encourages a circular cross section. For Ge the cross section of the resulting ingot is polygonal with rounded corners, but the surface tension is so strong in Si that the cross section is almost circular (Fig. 1-13).

In the fabrication of Si integrated circuits (Chapter 9) it is economical to use very large Si wafers,[†] so that many IC chips can be made simultaneously. As a result, considerable research and development have gone into methods for growing very large Si crystals. For example, Fig. 1-13 illustrates a 6-in.-diameter Si ingot and a wafer cut from such an ingot. Even larger wafer diameters are available for integrated circuit fabrication.

[†]*Wafers* are slices cut from a single-crystal ingot, generally using a diamond-tipped saw blade.

1.3.2   Zone Refining and Floating-Zone Growth

The use of a moving molten zone as in Figs. 1-11b and 1-14 (p. 16) can result in considerable purification over the starting material, particularly when several passes are made along the ingot. This process is called *zone refining*. Common techniques for melting the ingot include radiation of heat from a resistance heater, induction heating, and heating by electron bombardment. At the solidifying interface between the melt and the solid, there will be a certain distribution of impurities between the two phases. An important quantity that identifies this property is the *distribution coefficient* k_d, which is the ratio of the concentration of the impurity in the solid C_S to the concentration in the liquid C_L at equilibrium:

$$k_d = \frac{C_S}{C_L} \tag{1-2}$$

The distribution coefficient is a function of the material, the impurity, the temperature of the solid–liquid interface, and the growth rate. For an impurity with a distribution coefficient of one-half, the relative concentration of the impurity in the molten liquid to that in the refreezing solid is two to one. Thus the concentration of impurities in that portion of material that solidifies first is one-half the original concentration C_0. As the zone moves along the bar, however, impurities are driven along with the molten material until the concentration in the molten zone approaches C_0/k_d. At that point as many impurities enter the zone as leave it, and Eq. (1-2) is automatically satisfied with $C_S = C_0$ and $C_L = C_0/k_d$. After the first pass, a considerable region of the ingot exists with the original impurity concentration. If another pass is made in the same direction, however, impurities are again swept with the molten zone until the condition $C_L = 2C_0$ is reached. On the second pass this point occurs farther along the bar, and the purified region is longer. If repeated passes are made, the bar can be purified over much of its length. After many passes, most of the impurities have moved to the end of the bar, which can then be cut away, leaving a highly purified crystal. Of course, if the impurity distribution coefficient k_d is smaller than the $k_d = 0.5$ case, the condition $C_L = C_0/k_d$ is reached farther along the bar, and greater purification is obtained. Many important impurities in Si and Ge have very small distribution coefficients; these impurities can be removed effectively with only a few passes of the molten zone.

The distribution coefficient, which controls the zone refining process, is also important during any growth from a melt. This can be illustrated by an example involving Czochralski growth:

A Si crystal is to be grown by the Czochralski method, and it is desired that the ingot contain 10^{16} phosphorus atoms/cm^3.
    EXAMPLE 1-4

(a) What concentration of phosphorus atoms should the melt contain to give this impurity concentration in the crystal during the initial growth? For P in Si, $k_d = 0.35$.

(b) If the initial load of Si in the crucible is 5 kg, how many grams of phosphorus should be added? The atomic weight of phosphorus is 31.

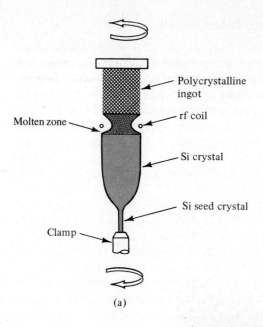

(a)

(b)

Figure 1-14
Floating-zone crystal growth: (a) schematic diagram of the growth process; (b) view through a furnace port of a Si crystal being grown by the floating-zone process. (Illustrations courtesy of Monsanto Company.)

SOLUTION

(a) Assume that $C_S = k_d C_L$ throughout the growth. Thus the initial concentration of P in the melt should be

$$\frac{10^{16}}{0.35} = 2.86 \times 10^{16} \text{ cm}^{-3}$$

(b) The P concentration is so small that the volume of melt can be calculated from the weight of Si. From Example 1-3 the density of Si is 2.33 g/cm³. In this example we will neglect the difference in density between solid and molten Si.

$$\frac{5000 \text{ g of Si}}{2.33 \text{ g/cm}^3} = 2146 \text{ cm}^3 \text{ of Si}$$

$$2.86 \times 10^{16} \text{ cm}^{-3} \times 2146 \text{ cm}^3 = 6.14 \times 10^{19} \text{ P atoms}$$

$$\frac{6.14 \times 10^{19} \text{ atoms} \times 31 \text{ g/mole}}{6.02 \times 10^{23} \text{ atoms/mole}} = 3.16 \times 10^{-3} \text{ g of P}$$

Since the P concentration in the growing crystal is only about one-third of that in the melt, Si is used up more rapidly than P in the growth. Thus the melt becomes richer in P as the growth proceeds, and the crystal is doped more heavily in the latter stages of growth. This assumes that k_d is not varied; a more uniformly doped ingot can be grown by varying the pull rate (and therefore k_d) appropriately.

■

If a seed crystal is included in the starting end of the ingot of Fig. 1-14, a crystal can be grown by the *floating-zone* method. In this case the molten zone does not touch the crucible walls. As a result, such crystals have lower oxygen contamination than do Czochralski crystals, in which oxygen is dissolved by the melt in contact with the silica (SiO_2) crucible.

One of the most important and versatile methods of crystal growth for device applications is the growth of a thin crystal layer on a wafer of a compatible crystal. The substrate crystal may be a wafer of the same material as the grown layer or a different material with a similar lattice structure. In this process the substrate serves as the seed crystal onto which the new crystalline material grows. The growing crystal layer maintains the crystal structure and orientation of the substrate. The technique of growing an oriented single-crystal layer on a substrate wafer is called *epitaxial growth,* or *epitaxy.* As we shall see in this section, epitaxial growth can be performed at temperatures considerably below the melting point of the substrate crystal. A variety of methods are used to provide the appropriate atoms to the surface of the growing layer. These methods include *chemical vapor deposition (CVD),*[†] growth from a melt (*liquid-phase epitaxy, LPE*), and evaporation of the elements in a vacuum (*molecular beam epitaxy, MBE*). With this wide range of epitaxial growth techniques, it is possible to grow a variety of crystals for device applications, having properties specifically designed for the electronic or optoelectronic device being made.

1.4
EPITAXIAL
GROWTH

1.4.1   Lattice Matching in Epitaxial Growth

When Si epitaxial layers are grown on Si substrates, there is a natural matching of the crystal lattice, and high-quality single crystal layers result. On the other hand, it is often desirable to obtain epitaxial layers that differ somewhat from the substrate. This can be accomplished easily if the lattice structure and lattice constant a match for the two materials. For example, GaAs and AlAs both have the zincblende structure, with a lattice constant of about 5.65 Å. As a re-

---

[†]The generic term *chemical vapor deposition* includes deposition of layers that may be polycrystalline or amorphous. When a CVD process results in a single-crystal epitaxial layer, a more specific term is *vapor-phase epitaxy (VPE)*.

sult, epitaxial layers of the ternary alloy AlGaAs can be grown on GaAs substrates with little lattice mismatch. Similarly, GaAs can be grown on Ge substrates (see Appendix III).

Since AlAs and GaAs have similar lattice constants, it is also true that the ternary alloy AlGaAs has essentially the same lattice constant over the entire range of compositions from AlAs to GaAs. As a result, one can choose the composition x of the ternary compound $Al_xGa_{1-x}As$ to fit the particular device requirement, and grow this composition on a GaAs wafer. The resulting epitaxial layer will be lattice-matched to the GaAs substrate.

Figure 1-15 illustrates the energy band gap E_g as a function of lattice constant a for several III–V ternary compounds as they are varied over their composition ranges. For example, as the ternary compound InGaAs is varied by choice of composition on the column III sublattice from InAs to GaAs, the band gap changes from 0.36 to 1.43 eV while the lattice constant of the crystal varies from 6.06 Å for InAs to 5.65 Å for GaAs. Clearly, we cannot grow this ternary compound over the entire composition range on a particular binary substrate, which has a fixed lattice constant. As Fig. 1-15 illustrates, however, it is possible to grow a specific composition of InGaAs on an InP substrate. The vertical (invariant lattice constant) line from InP to the InGaAs curve shows that a midrange ternary composition (actually, $In_{0.53}Ga_{0.47}As$) can be grown lattice-matched to an InP substrate. Similarly, a ternary InGaP alloy with about 50 percent Ga and 50 percent In on the column III sublattice can be grown lattice-matched to a GaAs substrate. To achieve a broader range of alloy compositions, grown lattice-matched on particular substrates, it is helpful to use quaternary alloys such as InGaAsP. The variation of compositions on both the column III and column V sublattices provides additional flexibility in choosing a particular band gap while providing lattice-matching to convenient binary substrates such as GaAs or InP.

In the case of GaAsP, the lattice constant is intermediate between that of GaAs and GaP, depending upon the composition. For example, GaAsP crystals used in red LEDs have 40 percent phosphorus and 60 percent arsenic on the column V sublattice. Since such a crystal cannot be grown directly on either a GaAs or a GaP substrate, it is necessary to gradually change the lattice constant as the crystal is grown. Using a GaAs or Ge wafer as a substrate, the growth is begun at a composition near GaAs. A region ~25 μm thick is grown while gradually introducing phosphorus until the desired As/P ratio is achieved. The desired epitaxial layer (e.g., 100 μm thick) is then grown on this graded layer. By this method epitaxial growth always occurs on a crystal of similar lattice constant. Although some crystal dislocations occur due to lattice strain in the graded region, such crystals are of high quality and can be used in LEDs.

In addition to the widespread use of lattice-matched epitaxial layers, the advanced epitaxial growth techniques described in the following sections allow the growth of very thin (~100Å) layers of lattice-mismatched crystals. If the mismatch is only a few percent and the layer is thin, the epitaxial layer grows with a lattice constant in compliance with that of the seed crystal. The resulting layer is in compression or tension along the surface plane as its lattice constant

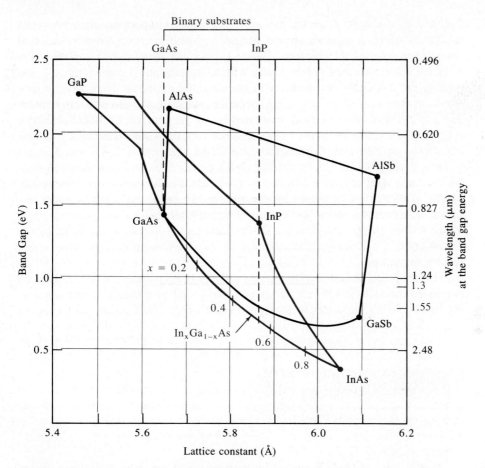

Relationship between band gap and lattice constant for alloys in the InGaAsP and AlGaAsSb systems. The dashed vertical lines show the lattice constants for the commercially available binary substrates GaAs and InP. For the marked example of $In_xGa_{1-x}As$, the ternary composition $x = 0.53$ can be grown lattice-matched on InP, since the lattice constants are the same. For quaternary alloys, the compositions on both the III and V sublattices can be varied to grow lattice-matched epitaxial layers along the dashed vertical lines between curves. For example, $In_xGa_{1-x}As_yP_{1-y}$ can be grown on InP substrates, with resulting band gaps ranging from 0.75 eV to 1.35 eV and with corresponding wavelengths for light emission from 1.65 μm to 0.9 μm.

Figure 1-15

adapts to the seed crystal. Such a layer is called *pseudomorphic* because it is not lattice-matched to the substrate without strain. Using thin alternating layers of slightly mismatched crystal layers, it is possible to grow a *strained-layer superlattice (SLS)* in which alternate layers are in tension and compression. The overall SLS lattice constant is an average of that of the two bulk materials.

1.4.2 Liquid-Phase Epitaxy

It is possible to grow crystals of many semiconductors from a liquid solution at temperatures well below their melting point. Since a mixture of the semiconductor with a second element may melt at a lower temperature than the semi-conductor itself, it is often an advantage to grow the crystal from solution at the temperature of the mixture. For example, the melting point of GaAs is 1238°C, whereas a mixture of GaAs with Ga metal has a considerably lower melting point (depending on the proportions of the mixture). Thus a GaAs seed crystal can be held in a Ga + GaAs solution, which is molten at a temperature low enough that the seed itself is not melted. If the solution is cooled slowly, a single-crystal GaAs layer grows on the seed. As the GaAs leaves the solution and grows on the parent crystal, the solution becomes richer in Ga and thus has a lower melting point. Upon further cooling, more GaAs leaves the solution, and the epitaxial crystal continues to grow. By this technique single crystals can be grown at temperatures low enough to eliminate many problems of im-purity introduction typical of growth at the crystal melting temperature. This method is particularly useful for III–V compounds in which Ga or In serves as the column III element, since these metals form solutions at conveniently low temperatures.

This method is called _liquid-phase epitaxy (LPE)_. In LPE the solution can be placed onto the wafer in a number of ways; one of the most direct methods involves holding the wafer in a graphite slider (Fig. 1-16) and moving it into a

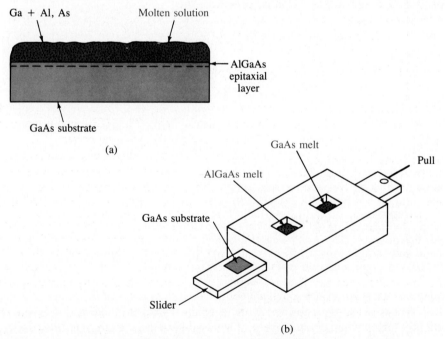

Figure 1-16   Liquid-phase epitaxial growth of AlGaAs and GaAs layers on a GaAs substrate: (a) cross section of the sample in contact with a Ga-rich melt containing Al and As; (b) carbon slider used to move the GaAs substrate between various melts. In this case, two pockets are provided, containing melts for AlGaAs and GaAs growth. The GaAs substrate on the slider is moved first into the AlGaAs growth chamber; after growth of this layer (shown in part a), the excess melt is wiped off as the slider moves the substrate to the next growth chamber.

pocket containing the melt. After growth, the epitaxial surface is wiped clean as it is moved to the next pocket, perhaps for further growth. In the fabrication of semiconductor lasers (Chapter 10), multiple layers of GaAs and AlGaAs can be obtained by using successive LPE growth in this type of slider system.

1.4.3   Vapor-Phase Epitaxy

The advantages of low temperature and high purity which characterize LPE growth can also be achieved by crystallization from the vapor phase. Crystalline layers can be grown onto a seed or substrate from a chemical vapor of the semiconductor material or from mixtures of chemical vapors containing the semiconductor. *Vapor-phase epitaxy (VPE)* is a particularly important source of semiconductor material for use in devices. Some compounds such as GaAs can be grown with better purity and crystal perfection by vapor epitaxy than by other methods. Furthermore, these techniques offer great flexibility in the actual fabrication of devices. When an epitaxial layer is grown on a substrate, it is relatively simple to obtain a sharp demarcation between the type of impurity doping in the substrate and in the grown layer. The advantages of this freedom to vary the impurity will be discussed in subsequent chapters. We point out here, however, that bipolar Si integrated-circuit devices (Chapter 9) are usually built in layers grown by VPE on Si wafers.

Epitaxial layers are generally grown on Si substrates by the controlled deposition of Si atoms onto the surface from a chemical vapor containing Si. In one method, a gas of silicon tetrachloride reacts with hydrogen gas to give Si and anhydrous HCl:

$$SiCl_4 + 2H_2 \longrightarrow Si + 4HCl \tag{1-3}$$

If this reaction occurs at the surface of a heated crystal, the Si atoms released in the reaction can be deposited as an epitaxial layer. The HCl remains gaseous at the reaction temperature and does not disturb the growing crystal.

This vapor epitaxy technique requires a chamber into which the gases can be introduced and a method for heating the Si wafers. Since the chemical reactions take place in this chamber, it is called a *reaction chamber* or, more simply, a *reactor*. Hydrogen gas is passed through a heated vessel in which $SiCl_4$ is evaporated; then the two gases are introduced into the reactor over the substrate crystal, along with other gases containing the desired doping impurities. The Si slice is placed on a graphite susceptor or some other material that can be heated to the reaction temperature with an rf heating coil. This method can be adapted to grow epitaxial layers of closely controlled impurity concentration on many Si slices simultaneously.

The reaction temperature for the hydrogen reduction of $SiCl_4$ is approximately 1250°C. Other reactions may be employed at somewhat lower temperatures, including the pyrolysis of silane (SiH_4) at 1000°C. Pyrolysis involves the breaking up of the silane at the reaction temperature:

$$SiH_4 \longrightarrow Si + 2H_2 \tag{1-4}$$

There are several advantages of this technique, including the fact that the lower reaction temperature reduces migration of impurities from the substrate to the growing epitaxial layer.

In some applications it is useful to grow thin Si layers on insulating substrates. For example, vapor-phase epitaxial techniques can be used to grow ∼1μm Si films on sapphire and other insulators. This application of VPE is discussed in Section 9.3.3.

Vapor-phase epitaxial growth is also important in the III–V compounds, such as GaAs, GaP, and the ternary alloy GaAsP, which is widely used in the fabrication of LEDs. Figure 1-17 illustrates schematically a vapor-phase reactor for the growth of these materials. Substrates are held at about 800°C on a rotating wafer holder while phosphine, arsine, and gallium chloride gases are mixed and passed over the samples. The GaCl is obtained by reacting anhydrous HCl with molten Ga within the reactor. Variation of the crystal composition for GaAsP can be controlled by altering the mixture of arsine and phosphine gases.

Another useful method for epitaxial growth of compound semiconductors is called *metal-organic vapor-phase epitaxy (MOVPE),* or *organometallic vapor-phase epitaxy (OMVPE).* For example, the organometallic compound trimethylgallium can be reacted with arsine to form GaAs and methane:

$$(CH_3)_3Ga + AsH_3 \longrightarrow GaAs + 3CH_4 \tag{1-5}$$

This reaction takes place at about 700°C, and epitaxial growth of high-quality GaAs layers can be obtained. Other compound semiconductors can also be grown by this method. For example, trimethylaluminum can be added to the gas mixture to grow AlGaAs. This growth method is widely used in the fabrication of a variety of devices, including solar cells and lasers. The convenient variability of the gas mixture allows the growth of multiple thin layers similar to those discussed below for molecular beam epitaxy.

Figure 1-17
Schematic diagram of a vapor-phase epitaxial (VPE) reactor used to grow GaAs, GaP, and the ternary compound GaAsP.

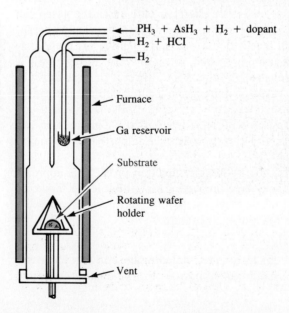

$PH_3 + AsH_3 + H_2 +$ dopant
$H_2 + HCl$
H_2

Furnace

Ga reservoir

Substrate

Rotating wafer holder

Vent

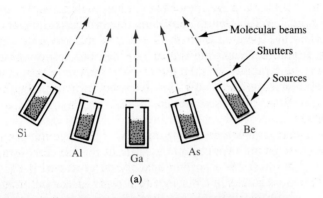

(a)

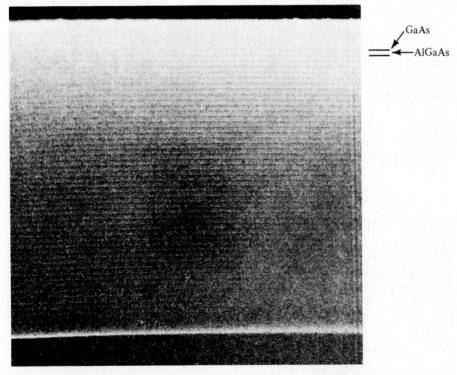

(b)

Crystal growth by molecular beam epitaxy (MBE): (a) evaporation cells inside a high-vacuum chamber directing beams of Al, Ga, As, and dopants onto a GaAs substrate; (b) scanning electron micro-graph of the cross section of an MBE-grown crystal having 100 alternating layers of GaAs (80 Å per layer) and AlGaAs (400 Å). (Photograph courtesy of AT&T Bell Laboratories.)

Figure 1-18

1.4.4   Molecular Beam Epitaxy

One of the most versatile techniques for growing epitaxial layers is called *molecular beam epitaxy (MBE)*. In this method the substrate is held in a high vacuum while molecular or atomic beams of the constituents impinge upon its surface (Fig. 1-18). For example, in the growth of AlGaAs layers on GaAs substrates, the Al, Ga, and As components, along with dopants, are heated in separate cylindrical cells. Collimated beams of these constituents escape into the vacuum and are directed onto the surface of the substrate. The rates at which these atomic beams strike the surface can be closely controlled, and growth of very high quality crystals results. The sample is held at a relatively low temperature (about 580°C for GaAs) in this growth procedure. Abrupt changes in doping or in crystal composition (e.g., changing from GaAs to AlGaAs) can be obtained by controlling shutters in front of the individual beams. Using slow growth rates (≤ 1 μm/h), it is possible to control the shutters to make composition changes on the scale of the lattice constant. For example, Fig. 1-18b illustrates a portion of a crystal grown with 100 alternating layers of GaAs and AlGaAs in a distance of 3.3 μm. Because of the high vacuum and close controls involved, MBE requires a rather sophisticated setup (Fig. 1-19). However, the versatility of this growth method makes it very attractive for many applications.

As MBE has developed in recent years, it has become common to replace some of the solid sources shown in Fig. 1-18 with gaseous chemical sources. This approach, called *chemical beam epitaxy,* or *gas-source MBE,* combines many of the advantages of MBE and VPE.

Figure 1-19
Molecular beam epitaxy facility at the University of Texas. In addition to MBE growth capability, this system includes numerous surface analytical instruments. (Photograph courtesy of Varian.)

PROBLEMS

1.1. Sodium chloride (NaCl) is a cubic crystal that differs from an sc in that alternating atoms are different; each Na is surrounded by six Cl nearest neighbors and vice versa in the three-dimensional lattice. Draw a two-dimensional NaCl lattice looking down a $\langle 100 \rangle$ direction and indicate a unit cell. Remember the unit cell must be repetitive upon displacement by the basis vectors.

1.2. (a) Label the planes illustrated in Fig. P1-2.
     (b) Draw equivalent $\langle 111 \rangle$, $\langle 100 \rangle$, $\langle 110 \rangle$ directions in a cubic lattice; use a unit cube for illustrating each set of equivalent directions.

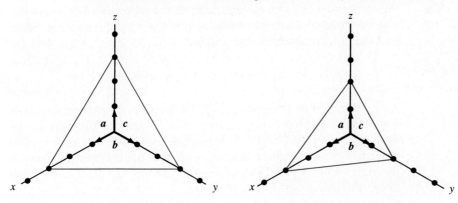

Figure P1-2

1.3. Show that the maximum fractions of the unit cell volume that can be filled by hard spheres in the sc, bcc, and diamond lattices are 0.52, 0.68, and 0.34, respectively.

1.4. (a) Find the number of Si atoms/cm^2 on the (100) surface of a Si wafer.
     (b) What is the distance (in Å) between nearest In neighbors in InP?

1.5. The ionic radii of Na$^+$ (atomic weight 23) and Cl$^-$ (atomic weight 35.5) are 1.0 and 1.8 Å, respectively. Treating the ions as hard spheres, calculate the density of NaCl. Compare this with the measured density of 2.17 g/cm^3. See Prob. 1.1.

1.6. Calculate the densities of Ge and InP from the lattice constants (Appendix III), atomic weights, and Avogadro's number. Compare the results with the densities given in Appendix III.

1.7. The atoms seen in Fig. 1-8b along a $\langle 100 \rangle$ direction of the diamond lattice are not all coplanar. Taking the top plane of colored atoms in Fig. 1-8a to be (0), the parallel plane $a/4$ down to be ($\frac{1}{4}$), the plane through the center to be $(\frac{1}{2})$, and the second plane of black atoms to be $(\frac{3}{4})$, label the plane of each atom in Fig. 1-8b.

1.8.   A Si crystal is to be pulled from the melt and doped with arsenic ($k_d = 0.3$). If the Si weighs 1 kg, how many grams of arsenic should be introduced to achieve 10^{15} cm^{-3} doping during the initial growth?

1.9.   In this problem we wish to calculate the distribution $C_s(x)$ of a particular impurity in a solid after one pass of a molten zone of length l. The distribution coefficient is k_d, and Eq. (1-2) is valid at x, the boundary between the molten zone and the solidifying crystal. Assume the geometry shown in Fig. P1-9; the concentration of the impurity throughout the molten zone is the same as $C_L(x)$, and the concentration to the right of the zone is the starting value C_0. Show that the distribution after one pass is

$$C_s(x) = C_0 - C_0(1 - k_d)e^{-k_d x/l}$$

Hint: Assume that the zone in Fig. P1-9 moves a distance Δx to the right. Write the following balance: The quantity of impurity in the zone after the shift $lAC_L(x + \Delta x)$ equals that before the shift, minus the quantity solidified at the left, plus that melted into the zone at the right. Obtain the standard derivative form and let Δx approach zero. The resulting differential equation can be solved easily for $C_L(x)$, using the boundary condition that $C_L(x = 0) = C_0$.

Figure P1-9

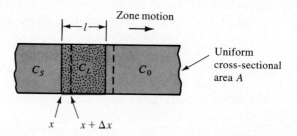

1.10.   We want to use zone refining to reduce the P ($k_d = 0.35$) concentration in a long Si ingot from an initial uniform concentration of C_0. The concentration after $i + 1$ passes can be expressed approximately as

$$C_{i+1}(x) = C_i(x) - (1 - k_d)C_i(x)e^{-k_d x/l}$$

Use this expression to calculate and plot the impurity concentration profile in the ingot after a number of passes. Plot normalized concentration C/C_0 vs. distance in zone lengths, x/l.

*1.11. Use the expression derived in Prob. 1-9 to calculate the impurity profile in the ingot after a single pass for a 1-m ingot with an initial concentration of 1×10^{17} cm^{-3} for a distribution coefficient of 0.1. Repeat, increasing k_d in increments of 0.1 to 0.9. Assume that the length of the molten zone is 5 cm. Plot the resulting impurity distributions and comment on the importance of the distribution coefficient in zone refining.

READING LIST

Beadle, W. E., J. C. C. Tsai, and R. D. Plummer, eds. *Quick Reference Manual for Silicon Integrated Circuit Technology*. New York: John Wiley-Interscience, 1986.

Brice, J. C. *Crystal Growth Processes*. New York: John Wiley, 1986.

Chang, L. L., and K. Ploog, eds. *Molecular Beam Epitaxy and Heterostructures*. Boston: Martinus Nijhoff Publishers, 1985.

Dupuis, R. D. "Metalorganic Chemical Vapor Deposition of III–V Semiconductors." *Science* 226, no. 4675 (June 1984): 623–29.

Green, M., J. E. E. Baglin, G. Y. Chin, H. W. Deckman, W. Mayo and D. Narasinham, eds. *Semiconductor-Based Heterostructures: Interfacial Structure and Stability*. Warrendale, Pa.: The Metallurgical Society, Inc., 1986.

Kittel, C. *Introduction to Solid State Physics*. 6th ed. New York: John Wiley, 1986.

Laudise, R. A. *Growth of Single Crystals*. Englewood Cliffs, N.J.: Prentice Hall, 1970.

Narayanamurti, V. "Crystalline Semiconductor Heterostructures." *Physics Today* 37, no. 10, (October 1984): 24–32.

Parker, E. H. C., ed. *The Technology and Physics of Molecular Beam Epitaxy*. New York: Plenum Press, 1985.

chapter 2

ATOMS AND ELECTRONS

Since this book is primarily an introduction to solid state devices, it would be preferable not to delay this discussion with subjects such as atomic theory, quantum mechanics, and electron models. However, the behavior of solid state devices is directly related to these subjects. For example, it would be difficult to understand how an electron is transported through a semiconductor device without some knowledge of the electron and its interaction with the crystal lattice. Therefore, in this chapter we shall investigate some of the important properties of electrons, with special emphasis on two points: (1) the electronic structure of atoms and (2) the interaction of atoms and electrons with excitation, such as the absorption and emission of light. By studying electron energies in an atom, we lay the foundation for understanding the influence of the lattice on electrons participating in current flow through a solid. Our discussions concerning the interaction of light with electrons form the basis for later descriptions of changes in the conductivity of a semiconductor with optical excitation, properties of light-sensitive devices, and lasers.

First, we shall investigate some of the experimental observations which led to the modern concept of the atom, and then we shall give a brief introduction to the theory of quantum mechanics. Several important concepts will emerge from this introduction: the electrons in atoms are restricted to certain energy levels by quantum rules; the electronic structure of atoms is determined from these quantum conditions; and this "quantization" defines certain allowable transitions involving absorption and emission of energy by the electrons.

The main effort of science is to describe what happens in nature, in as complete and concise a form as possible. In physics this effort involves observing natural phenomena, relating these observations to previously established theory, and finally establishing a physical model for the observations. The primary purpose of the model is to allow the information obtained in present observations to be used to understand new experiments. Therefore, the most useful models are expressed mathematically, so that quantitative explanations of new experiments can be made succinctly in terms of established principles. For example, we can explain the behavior of a spring-supported weight moving up and down periodically after an initial displacement, because the differential equations describing such a simple harmonic motion have been established and are understood by students of elementary physics. But the physical model upon which these equations of motion are based arises from serious study of natural phenomena such as gravitational force, the response of bodies to accelerating forces, the relationship of kinetic and potential energy, and the properties of springs. The mass and spring problem is relatively easy to solve because each of these properties of nature is well understood.

When a new physical phenomenon is observed, it is necessary to find out how it fits into the established models and "laws" of physics. In the vast majority of cases this involves a direct extension of the mathematics of well-established models to the particular conditions of the new problem. In fact, it is not uncommon for a scientist or engineer to predict that a new phenomenon should occur before it is actually observed, simply by a careful study and extension of existing models and laws. The beauty of science is that natural phenomena are not isolated events but are related to other events by a few analytically describable laws. However, it does happen occasionally that a set of observations cannot be described in terms of existing theories. In such cases it is necessary to develop models which are based as far as possible on existing laws, but which contain new aspects arising from the new phenomenon. Postulating new physical principles is a serious business, and it is done only when there is no possibility of explaining the observations with established theory. When new assumptions and models are made, their justification lies in the following question: "Does the model describe precisely the observations, and can reliable predictions be made based on the model?" The model is good or poor depending on the answer to this question.

In the 1920s it became necessary to develop a new theory to describe phenomena on the atomic scale. A long series of careful observations had been made that clearly indicated that many events involving electrons and atoms did not obey the classical laws of mechanics. It was necessary, therefore, to develop a new kind of mechanics to describe the behavior of particles on this small scale. This new approach, called *quantum mechanics,* describes atomic phenomena very well and also properly predicts the way in which electrons behave in solids — our primary interest here. Through the years, quantum mechanics has been so successful that now it stands beside the classical laws as a valid description of nature.

A special problem arises when students first encounter the theory of quantum mechanics. The problem is that quantum concepts are largely mathematical in nature and do not involve the "common sense" quality associated with classical mechanics. At first, many students find quantum concepts difficult, not so much because of the mathematics involved, but because they feel the concepts are somehow divorced from "reality." This is a reasonable reaction, since ideas which we consider to be real or intuitively satisfying are usually based on our own observation. Thus the classical laws of motion are easy to understand because we observe bodies in motion every day. On the other hand, we observe the effects of atoms and electrons only indirectly, and naturally we have very little feeling for what is happening on the atomic scale. It is necessary, therefore, to depend on the facility of the theory to predict experimental results rather than to attempt to force classical analogues onto the nonclassical phenomena of atoms and electrons.

Our approach in this chapter will be to investigate the important experimental observations that led to the quantum theory, and then to indicate how the theory accounts for these observations. Discussions of quantum theory must necessarily be largely qualitative in such a brief presentation, and those topics that are most important to solid state theory will be emphasized here. Several good references for further individual study are given at the end of this chapter.

**2.2
EXPERIMENTAL
OBSERVATIONS**

The experiments that led to the development of quantum theory were concerned with the nature of light and the relation of optical energy to the energies of electrons within atoms. These experiments supplied only indirect evidence of the nature of phenomena on the atomic scale; however, the cumulative results of a number of careful experiments showed clearly that a new theory was needed.

2.2.1   The Photoelectric Effect

An important observation by Planck indicated that radiation from a heated sample is emitted in discrete units of energy, called *quanta;* the energy units were described by $h\nu$, where ν is the frequency of the radiation, and h is a quantity now called Planck's constant ($h = 6.63 \times 10^{-34}$ J-s). Soon after Planck developed this hypothesis, Einstein interpreted an important experiment that clearly demonstrated the discrete nature (*quantization*) of light. This experiment involved absorption of optical energy by the electrons in a metal and the relationship between the amount of energy absorbed and the frequency of the light (Fig. 2-1). Let us suppose that monochromatic light is incident on the surface of a metal plate in a vacuum. The electrons in the metal absorb energy from the light, and some of the electrons receive enough energy to be ejected from the metal surface into the vacuum. This phenomenon is called the *photoelectric effect.* If the energy of the escaping electrons is measured, a plot can be made of the maximum energy as a function of the frequency ν of the incident light (Fig. 2-1b).

Monochromatic – Single color
INCident – Stricking

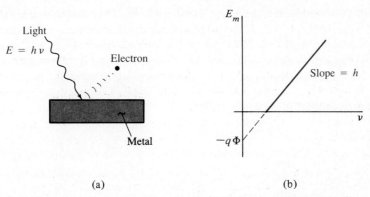

Figure 2-1
The photoelectric
effect: (a) electrons
are ejected from
the surface of a
metal when
exposed to light of
frequency v in a
vacuum; (b) plot of
the maximum
kinetic energy of
ejected electrons
vs. frequency of
the incoming light.

One simple way of finding the maximum energy of the ejected electrons is to place another plate above the one shown in Fig. 2-1a and then create an electric field between the two plates. The potential necessary to retard all electron flow between the plates gives the energy E_m (Prob. 2.1). For a particular frequency of light v incident on the sample, a maximum energy E_m is observed for the emitted electrons. The resulting plot of E_m vs. v is linear, with a slope equal to Planck's constant. The equation of the line shown in Fig. 2-1b is

$$E_m = hv - q\Phi \qquad (2\text{-}1)$$

where q is the magnitude of the electronic charge. The quantity Φ (volts) is a characteristic of the particular metal used. When Φ is multiplied by the electronic charge, an energy (joules) is obtained which represents the minimum energy required for an electron to escape from the metal into a vacuum. The energy $q\Phi$ is called the *work function* of the metal. These results indicate that the electrons receive an energy hv from the light and lose an amount of energy $q\Phi$ in escaping from the surface of the metal.

This experiment demonstrates clearly that Planck's hypothesis was correct — light energy is contained in discrete units rather than in a continuous distribution of energies. Other experiments also indicate that, in addition to the wave nature of light, the quantized units of light energy can be considered as localized packets of energy, called *photons*. Some experiments emphasize the wave nature of light, while other experiments reveal the discrete nature of photons. This duality is fundamental to quantum processes and does not imply an ambiguity in the theory.

2.2.2   Atomic Spectra

One of the most valuable experiments of modern physics is the analysis of absorption and emission of light by atoms. For example, an electric discharge can be created in a gas, so that the atoms begin to emit light with wavelengths characteristic of the gas. We see this effect in a neon sign, which is typically a glass tube filled with neon or a gas mixture, with electrodes for creating a discharge. If the intensity of the emitted light is measured as a function of wavelength, one finds a series of sharp lines rather than a continuous distribution of wavelengths. By the early 1900s the characteristic spectra for several atoms were well known. A portion of the measured emission spectrum for hydrogen is shown in Fig. 2-2, in which the vertical lines represent the positions of observed emission peaks on the wavelength scale. Wavelength (λ) is usually measured in angstroms (1 Å $= 10^{-10}$ m) and is related (in meters) to frequency by $\lambda = c/\nu$, where c is the speed of light (3×10^8 m/s). Photon energy $h\nu$ is then related to wavelength by

$$E = h\nu = \frac{hc}{\lambda} \tag{2-2}$$

Figure 2-2
Some important lines in the emission spectrum of hydrogen.

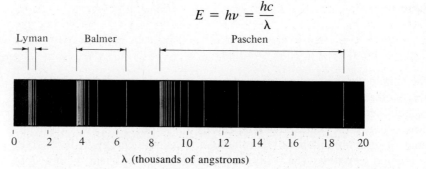

λ (thousands of angstroms)

The lines in Fig. 2-2 appear in several groups labeled the *Lyman, Balmer,* and *Paschen* series after their early investigators. Once the hydrogen spectrum was established, scientists noticed several interesting relationships among the lines. The various series in the spectrum were observed to follow certain empirical forms:

$$\text{Lyman:} \quad \nu = cR\left(\frac{1}{1^2} - \frac{1}{n^2}\right), \qquad n = 2, 3, 4, \ldots \tag{2-3a}$$

$$\text{Balmer:} \quad \nu = cR\left(\frac{1}{2^2} - \frac{1}{n^2}\right), \qquad n = 3, 4, 5, \ldots \tag{2-3b}$$

$$\text{Paschen:} \quad \nu = cR\left(\frac{1}{3^2} - \frac{1}{n^2}\right), \qquad n = 4, 5, 6, \ldots \tag{2-3c}$$

where R is a constant called the Rydberg constant ($R = 109{,}678$ cm^{-1}). If the photon energies $h\nu$ are plotted for successive values of the integer n, we notice that each energy can be obtained by taking sums and differences of other photon energies in the spectrum (Fig. 2-3). For example, E_{42} in the Balmer series

Figure 2-3
Relationships among photon energies in the hydrogen spectrum.

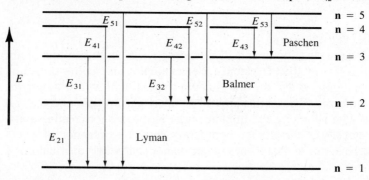

is the difference between E_{41} and E_{21} in the Lyman series. This relationship among the various series is called the *Ritz combination principle*. Naturally, these empirical observations stirred a great deal of interest in constructing a comprehensive theory for the origin of the photons given off by atoms.

The results of emission spectra experiments led Niels Bohr to construct a model for the hydrogen atom, based on the mathematics of planetary systems. If the electron in the hydrogen atom has a series of planetary-type orbits available to it, it can be excited to an outer orbit and then can fall to any one of the inner orbits, giving off energy corresponding to one of the lines of Fig. 2-3. To develop the model, Bohr made several postulates:

**2.3
THE BOHR MODEL**

1. Electrons exist in certain stable, circular orbits about the nucleus. This assumption implies that the orbiting electron does not give off radiation as classical electromagnetic theory would normally require of a charge experiencing angular acceleration; otherwise, the electron would not be stable in the orbit but would spiral into the nucleus as it lost energy by radiation.

2. The electron may shift to an orbit of higher or lower energy, thereby gaining or losing energy equal to the difference in the energy levels (by absorption or emission of a photon of energy $h\nu$).

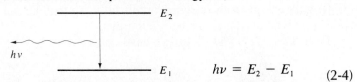

$$h\nu = E_2 - E_1 \qquad (2\text{-}4)$$

3. The angular momentum $\mathbf{p}_\theta$ of the electron in an orbit is always an integral multiple of Planck's constant divided by 2π ($h/2\pi$ is often abbreviated $\hbar$ for convenience). This assumption,

$$\mathbf{p}_\theta = \mathbf{n}\hbar, \qquad \mathbf{n} = 1, 2, 3, 4, \ldots \qquad (2\text{-}5)$$

is necessary to obtain the observed results of Fig. 2-3.

If we visualize the electron in a stable orbit of radius r about the proton of the hydrogen atom, we can equate the electrostatic force between the charges to the centripetal force:

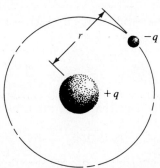

$$-\frac{q^2}{Kr^2} = -\frac{mv^2}{r} \qquad (2\text{-}6)$$

where $K = 4\pi\epsilon_0$ in MKS units, m is the mass of the electron, and v is its velocity. From assumption 3 we have

$$p_\theta = mvr = n\hbar \qquad (2\text{-}7)$$

Since n takes on integral values, r should be denoted by r_n to indicate the nth orbit. Then Eq. (2-7) can be written

$$m^2v^2 = \frac{n^2\hbar^2}{r_n^2} \qquad (2\text{-}8)$$

Substituting Eq. (2-8) into Eq. (2-6) we find that

$$\frac{q^2}{Kr_n^2} = \frac{1}{mr_n} \cdot \frac{n^2\hbar^2}{r_n^2} \qquad (2\text{-}9)$$

$$r_n = \frac{Kn^2\hbar^2}{mq^2} \qquad (2\text{-}10)$$

for the radius of the nth orbit of the electron. Now we must find the expression for the total energy of the electron in this orbit, so that we can calculate the energies involved in transitions between orbits.

From Eqs. (2-7) and (2-10) we have

$$v = \frac{n\hbar}{mr_n} \qquad (2\text{-}11)$$

$$v = \frac{n\hbar q^2}{Kn^2\hbar^2} = \frac{q^2}{Kn\hbar} \qquad (2\text{-}12)$$

Therefore, the kinetic energy of the electron is

$$\text{K.E.} = \frac{1}{2}mv^2 = \frac{mq^4}{2K^2n^2\hbar^2} \qquad (2\text{-}13)$$

The potential energy is the product of the electrostatic force and the distance between the charges:

$$\text{P.E.} = -\frac{q^2}{Kr_n} = -\frac{mq^4}{K^2n^2\hbar^2} \qquad (2\text{-}14)$$

Thus the total energy of the electron in the nth orbit is

$$E_n = \text{K.E.} + \text{P.E.} = -\frac{mq^4}{2K^2n^2\hbar^2} \qquad (2\text{-}15)$$

The critical test of the model is whether energy differences between orbits correspond to the observed photon energies of the hydrogen spectrum. The transitions between orbits corresponding to the Lyman, Balmer, and Paschen series are illustrated in Fig. 2-4. The energy difference between orbits $\mathbf{n}_1$ and $\mathbf{n}_2$ is given by

$$E_{n2} - E_{n1} = \frac{mq^4}{2K^2\hbar^2} \left(\frac{1}{\mathbf{n}_1^2} - \frac{1}{\mathbf{n}_2^2} \right) \qquad (2\text{-}16)$$

The frequency of light given off by a transition between these two orbits is

$$\nu_{21} = \left[\frac{mq^4}{2K^2\hbar^2 h} \right] \left(\frac{1}{\mathbf{n}_1^2} - \frac{1}{\mathbf{n}_2^2} \right) \qquad (2\text{-}17)$$

The factor in brackets is essentially the Rydberg constant R times the speed of light c (Prob. 2.4). A comparison of Eq. (2-17) with the experimental results summed up by Eq. (2-3) indicates that the Bohr theory provides a good model for electronic transitions within the hydrogen atom, as far as the early experimental evidence is concerned.

Whereas the Bohr model accurately describes the gross features of the hydrogen spectrum, it does not include many fine points. For example, experimental evidence indicates some splitting of levels in additon to the levels predicted by the theory. Also, difficulties arise in extending the model to atoms more complicated than hydrogen. Attempts were made to modify the Bohr model for more general cases, but it soon became obvious that a more comprehensive theory was needed. However, the partial success of the Bohr model was an important step toward the eventual development of the quantum theory. The concept that electrons are quantized in certain allowed energy levels, and the relationship of photon energy and transitions between levels had been established firmly by the Bohr theory.

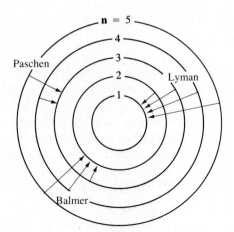

Figure 2-4
Electron orbits and transitions in the Bohr model of the hydrogen atom. Orbit spacing is not drawn to scale.

The principles of quantum mechanics were developed from two different points of view at about the same time (the late 1920s). One approach, developed by Heisenberg, utilizes the mathematics of matrices and is called *matrix mechanics*. Independently, Schrödinger developed an approach utilizing a wave equation, now called *wave mechanics*. These two mathematical formulations appear to be quite different. However, closer examination reveals that beyond the formalism, the basic principles of the two approaches are the same. It is possible to show, for example, that the results of matrix mechanics reduce to those of wave mechanics after mathematical manipulation. We shall concentrate here on the wave mechanics approach, since solutions to a few simple problems can be obtained with it, involving less mathematical discussion.

2.4.1 Probability and the Uncertainty Principle

It is impossible to describe with absolute precision events involving individual particles on the atomic scale. Instead, we must speak of the average values (*expectation values*) of position, momentum, and energy of a particle such as an electron. It is important to note, however, that the uncertainties revealed in quantum calculations are not based on some shortcoming of the theory. In fact, a major strength of the theory is that it describes the probabilistic nature of events involving atoms and electrons. The fact is that such quantities as the position and momentum of an electron *do not exist* apart from a particular uncertainty. The magnitude of this inherent uncertainty is described by the *Heisenberg uncertainty principle:*[†]

> In any measurement of the position and momentum of a particle, the uncertainties in the two measured quantities will be related by

$$(\Delta x)(\Delta p_x) \geq \hbar \qquad (2\text{-}18)$$

> Similarly, the uncertainties in an energy measurement will be related to the uncertainty in the time at which the measurement was made by

$$(\Delta E)(\Delta t) \geq \hbar \qquad (2\text{-}19)$$

These limitations indicate that simultaneous measurement of position and momentum or of energy and time are inherently inaccurate to some degree. Of course, Planck's constant h is a rather small number (6.63×10^{-34} J-s), and we are not concerned with this inaccuracy in the measurement of x and p_x for a truck, for example. On the other hand, measurements of the position of an electron and its speed are seriously limited by the uncertainty principle.

One implication of the uncertainty principle is that we cannot properly speak of *the* position of an electron, for example, but must look for the *proba-*

[†]This is often called the *principle of indeterminacy*.

bility of finding an electron at a certain position. Thus one of the important results of quantum mechanics is that a *probability density function* can be obtained for a particle in a certain environment, and this function can be used to find the expectation value of important quantities such as position, momentum, and energy. We are familiar with the methods for calculating discrete (single-valued) probabilities from common experience. For example, it is clear that the probability of drawing a particular card out of a random deck is $\frac{1}{52}$, and the probability that a tossed coin will come up heads is $\frac{1}{2}$. The techniques for making predictions when the probability varies are less familiar, however. In such cases it is common to define a probability density function which describes, for example, the probability of finding a particle within a certain volume. Given a probability density function $P(x)$ for a one-dimensional problem, the probability of finding the particle in a range from x to $x + dx$ is $P(x)\,dx$. Since the particle will be *somewhere*, this definition implies that

$$\int_{-\infty}^{\infty} P(x)\,dx = 1 \qquad (2\text{-}20)$$

if the function $P(x)$ is properly chosen. Equation (2-20) is implied by stating that the function $P(x)$ is *normalized* (i.e., the integral equals unity).

To find the average value of a function of x, we need only multiply the value of that function in each increment dx by the probability of finding the particle in that dx and sum over all x. Thus the average value of $f(x)$ is

$$\langle f(x) \rangle = \int_{-\infty}^{\infty} f(x)P(x)\,dx \qquad (2\text{-}21\text{a})$$

If the probability density function is not normalized, this equation should be written

$$\langle f(x) \rangle = \frac{\displaystyle\int_{-\infty}^{\infty} f(x)P(x)\,dx}{\displaystyle\int_{-\infty}^{\infty} P(x)\,dx} \qquad (2\text{-}21\text{b})$$

2.4.2   The Schrödinger Wave Equation

There are several ways to develop the wave equation by applying quantum concepts to various classical equations of mechanics. One of the simplest approaches is to consider a few basic postulates, develop the wave equation from them, and rely on the accuracy of the results to serve as a justification of the postulates. In more advanced texts these assumptions are dealt with in more convincing detail.

Basic Postulates

1. Each particle in a physical system is described by a wave function $\Psi(x, y, z, t)$. This function and its space derivative ($\partial\Psi/\partial x + \partial\Psi/\partial y + \partial\Psi/\partial z$) are continuous, finite, and single valued.

2. In dealing with classical quantities such as energy E and momentum p, we must relate these quantities with abstract quantum mechanical operators defined in the following way:

Classical variable	Quantum operator
x	x
$f(x)$	$f(x)$
$p(x)$	$\dfrac{\hbar}{j}\dfrac{\partial}{\partial x}$
E	$-\dfrac{\hbar}{j}\dfrac{\partial}{\partial t}$

and similarly for the other two directions.

3. The probability of finding a particle with wave function Ψ in the volume $dx\,dy\,dz$ is $\Psi^*\Psi\,dx\,dy\,dz$.[†] The product $\Psi^*\Psi$ is normalized according to Eq. (2-20) so that

$$\int_{-\infty}^{\infty} \Psi^*\Psi\,dx\,dy\,dz = 1$$

and the average value $\langle Q \rangle$ of any variable Q is calculated from the wave function by using the operator form Q_{op} defined in postulate 2:

$$\langle Q \rangle = \int_{-\infty}^{\infty} \Psi^* Q_{op} \Psi\,dx\,dy\,dz$$

Once we find the wave function Ψ for a particle, we can calculate its average position, energy, and momentum, within the limits of the uncertainty principle. Thus a major part of the effort in quantum calculations involves solving for Ψ within the conditions imposed by a particular physical system. We notice from assumption 3 that the probability density function is $\Psi^*\Psi$, or $|\Psi|^2$.

The classical equation for the energy of a particle can be written

$$\text{Kinetic energy + potential energy = total energy}$$

$$\frac{1}{2m}p^2 \quad + \quad V \quad = \quad E \tag{2-22}$$

In quantum mechanics we use the operator form for these variables (postulate 2); the operators are allowed to operate on the wave function Ψ. For a one-dimensional problem Eq. (2-22) becomes[‡]

$$-\frac{\hbar^2}{2m}\frac{\partial^2\Psi(x,t)}{\partial x^2} + V(x)\Psi(x,t) = -\frac{\hbar}{j}\frac{\partial\Psi(x,t)}{\partial t} \tag{2-23}$$

[†] Ψ^* is the *complex conjugate* of Ψ, obtained by reversing the sign on each j. Thus $(e^{jx})^* = e^{-jx}$.

[‡] The operational interpretation of $(\partial/\partial x)^2$ is the second derivative form $\partial^2/\partial x^2$; the square of j is -1.

which is the Schrödinger wave equation. In three dimensions the equation is

$$-\frac{\hbar^2}{2m}\nabla^2\Psi + V\Psi = -\frac{\hbar}{j}\frac{\partial\Psi}{\partial t} \qquad (2\text{-}24)$$

where $\nabla^2\Psi$ is

$$\frac{\partial^2\Psi}{\partial x^2} + \frac{\partial^2\Psi}{\partial y^2} + \frac{\partial^2\Psi}{\partial z^2}$$

The wave function Ψ in Eqs. (2-23) and (2-24) includes both space and time dependencies. It is common to calculate these dependencies separately and combine them later. Furthermore, many problems are time independent, and only the space variables are necessary. Thus we try to solve the wave equation by breaking it into two equations by the technique of separation of variables. Let $\Psi(x, t)$ be represented by the product $\psi(x)\phi(t)$. Using this product in Eq. (2-23) we obtain

$$-\frac{\hbar^2}{2m}\frac{\partial^2\psi(x)}{\partial x^2}\phi(t) + V(x)\psi(x)\phi(t) = -\frac{\hbar}{j}\psi(x)\frac{\partial\phi(t)}{\partial t} \qquad (2\text{-}25)$$

Now the variables can be separated (Prob. 2.8) to obtain the time-dependent equation in one dimension,

$$\frac{d\phi(t)}{dt} + \frac{jE}{\hbar}\phi(t) = 0 \qquad (2\text{-}26)$$

and the time-independent equation,

$$\frac{d^2\psi(x)}{dx^2} + \frac{2m}{\hbar^2}[E - V(x)]\psi(x) = 0 \qquad (2\text{-}27)$$

We can show that the separation constant E corresponds to the energy of the particle when particular solutions are obtained, such that a wave function ψ_n corresponds to a particle energy E_n.

These equations are the basis of wave mechanics. From them we can determine the wave functions for particles in various simple systems. For calculations involving electrons, the potential term $V(x)$ usually results from an electrostatic or magnetic field.

2.4.3   Potential Well Problem

It is quite difficult to find solutions to the Schrödinger equation for most realistic potential fields. One can solve the problem with some effort for the hydrogen atom, for example, but solutions for more complicated atoms are hard to obtain. There are several important problems, however, which illustrate the theory without complicated manipulation. The simplest problem is the potential energy well with infinite boundaries. Let us assume a particle is trapped in a potential well with $V(x)$ zero except at the boundaries $x = 0$ and L, where it is infinitely large (Fig. 2-5a)

$$V(x) = 0, \qquad 0 < x < L$$
$$V(x) = \infty, \qquad x = 0, L \tag{2-28}$$

Inside the well we set $V(x) = 0$ in Eq. (2-27)

$$\frac{d^2\psi(x)}{dx^2} + \frac{2m}{\hbar^2} E\psi(x) = 0, \qquad 0 < x < L \tag{2-29}$$

This is the wave equation for a free particle; it applies to the potential well problem in the region with no potential $V(x)$.

Possible solutions to Eq. (2-29) are $\sin kx$ and $\cos kx$, where k is $\sqrt{2mE}/\hbar$. In choosing a solution, however, we must examine the boundary conditions. The only allowable value of ψ at the walls is zero. Otherwise, there would be a nonzero $|\psi|^2$ outside the potential well, which is impossible because a particle cannot penetrate an infinite barrier. Therefore, we must choose only the sine solution and define k such that $\sin kx$ goes to zero at $x = L$.

$$\psi = A \sin kx, \qquad k = \frac{\sqrt{2mE}}{\hbar} \tag{2-30}$$

The constant A is the amplitude of the wave function and will be evaluated from the normalization condition (postulate 3). If ψ is to be zero at $x = L$, then k must be some integral multiple of π/L.

$$k = \frac{n\pi}{L}, \qquad n = 1, 2, 3, \ldots \tag{2-31}$$

From Eqs. (2-30) and (2-31) we can solve for the total energy E_n for each value of the integer n.

$$\frac{\sqrt{2mE_n}}{\hbar} = \frac{n\pi}{L} \tag{2-32}$$

$$E_n = \frac{n^2\pi^2\hbar^2}{2mL^2} \tag{2-33}$$

Figure 2-5
The problem of a particle in a potential well: (a) potential energy diagram; (b) wave functions in the first three quantum states; (c) probability density distribution for the second state.

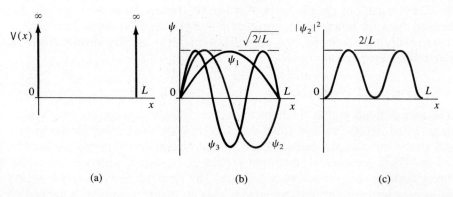

(a) (b) (c)

Thus for each allowable value of **n** the particle energy is described by Eq. (2-33). We notice that *the energy is quantized*. Only certain values of energy are allowed. The integer **n** is called a *quantum number;* the particular wave function ψ_n and corresponding energy state E_n describe the *quantum state* of the particle.

The quantized energy levels described by Eq. (2-33) appear in a variety of small-geometry structures encountered in semiconductor devices. We shall return to this potential well problem (often called the "particle in a box" problem) in later discussions.

The constant A is found from postulate 3.

$$\int_{-\infty}^{\infty} \psi^*\psi \, dx = \int_0^L A^2 \left(\sin \frac{\mathbf{n}\pi}{L} x\right)^2 dx = A^2 \frac{L}{2} \qquad (2\text{-}34)$$

Setting Eq. (2-34) equal to unity we obtain

$$A = \sqrt{\frac{2}{L}}, \qquad \psi_n = \sqrt{\frac{2}{L}} \sin \frac{\mathbf{n}\pi}{L} x \qquad (2\text{-}35)$$

The first three wave functions ψ_1, ψ_2, ψ_3, are sketched in Fig. 2-5b. The probability density function $\psi^*\psi$, or $|\psi|^2$, is sketched for ψ_2 in Fig. 2-5c.

2.4.4. Tunneling

The wave functions are relatively easy to obtain for the potential well with infinite walls, since the boundary conditions force ψ to zero at the walls. A slight modification of this problem illustrates a principle that is very important in some solid state devices — the quantum mechanical *tunneling* of an electron through a barrier of finite height and thickness. Let us consider the potential barrier of Fig. 2-6. If the barrier is not infinite, the boundary conditions do not

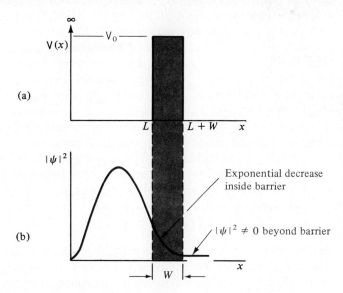

(a)

(b)

Figure 2-6
Quantum mechanical tunneling:
(a) potential barrier of height V_0 and thickness W;
(b) probability density for an electron with energy $E < V_0$, indicating a nonzero value of the wave function beyond the barrier.

force ψ to zero at the barrier. Instead, we must use the condition that ψ and its slope $d\psi/dx$ are continuous at each boundary of the barrier (postulate 1). Thus ψ must have a nonzero value within the barrier and also on the other side. Since ψ has a value to the right of the barrier, $\psi^*\psi$ exists there also, implying that there is some probability of finding the particle beyond the barrier. We notice that the particle does not go over the barrier; its total energy is assumed to be less than the barrier height V_0. The mechanism by which the particle "penetrates" the barrier is called tunneling. However, no classical analogue, including classical descriptions of tunneling through barriers, is appropriate for this effect. Quantum mechanical tunneling is intimately bound to the uncertainty principle. If the barrier is sufficiently thin, we cannot say with certainty that the particle exists only on one side. However, the wave function amplitude for the particle is reduced by the barrier as Fig. 2-6 indicates, so that by making the thickness W greater, we can reduce ψ on the right-hand side to the point that negligible tunneling occurs. Tunneling is important only over very small dimensions, but it can be of great importance in the conduction of electrons in solids, as we shall see in Chapters 5 and 6.

Recently, a novel electronic device called the resonant tunneling diode was developed. This device operates by tunneling electrons through "particle in a potential well" energy levels of the type described in Section 2.4.3. We discuss resonant tunneling in Chapter 12.

2.5 ATOMIC STRUCTURE AND THE PERIODIC TABLE

The Schrödinger equation describes accurately the interactions of particles with potential fields, such as electrons within atoms. Indeed, the modern understanding of atomic theory (the modern atomic *models*) comes from the wave equation and from Heisenberg's matrix mechanics. It should be pointed out, however, that the problem of solving the Schrödinger equation directly for complicated atoms is extremely difficult. In fact, only the hydrogen atom is generally solved directly; atoms of atomic number greater than one are usually handled by techniques involving approximations. Many atoms such as the alkali metals (Li, Na, etc.), which have a neutral core with a single electron in an outer orbit, can be treated by a rather simple extension of the hydrogen atom results. The hydrogen atom solution is also important in identifying the basic selection rules for describing allowed electron energy levels. These quantum mechanical results must coincide with the experimental spectra, and we expect the energy levels to include those predicted by the Bohr model. Without actually working through the mathematics for the hydrogen atom, in this section we shall investigate the energy level schemes dictated by the wave equation.

2.5.1 The Hydrogen Atom

Finding the wave functions for the hydrogen atom requires a solution of the Schrödinger equation in three dimensions for a coulombic potential field. Since the problem is spherically symmetric, the spherical coordinate system is used

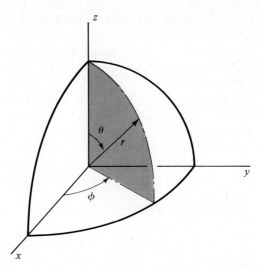

Figure 2-7
The spherical
coordinate system.

in the calculation (Fig. 2-7). The term $V(x, y, z)$ in Eq. (2-24) must be replaced by $V(r, \theta, \phi)$, representing the Coulomb potential which the electron experiences in the vicinity of the proton. The Coulomb potential varies only with r in spherical coordinates

$$V(r, \theta, \phi) = V(r) = -(4\pi\epsilon_0)^{-1}\frac{q^2}{r} \qquad (2\text{-}36)$$

as in Eq. (2-14).

When the separation of variables is made, the time-independent equation can be written as

$$\psi(r, \theta, \phi) = R(r)\Theta(\theta)\Phi(\phi) \qquad (2\text{-}37)$$

Thus the wave functions are found in three parts. Separate solutions must be obtained for the r-dependent equation, the θ-dependent equation, and the ϕ-dependent equation. After these three equations are solved, the total wave function ψ is obtained from the product.

As in the simple potential well problem, each of the three hydrogen atom equations gives a solution which is quantized. Thus we would expect a quantum number to be associated with *each* of the three parts of the wave equation. As an illustration, the ϕ-dependent equation obtained after separation of variables is

$$\frac{d^2\Phi}{d\phi^2} + \mathbf{m}^2\Phi = 0 \qquad (2\text{-}38)$$

where $\mathbf{m}$ is a quantum number. The solution to this equation is

$$\Phi_m(\phi) = Ae^{jm\phi} \qquad (2\text{-}39)$$

where A can be evaluated by the normalization condition, as before:

$$\int_0^{2\pi} \Phi_m^*(\phi)\Phi_m(\phi)\,d\phi = 1 \tag{2-40}$$

$$A^2 \int_0^{2\pi} e^{-jm\phi}e^{jm\phi}\,d\phi = A^2 \int_0^{2\pi} d\phi = 2\pi A^2 \tag{2-41}$$

Thus the value of A is

$$A = \frac{1}{\sqrt{2\pi}} \tag{2-42}$$

and the ϕ-dependent wave function is

$$\Phi_m(\phi) = \frac{1}{\sqrt{2\pi}}e^{jm\phi} \tag{2-43}$$

Since values of ϕ repeat every 2π radians, Φ should repeat also. This occurs if **m** is an integer, including negative integers and zero (Prob. 2.10). Thus the wave functions for the ϕ-dependent equation are quantized with the following selection rule for the quantum numbers:

$$\mathbf{m} = \ldots, -3, -2, -1, 0, +1, +2, +3, \ldots \tag{2-44}$$

By similar treatments, the functions $R(r)$ and $\Theta(\theta)$ can be obtained, each being quantized by its own selection rule. For the r-dependent equation, the quantum number **n** can be any positive integer (not zero), and for the θ-dependent equation the quantum number l can be zero or a positive integer. However, there are interrelationships among the equations which restrict the various quantum numbers used with a single wave function ψ_{nlm}:

$$\psi_{nlm}(r, \theta, \phi) = R_n(r)\Theta_l(\theta)\Phi_m(\phi) \tag{2-45}$$

These restrictions are summarized as follows:

$$\mathbf{n} = 1, 2, 3, \ldots \tag{2-46a}$$

$$l = 0, 1, 2, \ldots, (\mathbf{n} - 1) \tag{2-46b}$$

$$\mathbf{m} = -l, \ldots, -2, -1, 0, +1, +2, \ldots, +l \tag{2-46c}$$

In addition to the three quantum numbers arising from the three parts of the wave equation, there is an important quantization condition on the "spin" of the electron. Investigations of electron spin employ the theory of relativity as well as quantum mechanics; therefore, we shall simply state that the intrinsic angular momentum **s** of an electron with ψ_{nlm} specified is

$$\mathbf{s} = \pm\frac{\hbar}{2} \tag{2-47}$$

That is, in units of $\hbar$, the electron has a spin of $\frac{1}{2}$, and the angular momentum produced by this spin is positive or negative depending on whether the electron is "spin up" or "spin down." The important point for our discussion is that each allowed energy state of the electron in the hydrogen atom is uniquely described by four quantum numbers: **n**, l, **m**, and **s**.[†]

Using these four quantum numbers, we can identify the various states which the electron can occupy in a hydrogen atom. The number **n**, called the *principal* quantum number, specifies the "orbit" of the electron in Bohr terminology. Of course, the concept of orbit is replaced by probability density functions in quantum mechanical calculations. It is common to refer to states with a given principal quantum number as belonging to a *shell* rather than an orbit.

There is considerable fine structure in the energy levels about the Bohr orbits, due to the dictates of the other three quantum conditions. For example, an electron with **n** = 1 (the first Bohr orbit) can have only l = 0 and **m** = 0 according to Eq. (2-46), but there are two spin states allowed from Eq. (2-47). For **n** = 2, l can be 0 or 1, and **m** can be −1, 0, or +1. The various allowed combinations of quantum numbers appear in the first four columns of Table 2-1. From these combinations it is apparent that the electron in a hydrogen atom can occupy any one of a large number of excited states in addition to the lowest (*ground*) state ψ_{100}. Energy differences between the various states properly account for the observed lines in the hydrogen spectrum.

2.5.2 The Periodic Table

The quantum numbers discussed in Section 2.5.1 arise from solutions to the hydrogen atom problem. Thus the energies obtainable from the wave functions are unique to the hydrogen atom and cannot be extended to more complicated atoms without appropriate alterations. However, the quantum number selection rules are valid for more complicated structures, and we can use these rules to gain an understanding of the arrangement of atoms in the periodic table of chemical elements. Without these selection rules, it is difficult to understand why only two electrons fit into the first Bohr orbit of an atom, whereas eight electrons are allowed in the second orbit. After even the brief discussion of quantum numbers given above, we should be able to answer these questions with more insight.

Before discussing the periodic table, we must be aware of an important principle of quantum theory, the *Pauli exclusion principle*. This rule states that no two electrons in an interacting system[‡] can have the same set of quantum numbers **n**, l, **m**, **s**. In other words, only two electrons can have the same three quantum numbers **n**, l, **m**, and those two must have opposite spin. The importance of this principle cannot be overemphasized; it is basic to the electronic structure of all atoms in the periodic table. One implication of this principle is

---

[†] In many texts the numbers we have called **m** and **s** are referred to as $\mathbf{m}_l$ and $\mathbf{m}_s$, respectively.

[‡] An interacting system is one in which electron wave functions overlap — in this case an atom with two or more electrons.

Table 2-1. Quantum numbers to $n = 3$ and allowable states for the electron in a hydrogen atom: The first four columns show the various combinations of quantum numbers allowed by the selection rules of Eq. (2-46); the last two columns indicate the number of allowed states (combinations of n, l, m, and s) for each l (*subshell*) and n (shell, or Bohr orbit).

n	l	m	$s/\hbar$	Allowable states in subshell	Allowable states in complete shell
1	0	0	$\pm\frac{1}{2}$	2	2
2	0	0	$\pm\frac{1}{2}$	2	8
	1	-1	$\pm\frac{1}{2}$		
		0	$\pm\frac{1}{2}$	6	
		1	$\pm\frac{1}{2}$		
3	0	0	$\pm\frac{1}{2}$	2	18
	1	-1	$\pm\frac{1}{2}$		
		0	$\pm\frac{1}{2}$	6	
		1	$\pm\frac{1}{2}$		
	2	-2	$\pm\frac{1}{2}$		
		-1	$\pm\frac{1}{2}$		
		0	$\pm\frac{1}{2}$	10	
		1	$\pm\frac{1}{2}$		
		2	$\pm\frac{1}{2}$		

that by listing the various combinations of quantum numbers, we can determine into which shell each electron of a complicated atom fits, and how many electrons are allowed per shell. The quantum states summarized in Table 2-1 can be used to indicate the electronic configurations for atoms in the lowest energy state.

In the first electronic shell ($n = 1$), l can be only zero since the maximum value of l is always $n - 1$. Similarly, m can be only zero since m runs from the negative value of l to the positive value of l. Two electrons with opposite spin can fit in this ψ_{100} state; therefore, the first shell can have at most two electrons. For the helium atom (atomic number $Z = 2$) in the ground state, both electrons will be in the first Bohr orbit ($n = 1$), both will have $l = 0$ and $m = 0$, and they will have opposite spin. Of course, one or both of the He atom electrons can be excited to one of the higher energy states of Table 2-1 and subsequently relax to the ground state, giving off a photon characteristic of the He spectrum.

As Table 2-1 indicates, there can be two electrons in the $l = 0$ subshell, six electrons when $l = 1$, and ten electrons for $l = 2$. The electronic configurations of various atoms in the periodic table can be deduced from this list of

Table 2-2. Electronic configurations for atoms in the ground state.

Atomic number (Z)	Element	$1s$	$2s$	$2p$	$3s$	$3p$	$3d$	$4s$	$4p$	Shorthand notation
		n = 1	**2**		**3**			**4**		
		l = 0	**0**	**1**	**0**	**1**	**2**	**0**	**1**	
					Number of electrons					
1	H	1								$1s^1$
2	He	2								$1s^2$
3	Li		1							$1s^2\ 2s^1$
4	Be		2							$1s^2\ 2s^2$
5	B		2	1						$1s^2\ 2s^2\ 2p^1$
6	C	helium core,	2	2						$1s^2\ 2s^2\ 2p^2$
7	N	2 electrons	2	3						$1s^2\ 2s^2\ 2p^3$
8	O		2	4						$1s^2\ 2s^2\ 2p^4$
9	F		2	5						$1s^2\ 2s^2\ 2p^5$
10	Ne		2	6						$1s^2\ 2s^2\ 2p^6$
11	Na				1					[Ne] $3s^1$
12	Mg				2					$3s^2$
13	Al				2	1				$3s^2\ 3p^1$
14	Si	neon core,			2	2				$3s^2\ 3p^2$
15	P	10 electrons			2	3				$3s^2\ 3p^3$
16	S				2	4				$3s^2\ 3p^4$
17	Cl				2	5				$3s^2\ 3p^5$
18	Ar				2	6				$3s^2\ 3p^6$
19	K							1		[Ar] $4s^1$
20	Ca							2		$4s^2$
21	Sc						1	2		$3d^1\ 4s^2$
22	Ti						2	2		$3d^2\ 4s^2$
23	V						3	2		$3d^3\ 4s^2$
24	Cr						5	1		$3d^5\ 4s^1$
25	Mn						5	2		$3d^5\ 4s^2$
26	Fe						6	2		$3d^6\ 4s^2$
27	Co	argon core,					7	2		$3d^7\ 4s^2$
28	Ni	18 electrons					8	2		$3d^8\ 4s^2$
29	Cu						10	1		$3d^{10}\ 4s^1$
30	Zn						10	2		$3d^{10}\ 4s^2$
31	Ga						10	2	1	$3d^{10}\ 4s^2\ 4p^1$
32	Ge						10	2	2	$3d^{10}\ 4s^2\ 4p^2$
33	As						10	2	3	$3d^{10}\ 4s^2\ 4p^3$
34	Se						10	2	4	$3d^{10}\ 4s^2\ 4p^4$
35	Br						10	2	5	$3d^{10}\ 4s^2\ 4p^5$
36	Kr						10	2	6	$3d^{10}\ 4s^2\ 4p^6$

allowed states. The ground state electron structures for a number of atoms are listed in Table 2-2. There is a simple shorthand notation for electronic structures which is commonly used instead of such a table. The only new convention to remember in this notation is the naming of the l values:

$$l = 0, 1, 2, 3, 4, \ldots$$

$$s, p, d, f, g, \ldots$$

This convention was created by early spectroscopists who referred to the first four spectral groups as *s*harp, *p*rincipal, *d*iffuse, and *f*undamental. Alphabetical order is used beyond f. With this convention for l, we can write an electron state as follows:

$$\underset{(\mathbf{n} = 3)}{\overset{\text{6 electrons in the } 3p \text{ subshell}}{3p^6}} \quad (l = 1)$$

For example, the total electronic configuration for Si ($Z = 14$) in the ground state is

$$1s^2 2s^2 2p^6 3s^2 3p^2$$

We notice that Si has a closed Ne configuration (see Table 2-2) plus four electrons in an outer $\mathbf{n} = 3$ orbit ($3s^2 3p^2$). These are the four valence electrons of Si; two valence electrons are in an s state and two are in a p state. The Si electronic configuration can be written [Ne] $3s^2 3p^2$ for convenience, since the Ne configuration $1s^2 2s^2 2p^6$ forms a closed shell (typical of the inert elements). The column IV semiconductor Ge ($Z = 32$) has an electronic structure similar to Si, except that the four valence electrons are outside a closed $\mathbf{n} = 3$ shell. Thus the Ge configuration is [Ar] $3d^{10} 4s^2 4p^2$. In Chapter 3 we shall see that the four valence electrons of Ge and Si play very important roles in the bonding forces of the diamond lattice and in the conduction of charges through these semiconductors.

There are several cases in Table 2-2 that do not follow the most straightforward choice of quantum numbers. For example, we notice that in K ($Z = 19$) and Ca ($Z = 20$) the $4s$ state is filled before the $3d$ state; in Cr ($Z = 24$) and Cu ($Z = 29$) there is a transfer of an electron back to the $3d$ state. These exceptions, required by minimum energy considerations, are discussed more fully in most atomic physics texts.

PROBLEMS

2.1. (a) Sketch a simple vacuum tube device and the associated circuitry for measuring E_m in the photoelectric effect experiment. The electrodes can be placed in a sealed glass envelope.

(b) Sketch the photocurrent I vs. retarding voltage V that you would expect to measure for a given electrode material and configuration. Make the sketch for several intensities of light at a given wavelength.

(c) The work function of platinum is 4.09 eV. What retarding potential will be required to reduce the photocurrent to zero in a photoelectric experiment with Pt electrodes if the wavelength of incident light is 2440 Å? Remember that an energy of $q\Phi$ is lost by each electron in escaping the surface.

2.2. Calculate the Bohr radius in Å and energy in eV for $\mathbf{n} = 1, 2,$ and 3.

2.3. (a) Show that the various lines in the hydrogen spectrum can be expressed in angstroms as

$$\lambda(\mathring{A}) = \frac{911\mathbf{n}_1^2\mathbf{n}^2}{\mathbf{n}^2 - \mathbf{n}_1^2}$$

where $\mathbf{n}_1 = 1$ for the Lyman series, 2 for the Balmer series, and 3 for the Paschen series. The integer $\mathbf{n}$ is larger than $\mathbf{n}_1$.

(b) Calculate λ for the Lyman series to $\mathbf{n} = 5$, the Balmer series to $\mathbf{n} = 7$, and the Paschen series to $\mathbf{n} = 10$. Plot the results as in Fig. 2-2. What are the wavelength limits for each of the three series?

2.4. Show that the calculated Bohr expression for frequency of emitted light in the hydrogen spectrum, Eq. (2-17), corresponds to the experimental expressions, Eq. (2-3).

2.5. (a) The position of an electron is determined to within 1 Å. What is the minimum uncertainty in its momentum?

(b) An electron's energy is measured with an uncertainty of 1 eV. What is the minimum uncertainty in the time over which the measurement was made?

2.6. Show that Eq. (2-19) corresponds to Eq. (2-18) for a particle with velocity $\mathbf{v}$.

2.7. A sample of radioactive material undergoes decay such that the number of atoms $N(t)$ remaining in the unstable state at time t is related to the number N_0 at $t = 0$ by the relation $N(t) = N_0 \exp(-t/\tau)$. Show that τ is the average lifetime $\langle t \rangle$ of an atom in the unstable state before it spontaneously decays. Equation (2-21b) can be used with t substituted for x.

2.8. Separate the variables in Eq. (2-25) to obtain Eqs. (2-26) and (2-27). What is the solution to Eq. (2-26)? Assuming $\psi(x)$ has a particular time-independent value $\psi_n(x)$, use postulate 3 to show that the corresponding energy equals the separation constant E_n.

2.9. A free electron traveling in the x-direction can be described by a plane wave, with a wave function of the form $\psi_k(x) = Ae^{jk_x x}$, where $\mathbf{k}_x$ is a wave vector, or propagation constant. Use postulate 3 and the momentum operator to relate the electron momentum $\langle \mathbf{p}_x \rangle$ to $\mathbf{k}_x$.

2.10. Show that Eq. (2-44) is a proper choice for $\mathbf{m}$.

2.11. In forming NaCl, each Na gives up an electron to a Cl. What are the electronic configurations of the resulting ions?

READING LIST

Baym, G. *Lectures on Quantum Mechanics.* New York: Benjamin/Cummings, 1969.

Cohen-Tannoudji, C., B. Diu, and F. Laloe. *Quantum Mechanics.* New York: John Wiley, 1977.

Cohen, M. L., V. Heine, and J. C. Phillips. "The Quantum Mechanics of Materials." *Scientific American* 246, no. 6 (June 1982): 82–102.

Datta, S. *The Modular Series on Solid State Devices. Vol. VIII: Quantum Phenomena.* Reading, Mass.: Addison-Wesley, 1987.

Davydov, A. S. *Quantum Mechanics.* Elmsford, N.Y.: Pergamon Press, 1985.

Drummond, T. J., P. L. Gourley, and T. E. Zipperian. "Quantum-Tailored Solid-State Devices." *IEEE Spectrum* (June 1988): 33–37.

Feinberg, G. "Light." *Scientific American* 219, no. 3 (September 1968): 50–59.

Feynman, R. P., R. B. Leighton, and M. Sands. *The Feynman Lectures on Physics. Vol. III: Quantum Mechanics.* Reading, Mass.: Addison-Wesley, 1965.

Kittel, C. *Quantum Theory of Solids.* 2d ed. New York: John Wiley, 1987.

Liboff, R. M. *Introductory Quantum Mechanics.* San Francisco: Holden-Day, 1980.

Park, D. A. *Introduction to the Quantum Theory.* 2d ed. New York: McGraw-Hill, 1974.

Talley, H. E., and D. G. Daugherty. *Physical Principles of Semiconductor Devices.* Ames: Iowa State University Press, 1976.

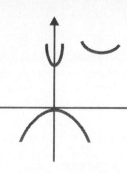

chapter 3

ENERGY BANDS AND CHARGE CARRIERS IN SEMICONDUCTORS

In this chapter we begin to discuss the specific mechanisms by which current flows in a solid. In examining these mechanisms we shall learn why some materials are good conductors of electric current, whereas others are poor conductors. We shall see how the conductivity of a semiconductor can be varied by changing the temperature or the number of impurities. These fundamental concepts of charge transport form the basis for later discussions of solid state device behavior.

In Chapter 2 we found that electrons are restricted to sets of discrete energy levels within atoms. Large gaps exist in the energy scale in which no energy states are available. In a similar fashion, electrons in solids are restricted to certain energies and are not allowed at other energies. The basic difference between the case of an electron in a solid and that of an electron in an isolated atom is that in the solid the electron has a *range,* or *band,* of available energies. The discrete energy levels of the isolated atom spread into bands of energies in the solid because in the solid the wave functions of electrons in neighboring atoms overlap, and an electron is not necessarily localized at a particular atom. Thus, for example, an electron in the outer orbit of one atom feels the influence of neighboring atoms, and its overall wave function is altered. Naturally, this influence affects the potential energy term and the boundary conditions in the Schrödinger equation, and we would expect to obtain different energies in the solution. Usually, the influence of neighboring atoms

3.1
BONDING FORCES AND ENERGY BANDS IN SOLIDS

on the energy levels of a particular atom can be treated as a small perturbation, giving rise to shifting and splitting of energy states into energy bands.

3.1.1 Bonding Forces in Solids

The interaction of electrons in neighboring atoms of a solid serves the very important function of holding the crystal together. For example, alkali halides such as NaCl are typified by *ionic bonding*. In the NaCl lattice, each Na atom is surrounded by six nearest neighbor Cl atoms, and vice versa. Four of the nearest neighbors are evident in the two-dimensional representation shown in Fig. 3-1. The electronic structure of Na ($Z = 11$) is [Ne]$3s^1$, and Cl ($Z = 17$) has the structure [Ne]$3s^2 3p^5$. In the lattice each Na atom gives up its outer $3s$ electron to a Cl atom, so that the crystal is made up of ions with the electronic structures of the inert atoms Ne and Ar (Ar has the electronic structure [Ne]$3s^2 3p^6$). However, the ions have net electric charges after the electron exchange. The Na$^+$ ion has a net positive charge, having lost an electron, and the Cl$^-$ ion has a net negative charge, having gained an electron.

Each Na$^+$ ion exerts an electrostatic attractive force upon its six Cl$^-$ neighbors, and vice versa. These coulombic forces pull the lattice together until a balance is reached with repulsive forces. A reasonably accurate calculation of the atomic spacing can be made by considering the ions as hard spheres being attracted together (Prob. 1.5).

An important observation in the NaCl structure is that all electrons are tightly bound to atoms. Once the electron exchanges have been made between the Na and Cl atoms to form the Na$^+$ and Cl$^-$ ions, the outer orbits of all atoms are completely filled. Since the ions have the closed-shell configurations of the inert atoms Ne and Ar, there are no loosely bound electrons to participate in current flow; as a result, NaCl is a good insulator.

Figure 3-1
The NaCl lattice: an example of ionic bonding.

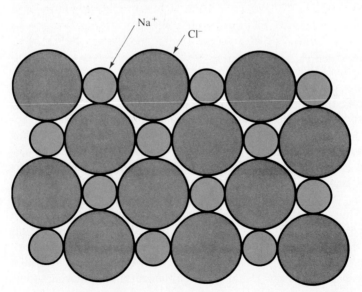

In a metal atom the outer electronic shell is only partially filled, usually by no more than three electrons. We have already noted that the alkali metals (e.g., Na) have only one electron in the outer orbit. This electron is loosely bound and is given up easily in ion formation. This accounts for the great chemical activity of the alkali metals, as well as for their high electrical conductivity. In the metal the outer electron of each alkali atom is contributed to the crystal as a whole, so that the solid is made up of ions with closed shells immersed in a sea of free electrons. The forces holding the lattice together arise from an interaction between the positive ion cores and the surrounding free electrons. This is one type of *metallic bonding*. Obviously, there are complicated differences in the bonding forces for various metals, as evidenced by the wide range of melting temperatures (234 K for Hg, 3643 K for W). However, the metals have the sea of electrons in common, and these electrons are free to move about the crystal under the influence of an electric field.

A third type of bonding is exhibited by the diamond lattice semiconductors. We recall that each atom in the Ge, Si, or C diamond lattice is surrounded by four nearest neighbors, each with four electrons in the outer orbit. In these crystals each atom shares its valence electrons with its four neighbors (Fig. 3-2). Bonding between nearest neighbor atoms is illustrated in the diamond lattice diagram of Fig. 1-9. The bonding forces arise from a quantum mechanical interaction between the shared electrons. This is known as *covalent bonding;* each electron pair constitutes a covalent bond. In the sharing process it is no longer relevant to ask which electron belongs to a particular atom — both belong to the bond. The two electrons are indistinguishable, except that they must have opposite spin to satisfy the Pauli exclusion principle. Covalent bonding is also found in certain molecules, such as H_2.

As in the case of the ionic crystals, no free electrons are available to the lattice in the covalent diamond structure of Fig. 3-2. By this reasoning Ge and Si should also be insulators. However, we have pictured an idealized lattice at 0 K in this figure. As we shall see in subsequent sections, an electron can be thermally or optically excited out of a covalent bond and thereby become free to participate in conduction. This is an important feature of semiconductors.

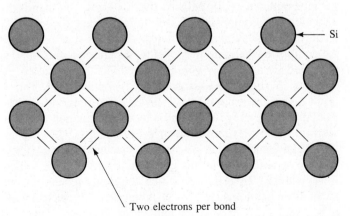

← Si

Figure 3-2
Covalent bonding in the Si crystal, viewed along a ⟨100⟩ direction (see also Figs. 1-8 and 1-9).

Two electrons per bond

Compound semiconductors such as GaAs have mixed bonding, in which both ionic and covalent bonding forces participate. Some ionic bonding is to be expected in a crystal such as GaAs because of the difference in placement of the Ga and As atoms in the periodic table. The ionic character of the bonding becomes more important as the atoms of the compound become further separated in the periodic table, as in the II–VI compounds.

3.1.2   Energy Bands

As isolated atoms are brought together to form a solid, various interactions occur between neighboring atoms, including those described in the preceding section. The forces of attraction and repulsion between atoms will find a balance at the proper interatomic spacing for the crystal. In the process, important changes occur in the electron energy level configurations, and these changes result in the varied electrical properties of solids.

Qualitatively, we can see that as atoms are brought together, the application of the Pauli exclusion principle becomes important. When two atoms are completely isolated from each other so that there is no interaction of electron wave functions between them, they can have identical electronic structures. As the spacing between the two atoms becomes smaller, however, electron wave functions begin to overlap. The exclusion principle dictates that no two electrons in a given interacting system may have the same quantum state; thus there must be a splitting of the discrete energy levels of the isolated atoms into new levels belonging to the pair rather than to individual atoms.

In a solid, many atoms are brought together, so that the split energy levels form essentially continuous *bands* of energies. As an example, Fig. 3-3 illustrates the imaginary formation of a diamond crystal from isolated carbon atoms. Each isolated carbon atom has an electronic structure $1s^2 2s^2 2p^2$ in the ground state. Each atom has available two $1s$ states, two $2s$ states, six $2p$ states, and higher states (see Tables 2-1 and 2-2). If we consider N atoms, there will be $2N$, $2N$, and $6N$ states of type $1s$, $2s$, and $2p$, respectively. As the interatomic spacing decreases, these energy levels split into bands, beginning with the outer ($\mathbf{n} = 2$) shell. As the "$2s$" and "$2p$" bands grow, they merge into a single band composed of a mixture of energy levels. This band of "$2s$–$2p$" levels contains $8N$ available states. As the distance between atoms approaches the equilibrium interatomic spacing of diamond, this band splits into two bands separated by an *energy gap* E_g. The upper band (called the *conduction band*) contains $4N$ states, as does the lower (*valence*) band. Thus, apart from the low-lying and tightly bound "$1s$" levels, the diamond crystal has two bands of available energy levels separated by an energy gap E_g wide, which contains no allowed energy levels for electrons to occupy. This gap is sometimes called a "forbidden band," since in a perfect crystal it contains no electron energy states.

We should pause at this point and count electrons. The lower "$1s$" band is filled with the $2N$ electrons which originally resided in the collective $1s$ states of the isolated atoms. However, there were $4N$ electrons in the original isolated $\mathbf{n} = 2$ shells ($2N$ in $2s$ states and $2N$ in $2p$ states). These $4N$ electrons must oc-

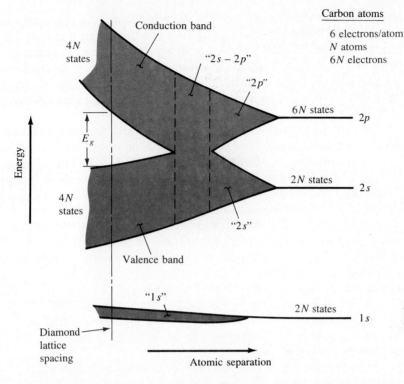

Figure 3-3
Formation of energy bands as a diamond crystal is formed by bringing together isolated carbon atoms. (From *Electrons and Holes in Semiconductors*, by W. Shockley, © 1950 by Litton Educational Publishing Co., Inc.; by permission of Van Nostrand Reinhold, Co. Inc.)

cupy states in the valence band or the conduction band in the crystal. At 0 K the electrons will occupy the lowest energy states available to them. In the case of the diamond crystal, there are exactly $4N$ states in the valence band available to the $4N$ electrons. Thus at 0 K, *every* state in the valence band will be filled, while the conduction band will be completely empty of electrons. As we shall see, this arrangement of completely filled and empty energy bands has an important effect on the electrical conductivity of the solid.

3.1.3 Metals, Semiconductors, and Insulators

Every solid has its own characteristic energy band structure. This variation in band structure is responsible for the wide range of electrical characteristics observed in various materials. The diamond band structure of Fig. 3-3, for example, can give a good picture of why carbon in the diamond lattice is a good insulator. To reach such a conclusion, we must consider the properties of completely filled and completely empty energy bands in the current conduction process.

Before discussing the mechanisms of current flow in solids further, we can observe here that for electrons to experience acceleration in an applied electric field, they must be able to move into new energy states. This implies there must be empty states (allowed energy states which are not already occupied by electrons) available to the electrons. For example, if relatively few electrons reside in an otherwise empty band, ample unoccupied states are available into which the electrons can move. On the other hand, the diamond structure is such that the valence band is completely filled with electrons at 0 K and the conduction band is empty. There can be no charge transport within the valence band, since no empty states are available into which electrons can move. There are no electrons in the conduction band, so no charge transport can take place there either. Thus carbon in the diamond structure has a high resistivity typical of insulators.

Semiconductor materials at 0 K have basically the same structure as insulators — a filled valence band separated from an empty conduction band by a band gap containing no allowed energy states (Fig. 3-4). The difference lies in the size of the band gap E_g, which is much smaller in semiconductors than in insulators. For example, the semiconductor Si has a band gap of about 1.1 eV compared with 5 eV for diamond. The relatively small band gaps of semiconductors (Appendix III) allow for excitation of electrons from the lower (valence) band to the upper (conduction) band by reasonable amounts of thermal or optical energy. For example, at room temperature a semiconductor with a 1-eV band gap will have a significant number of electrons excited thermally across the energy gap into the conduction band, whereas an insulator with $E_g = 10$ eV will have a negligible number of such excitations. Thus an important difference between semiconductors and insulators is that the number of electrons available for conduction can be increased greatly in semiconductors by thermal or optical energy.

In metals the bands either overlap or are only partially filled. Thus electrons and empty energy states are intermixed within the bands so that electrons can move freely under the influence of an electric field. As expected from the metallic band structures of Fig. 3-4, metals have a high electrical conductivity.

Figure 3-4
Typical band structures at 0 K.

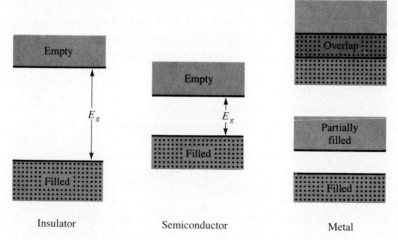

Insulator   Semiconductor   Metal

3.1.4 Direct and Indirect Semiconductors

The "thought experiment" of Section 3.1.2, in which isolated atoms were brought together to form a solid, is useful in pointing out the existence of energy bands and some of their properties. Other techniques are generally used, however, when quantitative calculations are made of band structures. In a typical calculation, a single electron is assumed to travel through a perfectly periodic lattice. The wave function of the electron is assumed to be in the form of a plane wave[†] moving, for example, in the x-direction with propagation constant $\mathbf{k}$, also called a *wave vector*. The space-dependent wave function for the electron is

$$\psi_{\mathbf{k}}(x) = U(\mathbf{k}_x, x)e^{jk_x x} \qquad (3\text{-}1)$$

where the function $U(\mathbf{k}_x, x)$ modulates the wave function according to the periodicity of the lattice.

In such a calculation, allowed values of energy can be plotted vs. the propagation constant $\mathbf{k}$. Since the periodicity of most lattices is different in various directions, the $(E, \mathbf{k})$ diagram must be plotted for the various crystal directions, and the full relationship between E and $\mathbf{k}$ is a complex surface which should be visualized in three dimensions.

The band structure of GaAs has a minimum in the conduction band and a maximum in the valence band for the same $\mathbf{k}$ value ($\mathbf{k} = 0$). On the other hand, Si has its valence band maximum at a different value of $\mathbf{k}$ than its conduction band minimum. Thus an electron making a smallest-energy transition from the conduction band to the valence band in GaAs can do so without a change in $\mathbf{k}$ value; on the other hand, a transition from the minimum point in the Si conduction band to the maximum point of the valence band requires some change in $\mathbf{k}$. Thus there are two classes of semiconductor energy bands: *direct* and *indirect* (Fig. 3-5). We can show that an indirect transition, involving a change in $\mathbf{k}$, requires a change of momentum for the electron.

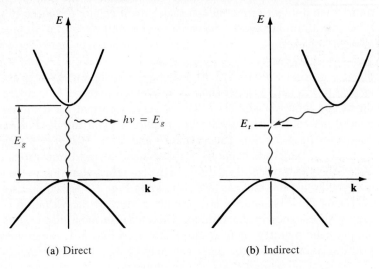

(a) Direct   (b) Indirect

Figure 3-5
Direct and indirect electron transitions in semiconductors: (a) direct transition with accompanying photon emission; (b) indirect transition via a defect level.

[†]Discussions of plane waves are available in most sophomore physics texts or in introductory electromagnetics texts.

EXAMPLE 3-1 Assuming that U is constant in Eq. (3-1) for an essentially free electron, show that the x-component of the electron momentum in the crystal is given by $\langle \mathbf{p}_x \rangle = \hbar \mathbf{k}_x$.

SOLUTION From Eq. (3-1),

$$\psi_\mathbf{k}(x) = U e^{j\mathbf{k}_x x}$$

Using Eq. (2-21b) and the momentum operator,

$$\langle \mathbf{p}_x \rangle = \frac{\displaystyle\int_{-\infty}^{\infty} U^2 e^{-j\mathbf{k}_x x} \frac{\hbar}{j} \frac{\partial}{\partial x} (e^{j\mathbf{k}_x x}) \, dx}{\displaystyle\int_{-\infty}^{\infty} U^2 \, dx}$$

$$= \frac{\hbar \mathbf{k}_x \displaystyle\int_{-\infty}^{\infty} U^2 \, dx}{\displaystyle\int_{-\infty}^{\infty} U^2 \, dx} = \hbar \mathbf{k}_x$$

This result implies that $(E, \mathbf{k})$ diagrams such as shown in Fig. 3-5 can be considered plots of electron energy vs. momentum, with a scaling factor $\hbar$.

The direct and indirect semiconductors are identified in Appendix III. In a direct semiconductor such as GaAs, an electron in the conduction band can fall to an empty state in the valence band, giving off the energy difference E_g as a photon of light. On the other hand, an electron in the conduction band minimum of an indirect semiconductor such as Si cannot fall directly to the valence band maximum but must undergo a momentum change as well as changing its energy. For example, it may go through some defect state (E_t) within the band gap. We shall discuss such defect states in Sections 4.2.1 and 4.3.2. In an indirect transition which involves a change in $\mathbf{k}$, the energy is generally given up as heat to the lattice rather than as an emitted photon. This difference between direct and indirect band structures is very important for deciding which semiconductors can be used in devices requiring light output. For example, semiconductor light emitters (Section 6.4) and lasers (Chapter 10) generally must be made of materials capable of direct band-to-band transitions or of indirect materials with vertical transitions between defect states.

Band diagrams such as those shown in Fig. 3-5 are cumbersome to draw in analyzing devices, and do not provide a view of the variation of electron energy with distance in the sample. Therefore, in most discussions we shall use simple band pictures such as those shown in Fig. 3-4, remembering that electron transitions across the band gap may be direct or indirect.

3.1.5 Variation of Energy Bands with Alloy Composition

As III–V ternary and quaternary alloys are varied over their composition ranges (see Sections 1.2.4 and 1.4.1), their band structures change. For example, Fig. 3-6 illustrates the band structure of GaAs and AlAs, and the way in which the bands change with composition x in the ternary compound $Al_xGa_{1-x}As$. The binary compound GaAs is a direct material, with a band gap of 1.43 eV at room temperature. For reference, we call the direct ($\mathbf{k} = 0$) conduction band minimum Γ. There are also two higher-lying indirect minima in the GaAs conduction band, but these are sufficiently far above Γ that few electrons reside there (we discuss an important exception in Chapter 12 in which high-field excitation of electrons into the indirect minima leads to the Gunn effect). We call the lowest-lying GaAs indirect minimum L and the other X. In AlAs the direct Γ minimum is much higher than the indirect X minimum, and this material is therefore indirect with a band gap of 2.16 eV at room temperature.

In the ternary alloy $Al_xGa_{1-x}As$ all of these conduction band minima move up relative to the valence band as the composition x varies from 0 (GaAs) to 1 (AlAs). However, the indirect minimum X moves up less than the others, and for compositions above about 38 percent Al this indirect minimum becomes the lowest-lying conduction band. Therefore, the ternary alloy AlGaAs is a direct semiconductor for Al compositions on the column III sublattice up to about 38 percent, and is an indirect semiconductor for higher Al mole fractions. The band gap energy E_g is shown in color on Fig. 3-6(c).

The variation of energy bands for the ternary alloy $GaAs_{1-x}P_x$ is generally similar to that of AlGaAs shown in Fig. 3-6. GaAsP is a direct semiconductor from GaAs to about $GaAs_{.55}P_{.45}$ and is indirect from this composition to GaP (see Fig. 6-19). This material is often used in visible LEDs.

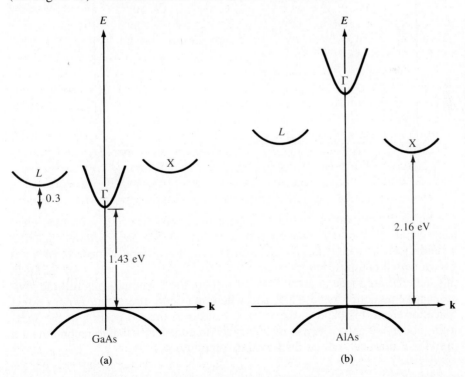

Figure 3-6
Variation of direct and indirect conduction bands in AlGaAs as a function of composition: (a) the $(E, \mathbf{k})$ diagram for GaAs, showing three minima in the conduction band; (b) AlAs band diagram;

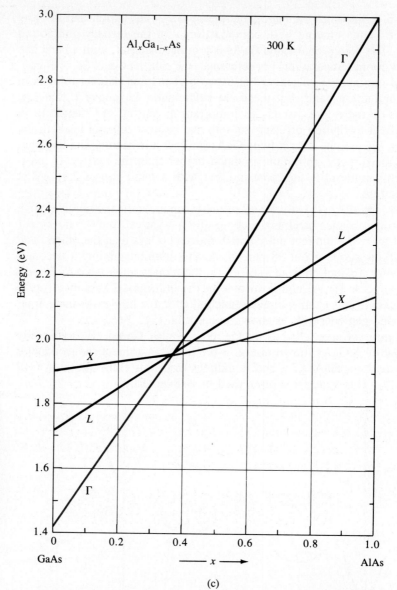

Figure 3-6 (cont.)
(c) positions of the three conduction band minima in $Al_xGa_{1-x}As$ as x varies over the range of compositions from GaAs ($x = 0$) to AlAs ($x = 1$). The smallest band gap, E_g (shown in color), follows the direct Γ band to $x = 0.38$, and then follows the indirect X band.

(c)

Since light emission is most efficient for direct materials, in which electrons can drop from the conduction band to the valence band without changing **k** (and therefore momentum), LEDs in GaAsP are generally made in material grown with a composition less than $x = 0.45$. For example, most red LEDs in this material are made at about $x = 0.4$, where the Γ minimum is still the lowest-lying conduction band edge, and where the photon resulting from a direct transition from this band to the valence band is in the red portion of the spectrum (about 1.9 eV). The use of impurities to enhance radiative recombination in indirect material will be discussed in Section 6.4.1.

The mechanism of current conduction is relatively easy to visualize in the case of a metal; the metal atoms are imbedded in a "sea" of relatively free electrons, and these electrons can move as a group under the influence of an electric field. This free electron view is oversimplified, but many important conduction properties of metals can be derived from just such a model. However, we cannot account for all of the electrical properties of semiconductors in this way. Since the semiconductor has a filled valence band and an empty conduction band at 0 K, we must consider the increase in conduction band electrons by thermal excitations across the band gap as the temperature is raised. In addition, after electrons are excited to the conduction band, the empty states left in the valence band can contribute to the conduction process. The introduction of impurities has an important effect on the energy band structure and on the availability of charge carriers. Thus there is considerable flexibility in controlling the electrical properties of semiconductors.

**3.2
CHARGE
CARRIERS IN
SEMICONDUCTORS**

3.2.1   Electrons and Holes

As the temperature of a semiconductor is raised from 0 K, some electrons in the valence band receive enough thermal energy to be excited across the band gap to the conduction band. The result is a material with some electrons in an otherwise empty conduction band and some unoccupied states in an otherwise filled valence band (Fig. 3-7).[†] For convenience, an empty state in the valence band is referred to as a *hole*. If the conduction band electron and the hole are created by the excitation of a valence band electron to the conduction band, they are called an *electron–hole pair* (abbreviated EHP).

After excitation to the conduction band, an electron is surrounded by a large number of unoccupied energy states. For example, the equilibrium number of electron–hole pairs in pure Si at room temperature is only about 10^{10} EHP/ cm^3, compared to the Si atom density of more than 10^{22} atoms/cm^3. Thus the

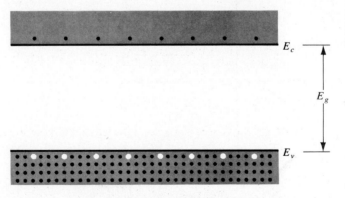

Figure 3-7
Electron–hole pairs in a semiconductor.

[†]In Fig. 3-7 and in subsequent discussions, we refer to the bottom of the conduction band as E_c and the top of the valence band as E_v.

few electrons in the conduction band are free to move about via the many available empty states.

The corresponding problem of charge transport in the valence band is somewhat more complicated. However, it is possible to show that the effects of current in a valence band containing holes can be accounted for by simply keeping track of the holes themselves.

In a filled band, all available energy states are occupied. For every electron moving with a given velocity, there is an equal and opposite electron motion elsewhere in the band. If we apply an electric field, the net current is zero because for every electron j moving with velocity $\mathbf{v}_j$ there is a corresponding electron j' with velocity $-\mathbf{v}_j$. Figure 3-8 illustrates this effect in terms of the electron energy vs. wave vector plot for the valence band. Since $\mathbf{k}$ is proportional to electron momentum, it is clear the two electrons have oppositely directed velocities. With N electrons/cm³ in the band we express the current density using a sum over all of the electron velocities, and including the charge $-q$ on each electron. In a unit volume,

$$J = (-q)\sum_i^N \mathbf{v}_i = 0 \qquad (\text{filled band}) \qquad (3\text{-}2a)$$

Now if we create a hole by removing the jth electron, the net current density in the valence band involves the sum over all velocities, minus the contribution of the electron we have removed.

$$J = (-q)\sum_i^N \mathbf{v}_i - (-q)\mathbf{v}_j \qquad (j\text{th } electron\ missing) \qquad (3\text{-}2b)$$

But the first term is zero, from Eq. (3-2a). Thus the net current is $+q\mathbf{v}_j$. In other words, the current contribution of the hole is equivalent to that of a positively charged particle with velocity $\mathbf{v}_j$, that of the missing electron. Of course,

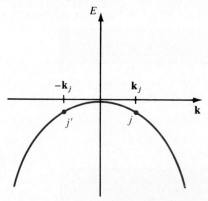

Figure 3-8   A valence band with all states filled, including states j and j', marked for discussion. The jth electron with wave vector $\mathbf{k}_j$ is matched by an electron at j' with the opposite wave vector $-\mathbf{k}_j$. There is no net current in the band unless an electron is removed. For example, if the jth electron is removed, the motion of the electron at j' is no longer compensated.

the charge transport is actually due to the motion of the now uncompensated electron (j'). Its current contribution $(-q)(-v_j)$ is equivalent to that of a positively charged particle with velocity $+v_j$. For simplicity, it is customary to treat empty states in the valence band as charge carriers with positive charge and positive mass.

In all the following discussions we shall concentrate on the electrons in the conduction band and on the holes in the valence band. We can account for the current flow in a semiconductor by the motion of these two types of charge carriers. We draw valence and conduction bands on an electron energy scale E, as in Fig. 3-8. However, we should remember that in the valence band, hole energy increases oppositely to electron energy, because the two carriers have opposite charge. Thus hole energy increases downward in Fig. 3-8 and holes, seeking the lowest energy state available, are generally found at the *top* of the valence band. In contrast, conduction band electrons are found at the bottom of the conduction band.

3.2.2 Effective Mass

The electrons in a crystal are not completely free, but instead interact with the periodic potential of the lattice. As a result, their "wave-particle" motion cannot be expected to be the same as for electrons in free space. Thus, in applying the usual equations of electrodynamics to charge carriers in a solid, we must use altered values of particle mass. In doing so, we account for most of the influences of the lattice, so that the electrons and holes can be treated as "almost free" carriers in most computations. The calculation of effective mass must take into account the shape of the energy bands in three-dimensional **k**-space, taking appropriate averages over the various energy bands.

Find the $(E, \mathbf{k})$ relationship for a free electron and relate it to the electron mass.   **EXAMPLE 3-2**

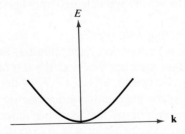

From Example 3-1, the electron momentum is $\mathbf{p} = m\mathbf{v} = \hbar\mathbf{k}$. Then   **SOLUTION**

$$E = \frac{1}{2}m\mathbf{v}^2 = \frac{1}{2}\frac{\mathbf{p}^2}{m} = \frac{\hbar^2}{2m}\mathbf{k}^2$$

Thus the electron energy is parabolic with wave vector $\mathbf{k}$. The electron mass is inversely related to the curvature (second derivative) of the $(E, \mathbf{k})$ relationship, since

$$\frac{d^2E}{d\mathbf{k}^2} = \frac{\hbar^2}{m}$$

Although electrons in solids are not free, most energy bands are close to parabolic at their minima (for conduction bands) or maxima (for valence bands). We can also approximate effective mass near those band extrema from the curvature of the band.

■

The effective mass of an electron in a band with a given $(E, \mathbf{k})$ relationship is given by

$$m* = \frac{\hbar^2}{d^2E/d\mathbf{k}^2} \tag{3-3}$$

Thus the curvature of the band determines the electron effective mass. For example, in Fig. 3-6a it is clear that the electron effective mass in GaAs is much smaller in the direct Γ conduction band (strong curvature) than in the L or X minima (weaker curvature, smaller value in the denominator of the $m*$ expression).

For a band centered at $\mathbf{k} = 0$ (such as the Γ band in GaAs), the $(E, \mathbf{k})$ relationship near the minimum is usually parabolic:

$$E = \frac{\hbar^2}{2m*}\mathbf{k}^2 + E_g \tag{3-4}$$

Comparing this relation to Eq. (3-3) indicates that the effective mass $m*$ is constant in a parabolic band. On the other hand, many conduction bands have complex $(E, \mathbf{k})$ relationships that depend on the direction of electron transport with respect to the principal crystal directions. In this case, the effective mass is a tensor quantity. However, we can use appropriate averages over such bands in most calculations.

A particularly interesting feature of Figs. 3-5 and 3-6 is that the curvature $d^2E/d\mathbf{k}^2$ is positive at the conduction band minima, but is negative at the valence band maxima. Thus, the electrons near the top of the valence band have *negative effective mass*, according to Eq. (3-3). Valence band electrons with negative charge and negative mass move in an electric field in the same direction as holes with positive charge and positive mass. As discussed in Section 3.2.1, we can fully account for charge transport in the valence band by considering hole motion.

In any calculation involving the mass of the charge carriers, we must use effective mass values for the particular material involved. Table 3-1 lists the effective masses for Ge, Si, and GaAs appropriate for one type of calculation. In this table and in all subsequent discussions, the electron effective mass is denoted by m_n^* and the hole effective mass by m_p^*. The n subscript indicates the electron as a negative charge carrier, and the p subscript indicates the hole as a positive charge carrier.

Table 3-1. Effective mass values for Ge, Si, and GaAs. The free electron rest mass is m_0. These are average values for effective mass, appropriate for use in the density of states calculation of Section 3.3.2; they take into account the number of equivalent bands in each material (4 equivalent conduction band valleys in Ge, 6 in Si). For Ge and Si, a different average is used for effective mass values in conductivity calculations.

	Ge	Si	GaAs
m_n^*	$0.55m_0$	$1.1m_0$	$0.067m_0$
m_p^*	$0.37m_0$	$0.56m_0$	$0.48m_0$

3.2.3   Intrinsic Material

A perfect semiconductor crystal with no impurities or lattice defects is called an *intrinsic* semiconductor. In such material there are no charge carriers at 0 K, since the valence band is filled with electrons and the conduction band is empty. At higher temperatures electron–hole pairs are generated as valence band electrons are excited thermally across the band gap to the conduction band. These EHPs are the only charge carriers in intrinsic material.

The generation of EHPs can be visualized in a qualitative way by considering the breaking of covalent bonds in the crystal lattice (Fig. 3-9). If one of the Si valence electrons is broken away from its position in the bonding structure such that it becomes free to move about in the lattice, a conduction electron is created and a broken bond (hole) is left behind. The energy required to break the bond is the band gap energy E_g. This model helps in visualizing the physical mechanism of EHP creation, but the energy band model is more productive for purposes of quantitative calculation. One important difficulty in the "broken bond" model is that the free electron and the hole seem deceptively localized in the lattice. Actually, the positions of the free electron and the hole are spread out over several lattice spacings and should be considered quantum mechanically by probability distributions (see Section 2.4).

Since the electrons and holes are created in pairs, the conduction band electron concentration n (electrons per cm^3) is equal to the concentration of holes in the valence band p (holes per cm^3). Each of these intrinsic carrier concentrations is commonly referred to as n_i. Thus *for intrinsic material*

$$n = p = n_i \qquad (3\text{-}5)$$

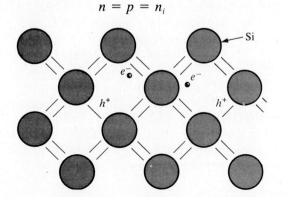

Figure 3-9
Electron–hole pairs in the covalent bonding model of the Si crystal.

e^- : Electron
h^+ : Hole

At a given temperature there is a certain concentration of electron–hole pairs n_i. Obviously, if a steady state carrier concentration is maintained, there must be *recombination* of EHPs at the same rate at which they are generated. Recombination occurs when an electron in the conduction band makes a transition (direct or indirect) to an empty state (hole) in the valence band, thus annihilating the pair. If we denote the generation rate of EHPs as g_i (EHP/cm³-s) and the recombination rate as r_i, equilibrium requires that

$$r_i = g_i \tag{3-6}$$

Each of these rates is temperature dependent. For example, $g_i(T)$ increases when the temperature is raised, and a new carrier concentration n_i is established such that the higher recombination rate $r_i(T)$ just balances generation. At any temperature, we can predict that the rate of recombination of electrons and holes r_i is proportional to the equilibrium concentration of electrons n_0 and the concentration of holes p_0:

$$r_i = \alpha_r n_0 p_0 = \alpha_r n_i^2 = g_i \tag{3-7}$$

The factor α_r is a constant of proportionality which depends on the particular mechanism by which recombination takes place. We shall discuss the calculation of n_i as a function of temperature in Section 3.3.3; recombination processes will be discussed in Chapter 4.

3.2.4 Extrinsic Material

In addition to the intrinsic carriers generated thermally, it is possible to create carriers in semiconductors by purposely introducing impurities into the crystal. This process, called *doping,* is the most common technique for varying the conductivity of semiconductors. By doping, a crystal can be altered so that it has a predominance of either electrons or holes. Thus there are two types of doped semiconductors, n-type (mostly electrons) and p-type (mostly holes). When a crystal is doped such that the equilibrium carrier concentrations n_0 and p_0 are different from the intrinsic carrier concentration n_i, the material is said to be *extrinsic*.

When impurities or lattice defects are introduced into an otherwise perfect crystal, additional levels are created in the energy band structure, usually within the band gap. For example, an impurity from column V of the periodic table (P, As, and Sb) introduces an energy level very near the conduction band in Ge or Si. This level is filled with electrons at 0 K, and very little thermal energy is required to excite these electrons to the conduction band (Fig. 3-10). Thus at about 50–100 K virtually all of the electrons in the impurity level are "donated" to the conduction band. Such an impurity level is called a *donor* level, and the column V impurities in Ge or Si are called donor impurities. From Fig. 3-10 we note that the material doped with donor impurities can have a considerable concentration of electrons in the conduction band, even when the temperature is too low for the intrinsic EHP concentration to be appreciable. Thus semiconductors doped with a significant number of donor atoms will have $n_0 \gg (n_i, p_0)$ at room temperature. This is n-type material.

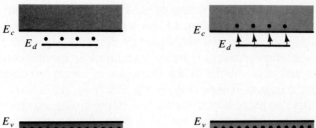

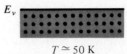

$T = 0\,K$ $T \simeq 50\,K$

Figure 3-10
Donation of electrons from a donor level to the conduction band.

Atoms from column III (B, Al, Ga, and In) introduce impurity levels in Ge or Si near the valence band. These levels are empty of electrons at 0 K (Fig. 3-11). At low temperatures, enough thermal energy is available to excite electrons from the valence band into the impurity level, leaving behind holes in the valence band. Since this type of impurity level "accepts" electrons from the valence band, it is called an *acceptor* level, and the column III impurities are acceptor impurities in Ge and Si. As Fig. 3-11 indicates, doping with acceptor impurities can create a semiconductor with a hole concentration p_0 much greater than the conduction band electron concentration n_0 (this is p-type material).

In the covalent bonding model, donor and acceptor atoms can be visualized as shown in Fig. 3-12. An Sb atom (column V) in the Si lattice has the four necessary valence electrons to complete the covalent bonds with the neighboring Si atoms, plus one extra electron. This fifth electron does not fit

Figure 3-11
Acceptance of valence band electrons by an acceptor level, and the resulting creation of holes.

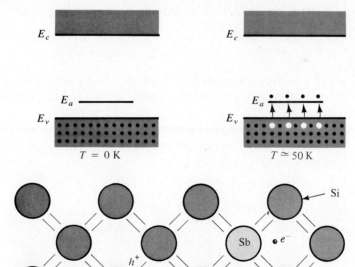

Figure 3-12
Donor and acceptor atoms in the covalent bonding model of a Si crystal.

into the bonding structure of the lattice and is therefore loosely bound to the Sb atom. A small amount of thermal energy enables this extra electron to overcome its coulombic binding to the impurity atom and be donated to the lattice as a whole. Thus it is free to participate in current conduction. This process is a qualitative model of the excitation of electrons out of a donor level and into the conduction band (Fig. 3-10). Similarly, the column III impurity Al has only three valence electrons to contribute to the covalent bonding (Fig. 3-12), thereby leaving one bond incomplete. With a small amount of thermal energy, this incomplete bond can be transferred to other atoms as the bonding electrons exchange positions. Again, the idea of an electron "hopping" from an adjacent bond into the incomplete bond at the Al site provides some physical insight into the behavior of an acceptor, but the model of Fig. 3-11 is preferable for most discussions.

We can calculate rather simply the approximate energy required to excite the fifth electron of a donor atom into the conduction band (the donor *binding energy*). Let us assume for rough calculations that the Sb atom of Fig. 3-12 has its four covalent bonding electrons rather tightly bound and the fifth "extra" electron loosely bound to the atom. We can approximate this situation by using the Bohr model results, considering the loosely bound electron as ranging about the tightly bound "core" electrons in a hydrogen-like orbit. From Eq. (2-15) the magnitude of the ground-state energy ($n = 1$) of such an electron is

$$E = \frac{mq^4}{2K^2\hbar^2} \tag{3-8}$$

The value of K must be modified from the free-space value $4\pi\epsilon_0$ used in the hydrogen atom problem to

$$K = 4\pi\epsilon_0\epsilon_r \tag{3-9}$$

where ϵ_r is the relative dielectric constant of the semiconductor material. In addition, we must use the effective mass m_n^* typical of the semiconductor.[†]

EXAMPLE 3-3   Calculate the approximate donor binding energy for Ge ($\epsilon_r = 16$, $m_n^* = 0.12m_0$).

SOLUTION   From Eq. (3-8) and Appendix II we have

$$E = \frac{m_n^* q^4}{8(\epsilon_0\epsilon_r)^2 h^2} = \frac{0.12(9.11 \times 10^{-31})(1.6 \times 10^{-19})^4}{8(8.85 \times 10^{-12} \times 16)^2(6.63 \times 10^{-34})^2}$$

$$= 1.02 \times 10^{-21} \text{ J} = 0.0064 \text{ eV}$$

[†]The "density of states" effective mass listed in Table 3-1 is not appropriate for this calculation. The value used here is an average of the effective mass in different crystallographic directions, called the "conductivity effective mass."

Thus the energy required to excite the donor electron from the **n** = 1 state to the free state (**n** = ∞) is ≃6 meV. This corresponds to the energy difference $E_c - E_d$ in Fig. 3-10 and is in very close agreement with actual measured values.

■

Generally, the column V donor levels lie approximately 0.01 eV below the conduction band in Ge, and the column III acceptor levels lie about 0.01 eV above the valence band. In Si the usual donor and acceptor levels lie about 0.03–0.06 eV from a band edge.

In III–V compounds, column VI impurities occupying column V sites serve as donors. For example, S, Se, and Te are donors in GaAs, since they substitute for As and provide an extra electron compared with the As atom. Similarly, impurities from column II (Be, Zn, Cd) substitute for column III atoms to form acceptors in the III–V compounds. A more ambiguous case arises when a III–V material is doped with Si or Ge, from column IV. These impurities are called *amphoteric,* meaning that Si or Ge can serve as donors or acceptors depending on whether they reside on the column III or column V sublattice of the crystal. In GaAs it is common for Si impurities to occupy Ga sites. Since the Si has an extra electron compared with the Ga it replaces, it serves as a donor. However, an excess of As vacancies arising during growth or processing of the GaAs can cause Si impurities to occupy As sites, where they serve as acceptors.

The importance of doping will become obvious when we discuss electronic devices made from junctions between p-type and n-type semiconductor material. The extent to which doping controls the electronic properties of semiconductors can be illustrated here by considering changes in the sample resistance which occur with doping. In Si, for example, the intrinsic carrier concentration n_i is about 10^{10} cm^{-3} at room temperature. If we dope Si with 10^{15} Sb atoms/cm^3, the conduction electron concentration changes by five orders of magnitude. The resistivity of Si changes from about 2×10^5 Ω-cm to 5 Ω-cm with this doping.

When a semiconductor is doped n-type or p-type, one type of carrier dominates. In the example given above, the conduction band electrons outnumber the holes in the valence band by many orders of magnitude. We refer to the small number of holes in n-type material as *minority carriers* and the relatively large number of conduction band electrons as *majority carriers*. Similarly, electrons are the minority carriers in p-type material, and holes are the majority carriers.

3.2.5   Electrons and Holes in Quantum Wells

We have discussed single-valued (*discrete*) energy levels in the band gap arising from doping, and a *continuum* of allowed states in the valence and conduction bands. A third possibility is the formation of discrete levels for electrons and holes as a result of quantum-mechanical confinement.

One of the most useful applications of MBE or OMVPE growth of multilayer compound semiconductors, as described in Section 1.4, is the fact that a continuous single crystal can be grown in which adjacent layers have different band gaps. For example, Fig. 3-13 shows the spatial variation in conduction and valence bands for a multilayer structure in which a very thin layer of GaAs is sandwiched between two layers of AlGaAs, which has a wider band gap than the GaAs. We will discuss the details of such *heterojunctions* (junctions between dissimilar materials) in Section 5.8. It is interesting to point out here, however, that a consequence of confining electrons and holes in a very thin layer is that these particles behave according to the *particle in a potential well* problem, with quantum states calculated in Section 2.4.3. Therefore, instead of having the continuum of states normally available in the conduction band, the conduction band electrons in the narrow-gap material are confined to discrete quantum states as described by Eq. (2-33), modified for effective mass and finite barrier height. Similarly, the states in the valence band available for holes are restricted to discrete levels in the quantum well. This is one of the clearest demonstrations of the quantum mechanical results discussed in Chapter 2. From a practical device point of view, the formation of discrete quantum states in the GaAs layer of Fig. 3-13 changes the energy at which photons can be emitted. An electron on one of the discrete conduction band states (E_1 in Fig. 3-13) can make a transition to an empty discrete valence band state in the GaAs quantum well (such as E_h), giving off a photon of energy $E_g + E_1 + E_h$, greater than the GaAs band gap. Semiconductor lasers have been made in which such a quantum well is used to raise the energy of the transition from the infrared, typical of GaAs, to the red portion of the spec-

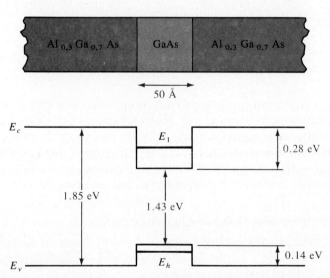

Figure 3-13   Energy band discontinuities for a thin layer of GaAs sandwiched between layers of wider band gap AlGaAs. In this case the GaAs region is so thin that quantum states are formed in the valence and conduction bands. Electrons in the GaAs conduction band reside on "particle in a potential well" states such as E_1 shown here, rather than in the usual conduction band states. Holes in the quantum well occupy similar discrete states, such as E_h.

trum. We will see other examples of quantum wells in semiconductor devices in later chapters, including applications of tunneling of electrons through such quantum states.

In calculating semiconductor electrical properties and analyzing device behavior, it is often necessary to know the number of charge carriers per cm^3 in the material. The majority carrier concentration is usually obvious in heavily doped material, since one majority carrier is obtained for each impurity atom (for the standard doping impurities). The concentration of minority carriers is not obvious, however, nor is the temperature dependence of the carrier concentrations.

**3.3
CARRIER
CONCENTRATIONS**

To obtain equations for the carrier concentrations we must investigate the distribution of carriers over the available energy states. This type of distribution is not difficult to calculate, but the derivation requires some background in statistical methods. Since we are primarily concerned here with the application of these results to semiconductor materials and devices, we shall accept the distribution function as given.

3.3.1   The Fermi Level

Electrons in solids obey *Fermi-Dirac* statistics.[†] In the development of this type of statistics, one must consider the indistinguishability of the electrons, their wave nature, and the Pauli exclusion principle. The rather simple result of these statistical arguments is that the distribution of electrons over a range of allowed energy levels at thermal equilibrium is

$$f(E) = \frac{1}{1 + e^{(E-E_F)/kT}}$$

(3-10)

where k is Boltzmann's constant ($k = 8.62 \times 10^{-5}$ eV/K $= 1.38 \times 10^{-23}$ J/K). The function $f(E)$, the *Fermi-Dirac distribution function*, gives the probability that an available energy state at E will be occupied by an electron at absolute temperature T. The quantity E_F is called the *Fermi level,* and it represents an important quantity in the analysis of semiconductor behavior. We no-

---

[†]Examples of other types of statistics are *Maxwell–Boltzmann* for classical particles (e.g., gas) and *Bose–Einstein* for photons. For two discrete energy levels E_2 and E_1 (with $E_2 > E_1$), classical gas atoms follow a Boltzmann distribution; the number n_2 of atoms in state E_2 is related to the number n_1 in E_1 at thermal equilibrium by

$$\frac{n_2}{n_1} = \frac{e^{-E_2/kT}}{e^{-E_1/kT}} = e^{-(E_2-E_1)/kT}$$

assuming the two levels have an equal number of states. The exponential term $\exp(-\Delta E/kT)$ is commonly called the *Boltzmann factor.* It appears also in the denominator of the Fermi–Dirac distribution function. We shall return to the Boltzmann distribution in Chapter 10 in discussions of the properties of lasers.

tice that, for an energy E equal to the Fermi level energy E_F, the occupation probability is

$$f(E_F) = [1 + e^{(E_F - E_F)/kT}]^{-1} = \frac{1}{1+1} = \frac{1}{2} \qquad (3\text{-}11)$$

Thus an energy state at the Fermi level has a probability of $\frac{1}{2}$ of being occupied by an electron.

A closer examination of $f(E)$ indicates that at 0 K the distribution takes the simple rectangular form shown in Fig. 3-14. With $T = 0$ in the denominator of the exponent, $f(E)$ is $1/(1 + 0) = 1$ when the exponent is negative ($E < E_F$), and is $1/(1 + \infty) = 0$ when the exponent is positive ($E > E_F$). This rectangular distribution implies that at 0 K every available energy state up to E_F is filled with electrons, and all states above E_F are empty.

At temperatures higher than 0 K, some probability exists for states above the Fermi level to be filled. For example, at $T = T_1$ in Fig. 3-14 there is some probability $f(E)$ that states above E_F are filled, and there is a corresponding probability $[1 - f(E)]$ that states below E_F are empty. The Fermi function is symmetrical about E_F for all temperatures (Prob. 3.3); that is, the probability $f(E_F + \Delta E)$ that a state ΔE above E_F is filled is the same as the probability $[1 - f(E_F - \Delta E)]$ that a state ΔE below E_F is empty. The symmetry of the distribution of empty and filled states about E_F makes the Fermi level a natural reference point in calculations of electron and hole concentrations in semiconductors.

In applying the Fermi-Dirac distribution to semiconductors, we must recall that $f(E)$ is the probability of occupancy of an *available* state at E. Thus if there is no available state at E (e.g., in the band gap of a semiconductor), there is no possibility of finding an electron there. We can best visualize the relation between $f(E)$ and the band structure by turning the $f(E)$ vs. E diagram on its side so that the E scale corresponds to the energies of the band diagram (Fig. 3-15). For intrinsic material we know that the concentration of holes in the valence band is equal to the concentration of electrons in the conduction band. Therefore, the Fermi level E_F must lie at the middle of the band gap in

Figure 3-14
The Fermi–Dirac distribution function.

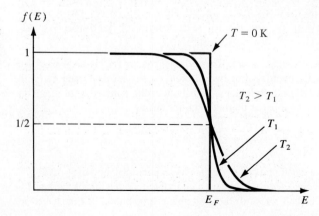

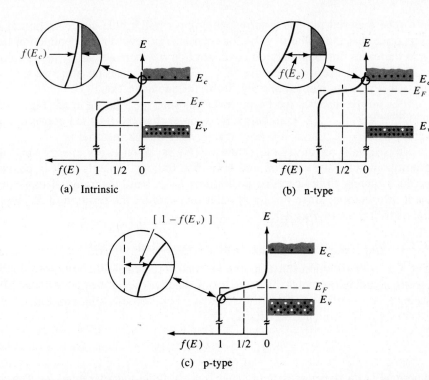

(a) Intrinsic

(b) n-type

(c) p-type

Figure 3-15
The Fermi
distribution
function applied to
semiconductors:
(a) intrinsic material;
(b) n-type material;
(c) p-type material.

intrinsic material.[†] Since $f(E)$ is symmetrical about E_F, the electron probability "tail" of $f(E)$ extending into the conduction band of Fig. 3-15a is symmetrical with the hole probability tail $[1 - f(E)]$ in the valence band. The distribution function has values within the band gap between E_v and E_c, but there are no energy states available, and no electron occupancy results from $f(E)$ in this range.

The tails in $f(E)$ are exaggerated in Fig. 3-15 for illustrative purposes. Actually, the probability values at E_v and E_c are quite small for intrinsic material at reasonable temperatures. For example, in Si at 300 K, $n_i = p_i \approx 10^{10}$ cm^{-3}, whereas the densities of available states at E_v and E_c are on the order of 10^{19} cm^{-3}. Thus the probability of occupancy $f(E)$ for an individual state in the conduction band and the hole probability $[1 - f(E)]$ for a state in the valence band are quite small. Because of the relatively large density of states in each band, small changes in $f(E)$ can result in significant changes in carrier concentration.

In n-type material there is a high concentration of electrons in the conduction band compared with the hole concentration in the valence band (recall Fig. 3-10). Thus in n-type material the distribution function $f(E)$ must lie above its intrinsic position on the energy scale (Fig. 3-15b). Since $f(E)$ retains its shape for a particular temperature, the larger concentration of electrons at E_c

---

[†]Actually the intrinsic E_F is displaced slightly from the middle of the gap, since the densities of available states in the valence and conduction bands are not equal (Section 3.3.2).

in n-type material implies a correspondingly smaller hole concentration at E_v. We notice that the value of $f(E)$ for each energy level in the conduction band (and therefore the total electron concentration n_0) increases as E_F moves closer to E_c. Thus the energy difference $(E_c - E_F)$ gives a measure of n; we shall express this relation mathematically in the following section.

For p-type material the Fermi level lies near the valence band (Fig. 3-15c) such that the $[1 - f(E)]$ tail below E_v is larger than the $f(E)$ tail above E_c. The value of $(E_F - E_v)$ indicates how strongly p-type the material is.

It is usually inconvenient to draw $f(E)$ vs. E on every energy band diagram to indicate the electron and hole distributions. Therefore, it is common practice merely to indicate the position of E_F in band diagrams. This is sufficient information, since for a particular temperature the position of E_F implies the distributions in Fig. 3-15.

3.3.2   Electron and Hole Concentrations at Equilibrium

The Fermi distribution function can be used to calculate the concentrations of electrons and holes in a semiconductor, if the densities of available states in the valence and conduction bands are known. For example, the concentration of electrons in the conduction band is

$$n_0 = \int_{E_c}^{\infty} f(E)N(E)\, dE \tag{3-12}$$

where $N(E)\, dE$ is the density of states (cm^{-3}) in the energy range dE. The subscript 0 used with the electron and hole concentration symbols (n_0, p_0) indicates equilibrium conditions. The number of electrons per unit volume in the energy range dE is the product of the density of states and the probability of occupancy $f(E)$. Thus the total electron concentration is the integral over the entire conduction band, as in Eq. (3-12).[†] The function $N(E)$ can be calculated by using quantum mechanics and the Pauli exclusion principle (Appendix IV).

It is shown in Appendix IV that $N(E)$ is proportional to $E^{1/2}$, so the density of states in the conduction band increases with electron energy. On the other hand, the Fermi function becomes extremely small for large energies. The result is that the product $f(E)N(E)$ decreases rapidly above E_c, and very few electrons occupy energy states far above the conduction band edge. Similarly, the probability of finding an empty state (hole) in the valence band $[1 - f(E)]$ decreases rapidly below E_v, and most holes occupy states near the top of the valence band. This effect is demonstrated in Fig. 3-16, which shows the density of available states, the Fermi function, and the resulting number of electrons and holes occupying available energy states in the conduction and valence bands at thermal equilibrium (i.e., with no excitations except thermal energy). For holes, increasing energy points down in Fig. 3-16, since the E scale refers to electron energy.

[†]The upper limit is actually improper in Eq. (3-12), since the conduction band does not extend to infinite energy. This is unimportant in the calculation of n_0, however, since $f(E)$ becomes negligibly small for large values of E. Most electrons occupy states near the bottom of the conduction band at equilibrium.

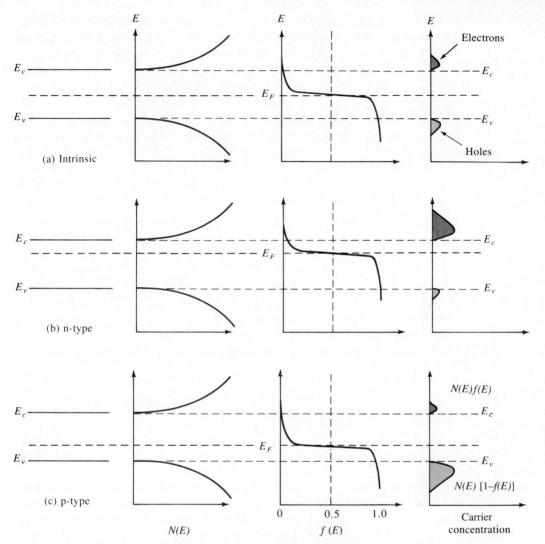

Schematic band diagram, density of states, Fermi–Dirac distribution, and the carrier concentrations for (a) intrinsic, (b) n-type, and (c) p-type semiconductors at thermal equilibrium.   **Figure 3-16**

The result of the integration of Eq. (3-12) is the same as that obtained if we represent all of the distributed electron states in the conduction band by an *effective density of states* N_c located at the conduction band edge E_c. Therefore, the conduction band electron concentration is simply the effective density of states at E_c times the probability of occupancy at E_c:[†]

$$n_0 = N_c f(E_c) \qquad (3\text{-}13)$$

[†]The simple expression for n_0 obtained in Eq. (3-13) is the direct result of integrating Eq. (3-12), as in Appendix IV. Equations (3-15) and (3-19) properly include the effects of the conduction and valence bands through the density-of-states terms.

In this expression we assume the Fermi level E_F lies at least several kT below the conduction band. Then the exponential term is large compared with unity, and the Fermi function $f(E_c)$ can be simplified as

$$f(E_c) = \frac{1}{1 + e^{(E_c - E_F)/kT}} \simeq e^{-(E_c - E_F)/kT} \qquad (3\text{-}14)$$

Since kT at room temperature is only 0.026 eV, this is generally a good approximation. For this condition the concentration of electrons in the conduction band is

$$\boxed{n_0 = N_c e^{-(E_c - E_F)/kT}} \qquad (3\text{-}15)$$

The effective density of states N_c is shown in Appendix IV to be

$$N_c = 2\left(\frac{2\pi m_n^* kT}{h^2}\right)^{3/2} \qquad (3\text{-}16)$$

Since the quantitites in Eq. (3-16) are known, values of N_c can be tabulated as a function of temperature. As Eq. (3-15) indicates, the electron concentration increases as E_F moves closer to the conduction band. This is the result we would predict from Fig. 3-15b.

By similar arguments, the concentration of holes in the valence band is

$$p_0 = N_v[1 - f(E_v)] \qquad (3\text{-}17)$$

where N_v is the effective density of states in the valence band. The probability of finding an empty state at E_v is

$$1 - f(E_v) = 1 - \frac{1}{1 + e^{(E_v - E_F)/kT}} \simeq e^{-(E_F - E_v)/kT} \qquad (3\text{-}18)$$

for E_F larger than E_v by several kT. From these equations, the concentration of holes in the valence band is

$$\boxed{p_0 = N_v e^{-(E_F - E_v)/kT}} \qquad (3\text{-}19)$$

The effective density of states in the valence band reduced to the band edge is

$$N_v = 2\left(\frac{2\pi m_p^* kT}{h^2}\right)^{3/2} \qquad (3\text{-}20)$$

As expected from Fig. 3-15c, Eq. (3-19) predicts that the hole concentration increases as E_F moves closer to the valence band.

The electron and hole concentrations predicted by Eqs. (3-15) and (3-19) are valid whether the material is intrinsic or doped, provided thermal equilibrium is maintained. Thus *for intrinsic material, E_F lies at some intrinsic level E_i near the middle of the band gap* (Fig. 3-15a), and the intrinsic electron and hole concentrations are

$$n_i = N_c e^{-(E_c - E_i)/kT}, \qquad p_i = N_v e^{-(E_i - E_v)/kT} \qquad (3\text{-}21)$$

The product of n_0 and p_0 at equilibrium is a constant for a particular material and temperature, even if the doping is varied:

$$n_0 p_0 = (N_c e^{-(E_c - E_F)/kT})(N_v e^{-(E_F - E_v)/kT}) = N_c N_v e^{-(E_c - E_v)/kT}$$

$$= N_c N_v e^{-E_g/kT} \tag{3-22a}$$

$$n_i p_i = (N_c e^{-(E_c - E_i)/kT})(N_v e^{-(E_i - E_v)/kT}) = N_c N_v e^{-E_g/kT} \tag{3-22b}$$

The intrinsic electron and hole concentrations are equal (since the carriers are created in pairs), $n_i = p_i$; thus the intrinsic concentration is

$$n_i = \sqrt{N_c N_v}\, e^{-E_g/2kT} \tag{3-23}$$

The constant product of electron and hole concentrations in Eq. (3-22) can be written conveniently as

$$\boxed{n_0 p_0 = n_i^2} \tag{3-24}$$

This is an important relation, and we shall use it extensively in later calculations. The intrinsic concentration for Si at room temperature is approximately $n_i = 1.5 \times 10^{10}$ cm^{-3}.

Comparing Eqs. (3-21) and (3-23), we note that the intrinsic level E_i is the middle of the band gap ($E_c - E_i = E_g/2$), if the effective densities of states N_c and N_v are equal. There is usually some difference in effective mass for electrons and holes, however, and N_c and N_v are slightly different as Eqs. (3-16) and (3-20) indicate. Thus the intrinsic level E_i is displaced from the middle of the band gap, although the deviation is usually slight (Prob. 3.8).

Another convenient way of writing Eqs. (3-15) and (3-19) is

$$\boxed{\begin{aligned} n_0 &= n_i e^{(E_F - E_i)/kT} \\ p_0 &= n_i e^{(E_i - E_F)/kT} \end{aligned}} \qquad \begin{aligned} &\text{(3-25a)} \\ &\text{(3-25b)} \end{aligned}$$

obtained by the application of Eq. (3-21). This form of the equations indicates directly that the electron concentration is n_i when E_F is at the intrinsic level E_i, and that n_0 increases exponentially as the Fermi level moves away from E_i toward the conduction band. Similarly, the hole concentration p_0 varies from n_i to larger values as E_F moves from E_i toward the valence band. Since these equations reveal the qualitative features of carrier concentration so directly, they are particularly convenient to remember.

A Si sample is doped with 10^{17} As atoms/cm^3. What is the equilibrium hole concentration p_0 at 300 K? Where is E_F relative to E_i?   **EXAMPLE 3-4**

Since $N_d \gg n_i$, we can approximate $n_0 = N_d$ and   **SOLUTION**

$$p_0 = \frac{n_i^2}{n_0} = \frac{2.25 \times 10^{20}}{10^{17}} = 2.25 \times 10^3 \text{ cm}^{-3}$$

From Eq. (3-25a), we have

$$E_F - E_i = kT \ln\frac{n_0}{n_i} = 0.0259 \ln\frac{10^{17}}{1.5 \times 10^{10}} = 0.407 \text{ eV}$$

The resulting band diagram is:

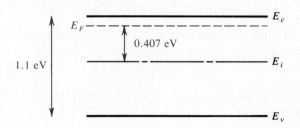

3.3.3   Temperature Dependence of Carrier Concentrations

The variation of carrier concentration with temperature is indicated by Eq. (3-25). Initially, the variation of n_0 and p_0 with T seems relatively straightforward in these relations. The problem is complicated, however, by the fact that n_i has a strong temperature dependence [Eq. (3-23)] and that E_F can also vary with temperature. Let us begin by examining the intrinsic carrier concentration. By combining Eqs. (3-23), (3-16), and (3-20) we obtain

$$n_i(T) = 2\left(\frac{2\pi kT}{h^2}\right)^{3/2} (m_n^* m_p^*)^{3/4} e^{-E_g/2kT} \qquad (3-26)$$

The exponential temperature dependence dominates $n_i(T)$, and a plot of $\ln n_i$ vs. $10^3/T$ appears almost linear (Fig. 3-17).[†] In this figure we neglect variations due to the $T^{3/2}$ dependence of the density-of-states function and the fact that E_g varies somewhat with temperature.[‡] The value of n_i at any temperature is a definite number for a given semiconductor, and is known for most materials. Thus we can take n_i as given in calculating n_0 or p_0 from Eq. (3-25).[§]

With n_i and T given, the unknowns in Eq. (3-25) are the carrier concentrations and the Fermi level position relative to E_i. One of these two quantities must be given if the other is to be found. If the carrier concentration is held at a certain value, as in heavily doped extrinsic material, E_F can be obtained from

---

[†]When plotting quantities such as carrier concentration, which involve a Boltzmann factor, it is common to use an inverse temperature scale. This allows terms which are exponential in $1/T$ to appear linear in the semilogarithmic plot. When reading such graphs, remember that temperature increases from right to left.

[‡]For Si the band gap E_g varies from about 1.11 eV at 300 K to about 1.16 eV at 0 K.

[§]Care must be taken to use consistent units in these calculations. For example, if an energy such as E_g is expressed in electron volts (eV), it should be multiplied by q (1.6×10^{-19} C) to convert to joules if k is in J/K; alternatively, E_g can be kept in eV and the value of k in eV/K can be used. At 300 K we can use $kT = 0.0259$ eV and E_g in eV.

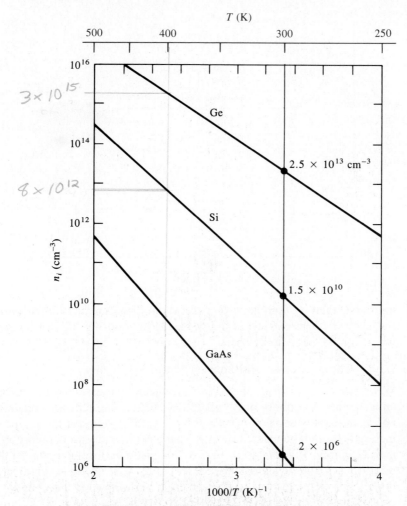

Figure 3-17
Intrinsic carrier concentration for Ge, Si, and GaAs as a function of inverse temperature. The room temperature values are marked for reference.

Eq. (3-25). The temperature dependence of electron concentration in a doped semiconductor can be visualized as shown in Fig. 3-18. In this example, Si is doped n-type with a donor concentration N_d of 10^{15} cm^{-3}. At very low temperatures (large $1/T$), negligible intrinsic EHPs exist, and the donor electrons are bound to the donor atoms. As the temperature is raised, these electrons are donated to the conduction band, and at about 100 K ($1000/T = 10$) all the donor atoms are ionized. This temperature range is called the *ionization* region. Once the donors are ionized, the conduction band electron concentration is $n_0 \simeq N_d = 10^{15}$ cm^{-3}, since one electron is obtained for each donor atom. When every available extrinsic electron has been transferred to the conduction band, n_0 is virtually constant with temperature until the concentration of intrinsic carriers n_i becomes comparable to the extrinsic concentration N_d. Finally, at higher temperatures n_i is much greater than N_d, and the intrinsic carriers dominate. In most devices it is desirable to control the carrier concentration by dop-

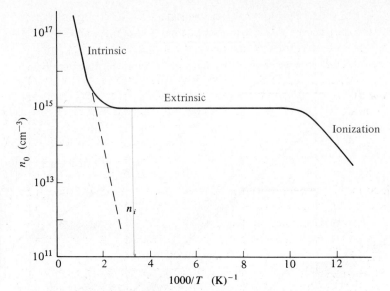

Figure 3-18
Carrier
concentration
vs. inverse
temperature for Si
doped with 10^{15}
donors/cm^3.

ing rather than by thermal EHP generation. Thus one usually dopes the material such that the extrinsic range extends beyond the highest temperature at which the device is to be used.

3.3.4   Compensation and Space Charge Neutrality

When the concept of doping was introduced, we assumed the material contained either N_d donors or N_a acceptors, so that the extrinsic majority carrier concentrations were $n_0 \simeq N_d$ or $p_0 \simeq N_a$, respectively, for the n-type or p-type material. It often happens, however, that a semiconductor contains both donors and acceptors. For example, Fig. 3-19 illustrates a semiconductor for which both donors and acceptors are present, but $N_d > N_a$. The predominance of donors makes the material n-type, and the Fermi level is therefore in the upper part of the band gap. Since E_F is well above the acceptor level E_a, this level is essentially filled with electrons. However, with E_F above E_i, we cannot expect

Figure 3-19
Compensation in
an n-type
semiconductor
$(N_d > N_a)$.

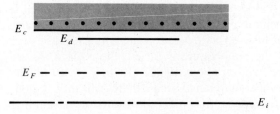

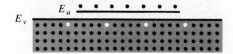

a hole concentration in the valence band commensurate with the acceptor concentration. In fact, the filling of the E_a states occurs at the expense of the donated conduction band electrons. The mechanism can be visualized as follows: Assume an acceptor state is filled with a valence band electron as described in Fig. 3-11, with a hole resulting in the valence band. This hole is then filled by recombination with one of the conduction band electrons. Extending this logic to all the acceptor atoms, we expect the resultant concentration of electrons in the conduction band to be $N_d - N_a$ instead of the total N_d. This process is called *compensation*. By this process it is possible to begin with an n-type semiconductor and add acceptors until $N_a = N_d$ and no donated electrons remain in the conduction band. In such compensated material, $n_0 = n_i = p_0$ and intrinsic conduction is obtained. With further acceptor doping the semiconductor becomes p-type with a hole concentration of essentially $N_a - N_d$.

The exact relationship among the electron, hole, donor, and acceptor concentrations can be obtained by considering the requirements for *space charge neutrality*. If the material is to remain electrostatically neutral, the sum of the positive charges (holes and ionized donor atoms) must balance the sum of the negative charges (electrons and ionized acceptor atoms):

$$p_0 + N_d^+ = n_0 + N_a^- \tag{3-27}$$

Thus in Fig. 3-19 the net electron concentration in the conduction band is

$$n_0 = p_0 + (N_d^+ - N_a^-) \tag{3-28}$$

If the material is doped n-type ($n_0 \gg p_0$) and all the impurities are ionized, we can approximate Eq. (3-28) by $n_0 \simeq N_d - N_a$.

Since the intrinsic semiconductor itself is electrostatically neutral and the doping atoms we add are also neutral, the requirement of Eq. (3-27) must be maintained at equilibrium. The electron and hole concentrations and the Fermi level adjust such that Eqs. (3-27) and (3-25) are satisfied.

Knowledge of carrier concentrations in a solid is necessary for calculating current flow in the presence of electric or magnetic fields. In addition to the values of n and p, we must be able to take into account the collisions of the charge carriers with the lattice and with the impurities. These processes will affect the ease with which electrons and holes can flow through the crystal, that is, their *mobility* within the solid. As should be expected, these collision and scattering processes depend on temperature, which affects the thermal motion of the lattice atoms and the velocity of the carriers.

**3.4
DRIFT OF CARRIERS
IN ELECTRIC AND
MAGNETIC FIELDS**

3.4.1   Conductivity and Mobility

The charge carriers in a solid are in constant motion, even at thermal equilibrium. At room temperature, for example, the thermal motion of an individual electron may be visualized as random scattering from lattice atoms, impurities, other electrons, and defects (Fig. 3-20). Since the scattering is ran-

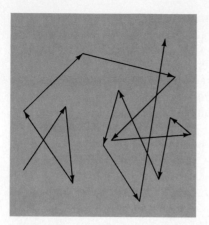

Figure 3-20
Thermal motion of
an electron in a
solid.

dom, there is no net motion of the group of n electrons/cm^3 over any period of time. This is not true of an individual electron, of course. The probability of the electron in Fig. 3-20 returning to its starting point after some time t is negligibly small. However, if a large number of electrons is considered (e.g., 10^{16} cm^{-3} in an n-type semiconductor), there will be no preferred direction of motion for the group of electrons and no net current flow.

If an electric field $\mathscr{E}_x$ is applied in the x-direction, each electron experiences a net force $- q\mathscr{E}_x$ from the field. This force may be insufficient to alter appreciably the random path of an individual electron; the effect when averaged over all the electrons, however, is a net motion of the group in the $-x$-direction. If $\mathbf{p}_x$ is the x-component of the total momentum of the group, the force of the field on the n electrons/cm^3 is

$$-nq\mathscr{E}_x = \frac{d\mathbf{p}_x}{dt}\bigg|_{\text{field}} \qquad (3\text{-}29)$$

Initially, Eq. (3-29) seems to indicate a continuous acceleration of the electrons in the $-x$-direction. This is not the case, however, because the net acceleration of Eq. (3-29) is just balanced in steady state by the decelerations of the collision processes. Thus while the steady field $\mathscr{E}_x$ does produce a net momentum $\mathbf{p}_{-x}$, the net rate of change of momentum when collisions are included must be zero in the case of steady state current flow.

To find the total rate of momentum change from collisions, we must investigate the collision probabilities more closely. If the collisions are truly random, there will be a constant probability of collision at any time for each electron. Let us consider a group of N_0 electrons at time $t = 0$ and define $N(t)$ as the number of electrons that *have not* undergone a collision by time t. The rate of decrease in $N(t)$ at any time t is proportional to the number left unscattered at t,

$$-\frac{dN(t)}{dt} = \frac{1}{\bar{t}}N(t) \qquad (3\text{-}30)$$

where $\bar{t}^{-1}$ is a constant of proportionality.

The solution to Eq. (3-30) is an exponential function

$$N(t) = N_0 e^{-t/\bar{t}} \tag{3-31}$$

and $\bar{t}$ represents the mean time between scattering events,[†] called the *mean free time*. The probability that any electron has a collision in the time interval dt is $dt/\bar{t}$. Thus the differential change in $\mathbf{p}_x$ due to collisions in time dt is

$$d\mathbf{p}_x = -\mathbf{p}_x \frac{dt}{\bar{t}} \tag{3-32}$$

The rate of change of $\mathbf{p}_x$ due to the decelerating effect of collisions is

$$\left. \frac{d\mathbf{p}_x}{dt} \right|_{\text{collisions}} = -\frac{\mathbf{p}_x}{\bar{t}} \tag{3-33}$$

The sum of acceleration and deceleration effects must be zero for steady state. Taking the sum of Eqs. (3-29) and (3-33), we have

$$-\frac{\mathbf{p}_x}{\bar{t}} - nq\mathscr{E}_x = 0 \tag{3-34}$$

The average momentum per electron is

$$\langle \mathbf{p}_x \rangle = \frac{\mathbf{p}_x}{n} = -q\bar{t}\mathscr{E}_x \tag{3-35}$$

where the angular brackets indicate an average over the entire group of electrons. As expected for steady state, Eq. (3-35) indicates that the electrons have *on the average* a constant net velocity in the negative x-direction:

$$\langle \mathbf{v}_x \rangle = \frac{\langle \mathbf{p}_x \rangle}{m_n^*} = -\frac{q\bar{t}}{m_n^*}\mathscr{E}_x \tag{3-36}$$

Actually, the individual electrons move in many directions by thermal motion during a given time period, but Eq. (3-36) tells us the *net drift* of an average electron in response to the electric field. The drift speed described by Eq. (3-36) is usually much smaller than the random speed due to thermal motion v_{th}.

The current density resulting from this net drift is just the number of electrons crossing a unit area per unit time ($n\langle \mathbf{v}_x \rangle$) multiplied by the charge on the electron ($-q$):

$$\boxed{\begin{array}{c} J_x = -qn\langle \mathbf{v}_x \rangle \\[2mm] \dfrac{\text{ampere}}{\text{cm}^2} = \dfrac{\text{coulomb}}{\text{electron}} \cdot \dfrac{\text{electrons}}{\text{cm}^3} \cdot \dfrac{\text{cm}}{\text{s}} \end{array}} \tag{3-37}$$

[†] Equations (3-30) and (3-31) are typical of events dominated by random processes, and the forms of these equations occur often in many branches of physics and engineering. For example, in the radioactive decay of unstable nuclear isotopes, N_0 nuclides decay exponentially with a mean lifetime $\bar{t}$. Other examples will be found in this text, including the absorption of light in a semiconductor and the recombination of excess EHPs.

Using Eq. (3-36) for the average velocity, we obtain

$$J_x = \frac{nq^2\bar{t}}{m_n^*}\mathscr{E}_x \tag{3-38}$$

Thus the current density is proportional to the electric field, as we expect from Ohm's law:

$$J_x = \sigma\mathscr{E}_x, \qquad \text{where } \sigma \equiv \frac{nq^2\bar{t}}{m_n^*} \tag{3-39}$$

The conductivity $\sigma(\Omega\text{-cm})^{-1}$ can be written

$$\sigma = qn\mu_n, \qquad \text{where } \mu_n \equiv \frac{q\bar{t}}{m_n^*} \tag{3-40}$$

The quantity μ_n, called the *electron mobility*, describes the ease with which electrons drift in the material. Mobility is a very important quantity in characterizing semiconductor materials and in device development.

The mobility defined in Eq. (3-40) can be expressed as the average particle drift velocity per unit electric field. Comparing Eqs. (3-36) and (3-40), we have

$$\mu_n = -\frac{\langle \mathsf{v}_x \rangle}{\mathscr{E}_x} \tag{3-41}$$

The units of mobility are $(\text{cm/s})/(\text{V/cm}) = \text{cm}^2/\text{V-s}$, as Eq. (3-41) suggests. The minus sign in the definition results in a positive value of mobility, since electrons drift opposite to the field.

The current density can be written in terms of mobility as

$$J_x = qn\mu_n\mathscr{E}_x \tag{3-42}$$

This derivation has been based on the assumption that the current is carried primarily by electrons. For hole conduction we change n to p, $-q$ to $+q$, and μ_n to μ_p, where $\mu_p = +\langle \mathsf{v}_x \rangle/\mathscr{E}_x$ is the mobility for holes. If both electrons and holes participate, we must modify Eq. (3-42) to

$$\boxed{J_x = q(n\mu_n + p\mu_p)\mathscr{E}_x = \sigma\mathscr{E}_x} \tag{3-43}$$

Values of μ_n and μ_p are given for many of the common semiconductor materials in Appendix III. According to Eq. (3-40), the parameters determining mobility are m^* and mean free time $\bar{t}$. Effective mass is a property of the material's band structure, as described by Eq. (3-3). Thus we expect m_n^* to be small in the strongly curved Γ minimum of the GaAs conduction band (Fig. 3-6), with the result that μ_n is very high. In a more gradually curved band, a larger m^* in the denominator of Eq. (3-40) leads to a smaller value of mobility. It is reasonable to expect that lighter particles are more mobile than heavier particles (which is satisfying, since the common-sense value of effective mass is not always apparent). The other parameter determining mobility is the mean time between scattering events, $\bar{t}$. In Section 3.4.3 we shall see that this is determined primarily by temperature and impurity concentration in the semiconductor.

3.4.2   Drift and Resistance

Let us look more closely at the drift of electrons and holes. If the semiconductor bar of Fig. 3-21 contains both types of carrier, Eq. (3-43) gives the conductivity of the material. The resistance of the bar is then

$$R = \frac{\rho L}{wt} = \frac{L}{wt}\frac{1}{\sigma} \tag{3-44}$$

where ρ is the resistivity (Ω-cm). The physical mechanism of carrier drift requires that the holes in the bar move as a group in the direction of the electric field and that the electrons move as a group in the opposite direction. Both the electron and the hole components of current are in the direction of the $\mathscr{E}$ field, since conventional current is positive in the direction of hole flow and opposite to the direction of electron flow. The drift current described by Eq. (3-43) is constant throughout the bar. A valid question arises, therefore, concerning the nature of the electron and hole flow at the contacts and in the external circuit. We should specify that the contacts to the bar of Fig. 3-21 are *ohmic*, meaning that they are perfect sources and sinks of both carrier types and have no special tendency to inject or collect either electrons or holes.

If we consider that current is carried around the external circuit by electrons, there is no problem in visualizing electrons flowing into the bar at one end and out at the other (always opposite to I). Thus for every electron leaving the left end ($x = 0$) of the bar in Fig. 3-21, there is a corresponding electron entering at $x = L$, so that the electron concentration in the bar remains constant at n. But what happens to the holes at the contacts? As a hole reaches the ohmic contact at $x = L$, it recombines with an electron, which must be supplied through the external circuit. As this hole disappears, a corresponding hole must appear at $x = 0$ to maintain space charge neutrality. It is reasonable to consider the source of this hole as the generation of an EHP at $x = 0$, with the hole flowing into the bar and the electron flowing into the external circuit.

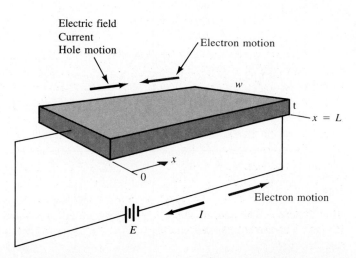

Electric field
Current
Hole motion

Electron motion

Electron motion

w

t

$x = L$

x

0

I

E

Figure 3-21
Drift of electrons and holes in a semiconductor bar.

EXAMPLE 3-5   Find the resistivity of intrinsic Ge at 300 K.

SOLUTION   From Appendix III, $\mu_n = 3900$ and $\mu_p = 1900$ cm^2/V-s for intrinsic Ge. Thus, since $n_0 = p_0 = n_i$,

$$\sigma_i = q(\mu_n + \mu_p)n_i = 1.6 \times 10^{-19}(5800)\,(2.5 \times 10^{13})$$

$$= 2.32 \times 10^{-2}\,(\Omega\text{-cm})^{-1}$$

$$\rho_i = \sigma_i^{-1} = 43\ \Omega\text{-cm}$$

which agrees with the experimental value in Appendix III.

3.4.3   Effects of Temperature and Doping on Mobility

The two basic types of scattering mechanisms that influence electron and hole mobility are *lattice scattering* and *impurity scattering*. In lattice scattering a carrier moving through the crystal is scattered by a vibration of the lattice, resulting from the temperature.[†] The frequency of such scattering events increases as the temperature increases, since the thermal agitation of the lattice becomes greater. Therefore, we should expect the mobility to decrease as the sample is heated (Fig. 3-22). On the other hand, scattering from crystal defects such as ionized impurities becomes the dominant mechanism at low temperatures. Since the atoms of the cooler lattice are less agitated, lattice scattering is less important; however, the thermal motion of the carriers is also slower.

Figure 3-22
Approximate temperature dependence of mobility with both lattice and impurity scattering.

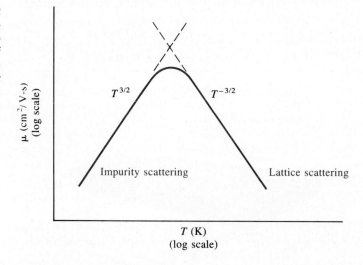

[†]Collective vibrations of atoms in the crystal are called *phonons*. Thus lattice scattering is also known as *phonon scattering*.

Since a slowly moving carrier is likely to be scattered more strongly by an in-teraction with a charged ion than is a carrier with greater momentum, impurity scattering events cause a decrease in mobility with decreasing temperature. As Fig. 3-22 indicates, the approximate temperature dependencies are $T^{-3/2}$ for lattice scattering and $T^{3/2}$ for impurity scattering. Since the scattering proba-bility of Eq. (3-32) is inversely proportional to the mean free time and there-fore to mobility, the mobilities due to two or more scattering mechanisms add inversely:

$$\frac{1}{\mu} = \frac{1}{\mu_1} + \frac{1}{\mu_2} + \cdots \qquad (3-45)$$

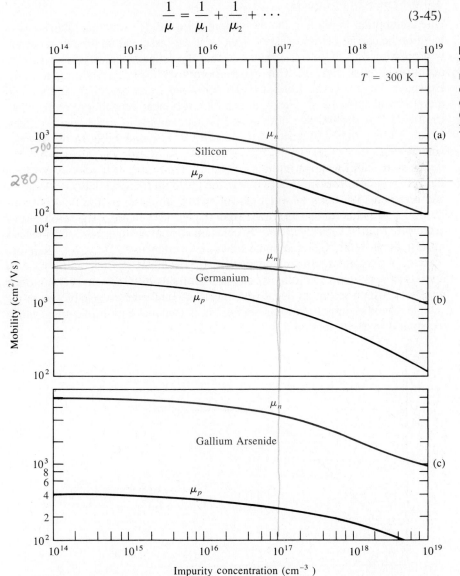

Figure 3-23
Variation of mobility with doping impurity concentration for Ge, Si, and GaAs at 300 K.

As a result, the mechanism causing the lowest mobility value dominates, as shown in Fig. 3-22.

As the concentration of impurities increases, the effects of impurity scattering are felt at higher temperatures. For example, the electron mobility μ_n of intrinsic silicon at 300 K is 1350 cm^2/(V-s). With a donor doping concentration of 10^{17} cm^{-3}, however, μ_n is 700 cm^2/(V-s). Thus the presence of the 10^{17} ionized donors/cm^3 introduces a significant amount of impurity scattering. This effect is illustrated in Fig. 3-23, which shows the variation of mobility with doping concentration at room temperature.

3.4.4 High-Field Effects

One assumption implied in the derivation of Eq. (3-39) was that Ohm's law is valid in the carrier drift processes. That is, it was assumed that the drift current is proportional to the electric field and that the proportionality constant (σ) is not a function of field $\mathscr{E}$. This assumption is valid over a wide range of $\mathscr{E}$. However, large electric fields ($>10^3$V/cm) can cause the drift velocity and therefore the current $J = -qn\mathbf{v}_d$ to exhibit a sublinear dependence on the electric field. This dependence of σ upon $\mathscr{E}$ is an example of a *hot carrier* effect, which implies that the carrier drift velocity $\mathbf{v}_d$ is comparable to the thermal velocity $\mathbf{v}_{th}$.

In many cases an upper limit is reached for the carrier drift velocity in a high field (Fig. 3-24). This limit occurs near the mean thermal velocity ($\simeq 10^7$ cm/s) and represents the point at which added energy imparted by the field is transferred to the lattice rather than increasing the carrier velocity. The result of this *scattering limited velocity* is a fairly constant current at high field. This behavior is typical of Si, Ge, and some other semiconductors. However, there are other important effects in some materials; for example, in Chapter 12 we shall discuss a *decrease* in electron velocity at high fields for GaAs and certain other materials, which results in negative conductivity and current instabilities in the sample. Another important high-field effect is avalanche multiplication, which we shall discuss in Section 5.4.2.

Figure 3-24
Saturation of electron drift velocity at high electric fields for Si.

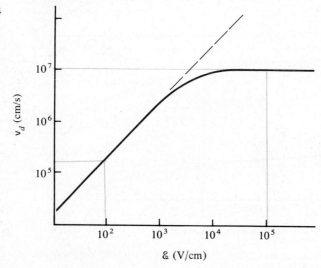

3.4.5 The Hall Effect

If a magnetic field is applied perpendicular to the direction in which holes drift in a p-type bar, the path of the holes tends to be deflected (Fig. 3-25). Using vector notation, the total force on a single hole due to the electric and magnetic fields is

$$\mathbf{F} = q(\mathscr{E} + \mathbf{v} \times \mathscr{B}) \qquad (3\text{-}46)$$

In the y-direction the force is

$$F_y = q(\mathscr{E}_y - \mathbf{v}_x \mathscr{B}_z) \qquad (3\text{-}47)$$

The important result of Eq. (3-47) is that unless an electric field $\mathscr{E}_y$ is established along the width of the bar, each hole will experience a net force (and therefore an acceleration) in the $-y$-direction due to the $q\mathbf{v}_x\mathscr{B}_z$ product. Therefore, to maintain a steady state flow of holes down the length of the bar, the electric field $\mathscr{E}_y$ must just balance the product $\mathbf{v}_x\mathscr{B}_z$:

$$\mathscr{E}_y = \mathbf{v}_x \mathscr{B}_z \qquad (3\text{-}48)$$

so that the net force F_y is zero. Physically, this electric field is set up when the magnetic field shifts the hole distribution slightly in the $-y$-direction. Once the electric field $\mathscr{E}_y$ becomes as large as $\mathbf{v}_x\mathscr{B}_z$, no net lateral force is experienced by the holes as they drift along the bar. The establishment of the electric field $\mathscr{E}_y$ is known as the *Hall effect*, and the resulting voltage $V_{AB} = \mathscr{E}_y w$ is called

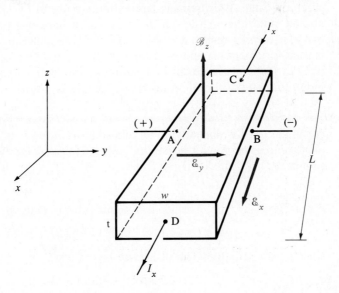

Figure 3-25
The Hall effect.

the *Hall voltage*. If we use the expression derived in Eq. (3-37) for the drift velocity (using $+q$ and p_0 for holes), the field $\mathscr{E}_y$ becomes

$$\mathscr{E}_y = \frac{J_x}{qp_0}\mathscr{B}_z = R_H J_x \mathscr{B}_z, \qquad R_H \equiv \frac{1}{qp_0} \tag{3-49}$$

Thus the Hall field is proportional to the product of the current density and the magnetic flux density. The proportionality constant $R_H = (qp_0)^{-1}$ is called the *Hall coefficient*. A measurement of the Hall voltage for a known current and magnetic field yields a value for the hole concentration p_0

$$p_0 = \frac{1}{qR_H} = \frac{J_x \mathscr{B}_z}{q\mathscr{E}_y} = \frac{(I_x/wt)\mathscr{B}_z}{q(V_{AB}/w)} = \frac{I_x \mathscr{B}_z}{qtV_{AB}} \tag{3-50}$$

Since all of the quantities in the right-hand side of Eq. (3-50) can be measured, the Hall effect can be used to give quite accurate values for carrier concentration.

If a measurement of resistance R is made, the sample resistivity ρ can be calculated:

$$\rho(\Omega\text{-cm}) = \frac{Rwt}{L} = \frac{V_{CD}/I_x}{L/wt} \tag{3-51}$$

Since the conductivity $\sigma = 1/\rho$ is given by $q\mu_p p_0$, the mobility is simply the ratio of the Hall coefficient and the resistivity:

$$\mu_p = \frac{\sigma}{qp_0} = \frac{1/\rho}{q(1/qR_H)} = \frac{R_H}{\rho} \tag{3-52}$$

Measurements of the Hall coefficient and the resistivity over a range of temperatures yield plots of majority carrier concentration and mobility vs. temperature. Such measurements are extremely useful in the analysis of semiconductor materials. Although the discussion here has been related to p-type material, similar results are obtained for n-type material. A negative value of q is used for electrons, and the Hall voltage V_{AB} and Hall coefficient R_H are negative. In fact, measurement of the sign of the Hall voltage is a common technique for determining if an unknown sample is p-type or n-type.

EXAMPLE 3-6   A sample of Si is doped with 10^{17} phosphorus atoms/cm³. What would you expect to measure for its resistivity? What Hall voltage would you expect in a sample 100 μm thick if $I_x = 1$ mA and $\mathscr{B}_z = 1$ kG $= 10^{-5}$ Wb/cm²?

SOLUTION   From Fig. 3-23, the mobility is 700 cm²/(V-s). Thus the conductivity is

$$\sigma = q\mu_n n_0 = (1.6 \times 10^{-19})(700)(10^{17}) = 11.2(\Omega\text{-cm})^{-1}$$

since p_0 is negligible. The resistivity is

$$\rho = \sigma^{-1} = 0.0893 \ \Omega\text{-cm}$$

The Hall coefficient is

$$R_H = -(qn_0)^{-1} = -62.5 \text{ cm}^3/\text{C}$$

from Eq. (3-49), or we could use Eq. (3-52). The Hall voltage is

$$V_{AB} = \frac{I_x \mathscr{B}_z}{t} R_H = \frac{(10^{-3} \text{ A})(10^{-5} \text{ Wb/cm}^2)}{10^{-2} \text{ cm}}(-62.5 \text{ cm}^3/\text{C}) = -62.5 \text{ } \mu\text{V}$$

■

In this chapter we have discussed homogeneous semiconductors, without variations in doping and without junctions between dissimilar materials. In the following chapters we will be considering cases in which nonuniform doping occurs in a given semiconductor, or junctions occur between different semiconductors or a semiconductor and a metal. These cases are crucial to the various types of electronic and optoelectronic devices made in semiconductors. In anticipation of those discussions, an important concept should be established here regarding the demands of equilibrium. That concept can be summarized by noting that *no discontinuity or gradient can arise in the equilibrium Fermi level E_F.*

To demonstrate this assertion, let us consider two materials in intimate contact such that electrons can move between the two (Fig. 3-26). These may be, for example, dissimilar semiconductors, n- and p-type regions, a metal and a semiconductor, or simply two adjacent regions of a nonuniformly doped semiconductor. Each material is described by a Fermi–Dirac distribution function and some distribution of available energy states that electrons can occupy.

There is no current, and therefore no net charge transport, at thermal equilibrium. There is also no net transfer of energy. Therefore, for each energy E in Fig. 3-26 any transfer of electrons from material 1 to material 2 must be exactly balanced by the opposite transfer of electrons from 2 to 1. We will let the density of states at energy E in material 1 be called $N_1(E)$ and in material 2 we

3.5
INVARIANCE OF
THE FERMI LEVEL
AT EQUILIBRIUM

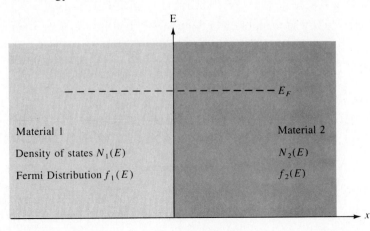

E

Material 1
Density of states $N_1(E)$
Fermi Distribution $f_1(E)$

Material 2
$N_2(E)$
$f_2(E)$

E_F

x

Figure 3-26
Two materials in intimate contact at equilibrium. Since the net motion of electrons is zero, the equilibrium Fermi level must be constant throughout.

will call it $N_2(E)$. At energy E the rate of transfer of electrons from 1 to 2 is proportional to the number of filled states at E in material 1 times the number of empty states at E in material 2:

$$\text{rate from 1 to 2} \propto N_1(E)f_1(E) \cdot N_2(E)[1 - f_2(E)] \qquad (3\text{-}53)$$

where $f(E)$ is the probability of a state being filled at E in each material, i.e., the Fermi–Dirac distribution function given by Eq. (3-10). Similarly,

$$\text{rate from 2 to 1} \propto N_2(E)f_2(E) \cdot N_1(E)[1 - f_1(E)] \qquad (3\text{-}54)$$

At equilibrium these must be equal:

$$N_1(E)f_1(E) \cdot N_2(E)[1 - f_2(E)] = N_2(E)f_2(E) \cdot N_1(E)[1 - f_1(E)] \qquad (3\text{-}55)$$

Rearranging terms, we have, at energy E,

$$N_1 f_1 N_2 - N_1 f_1 N_2 f_2 = N_2 f_2 N_1 - N_2 f_2 N_1 f_1 \qquad (3\text{-}56)$$

which results in

$$f_1(E) = f_2(E), \quad \text{that is,} \quad [1 + e^{(E-E_{F1})/kT}]^{-1} = [1 + e^{(E-E_{F2})/kT}]^{-1} \qquad (3\text{-}57)$$

Therefore, we conclude that $E_{F1} = E_{F2}$. That is, there is no discontinuity in the equilibrium Fermi level. More generally, we can state that the Fermi level at equilibrium must be constant throughout materials in intimate contact. One way of stating this is that no gradient exists in the Fermi level at equilibrium:

$$\boxed{\frac{dE_F}{dx} = 0} \qquad (3\text{-}58)$$

We will make considerable use of this result in the chapters to follow.

PROBLEMS

3.1. It was mentioned in Section 3.2 that the covalent bonding model gives a false impression of the localization of carriers. As an illustration, calculate the radius of the electron orbit around the donor in Fig. 3-12, assuming a ground state hydrogen-like orbit in Si. Compare with the Si lattice constant. Use $m_n^* = 0.26m_0$ for Si.

3.2. Calculate values for the Fermi function $f(E)$ at 300 K and plot vs. energy in eV as in Fig. 3-14. Choose $E_F = 1$ eV and make the calculated points closer together near the Fermi level to obtain a smooth curve. Notice that $f(E)$ varies quite rapidly within a few kT of E_F.

3.3. Show that the probability that a state ΔE above the Fermi level E_F is filled equals the probability that a state ΔE below E_F is empty.

3.4. Given m_n^* and m_p^* from Table 3-1, calculate the effective densities of states N_c and N_v for Si at 300 K (assume m_n^* and m_p^* do not vary with temperature). Calculate the intrinsic carrier concentration and compare this with the value given in Fig. 3-17. Expect some error to arise from the simplifying assumptions.

3.5. Show that Eqs. (3-25) result from Eqs. (3-15) and (3-19). If $n_0 = 5 \times 10^{16}$ cm^{-3}, where is the Fermi level relative to E_i in Si at 300 K?

3.6. Calculate the band gap of Si from Eq. (3-23) and the plot of n_i vs. $1000/T$ (Fig. 3-17). *Hint:* the slope cannot be measured directly from a semilogarithmic plot; read the values from two points on the plot and take the natural logarithm as needed for the solution.

3.7. (a) Explain why holes are found at the *top* of the valence band, whereas electrons are found at the *bottom* of the conduction band.
   (b) Explain why Si doped with 10^{14} cm^{-3} Sb is n-type at 400 K but similarly doped Ge is not.

3.8. Derive an expression relating the intrinsic level E_i to the center of the band gap $E_g/2$. Calculate the displacement of E_i from $E_g/2$ for Si at 300 K, assuming the effective mass values in Table 3-1 are valid at this temperature.

3.9. (a) A Si sample is doped with 10^{17} boron atoms/cm^3. What is the electron concentration n_0 at 300 K? What is the resistivity?
   (b) A Ge sample is doped with 3×10^{13} Sb atoms/cm^3. Using the requirements of space charge neutrality, calculate the electron concentration n_0 at 300 K.

3.10. A Si sample is doped with 6×10^{15} cm^{-3} donors and 2×10^{15} cm^{-3} acceptors. Find the position of the Fermi level with respect to E_i at 300 K. What is the value and sign of the Hall coefficient?

3.11. (a) Show that the minimum conductivity of a semiconductor sample occurs when $n_0 = n_i \sqrt{\mu_p/\mu_n}$. *Hint:* begin with Eq. (3-43) and apply Eq. (3-24).
   (b) What is the expression for the minimum conductivity σ_{min}?
   (c) Calculate σ_{min} for Si at 300 K and compare with the intrinsic conductivity.

3.12. (a) A Si bar 0.1 cm long and 100 μm^2 in cross-sectional area is doped with 10^{17} cm^{-3} antimony. Find the current at 300 K with 10V applied. Repeat for a Si bar 1 μm long.
   (b) How long does it take an average electron to drift 1 μm in pure Si at an electric field of 100 V/cm? Repeat for 10^5 V/cm.

3.13. (a) A Si sample is doped with 10^{15} cm^{-3} boron atoms, and a certain number of shallow donors. The Fermi level is 0.33 eV above E_i at 300 K. What is the donor concentration N_d?

   (b) A Si sample contains 10^{16} cm^{-3} In acceptor atoms and a certain number of shallow donors. The In acceptor level is 0.16 eV above E_v, and E_F is 0.2 eV above E_v at 300 K. How many (cm^{-3}) In atoms are unionized (i.e., neutral)?

3.14. In soldering wires to a sample such as that shown in Fig. 3-25, it is difficult to align the Hall probes A and B precisely. If B is displaced slightly down the length of the bar from A, an erroneous Hall voltage results. Show that the true Hall voltage V_H can be obtained from two measurements of V_{AB}, with the magnetic field first in the $+z$-direction and then in the $-z$-direction.

3.15. A Ge sample is properly contacted and oriented in a 5-kG magnetic field as in Fig. 3-25. The current is 4 mA. The sample dimensions are $w = 0.25$ mm, $t = 50$ μm, and $L = 2.5$ mm. The following data are taken: $V_{AB} = -2.5$ mV and $V_{CD} = 170$ mV. Find the type and concentration of the majority carrier and its mobility. *Note:* 1 kG $= 10^{-5}$ Wb/cm^2. Comment on the mobility value.

*3.16. Use Eq. (3-45) to calculate and plot the mobility vs. temperature $\mu(T)$ from 10 K to 500 K for Si doped with $N_d = 10^{14}$, 10^{16}, and 10^{18} donors cm^{-3}. Consider the mobility to be determined by impurity and phonon (lattice) scattering. Impurity scattering limited mobility can be described by

$$\mu_I = 3.29 \times 10^{15} \frac{\epsilon_r^2 T^{3/2}}{N_d^+ (m_n^*/m_0)^{1/2}\left[\ln(1 + z) - \dfrac{z}{1 + z}\right]}$$

where

$$z = 1.3 \times 10^{13} \epsilon_r T^2 (m_n^*/m_0) (N_d^+)^{-1}$$

Assume that the ionized impurity concentration N_d^+ is equal to N_d at all temperatures.

The conductivity effective mass m_n^* for Si is 0.26 m_0. Acoustic phonon (lattice) scattering limited mobility can be described by

$$\mu_{AC} = 1.18 \times 10^{-5} c_1 (m_n^*/m_0)^{-5/2} T^{-3/2} (E_{AC})^{-2}$$

where the stiffness (c_1) is given by

$$c_1 = 1.9 \times 10^{12} \text{ dyn cm}^{-2} \text{ for Si}$$

and the conduction band acoustic deformation potential (E_{AC}) is

$$E_{AC} = 9.5 \text{ eV for Si}$$

*3.17. Rework Prob. 3.16 considering carrier freeze-out onto donors at low T. That is, consider

$$N_d^+ = \frac{N_d}{1 + \exp(E_d/kT)}$$

as the ionized impurity concentration. Consider the donor ionization energy (E_d) to be 45 meV for Si.

READING LIST

Ashcroft, N. W., and N. D. Mermin. *Solid State Physics.* New York: Holt, Rinehart & Winston, 1976.

Blakemore, J. S. *Semiconductor Statistics.* New York: Dover Publications, 1987.

Capasso, F. "Band-Gap Engineering: From Physics and Materials to New Semiconductor Devices," *Science* 235, (January 9, 1987): 172–6.

Drummond, T. J., P. L. Gourley, and T. E. Zipperian. "Quantum-Tailored Solid-State Devices." *IEEE Spectrum* (June 1988): 33–37.

Hess, K. *Advanced Theory of Semiconductor Devices.* Englewood Cliffs, N.J.: Prentice Hall, 1988.

Hess, K., and N. Holonyak, Jr. "Hot Electrons in Layered Semiconductors." *Physics Today* 33, no. 10 (October 1980): 40–47.

Kittel, C. *Introduction to Solid State Physics.* 6th ed. New York: John Wiley, 1986.

Pierret, R. F. *The Modular Series on Solid State Devices. Vol. I: Semiconductor Fundamentals,* 2nd ed. Reading, Mass.: Addison-Wesley, 1988.

Pierret, R. F. *The Modular Series on Solid State Devices. Vol. VI: Advanced Semiconductor Fundamentals.* Reading, Mass.: Addison-Wesley, 1987.

Pulfrey, D. L., and N. G. Tarr. *Introduction to Microelectronic Devices.* Englewood Cliffs, N.J.: Prentice Hall, 1989.

Van Der Ziel, A. *Solid State Physical Electronics.* 3d ed. Englewood Cliffs, N.J.: Prentice Hall, 1976.

Wang, S. *Fundamentals of Semiconductor Theory and Device Physics.* Englewood Cliffs, N.J.: Prentice Hall, 1989.

Wolfe, C. M., N. Holonyak, and G. E. Stillman, *Physical Properties of Semiconductors.* Englewood Cliffs, N.J.: Prentice Hall, 1989.

chapter 4

EXCESS CARRIERS
IN SEMICONDUCTORS

Most semiconductor devices operate by the creation of charge carriers in excess of the thermal equilibrium values. These excess carriers can be created by optical excitation or electron bombardment, or as we shall see in Chapter 5, they can be injected across a forward-biased p-n junction. However the excess carriers arise, they can dominate the conduction processes in the semiconductor material. In this chapter we shall investigate the creation of excess carriers by optical absorption and the resulting properties of photoluminescence and photoconductivity. We shall study more closely the mechanism of electron–hole pair recombination and the effects of carrier trapping. Finally, we shall discuss the diffusion of excess carriers due to a carrier gradient, which serves as a basic mechanism of current conduction along with the mechanism of drift in an electric field.

4.1
OPTICAL
ABSORPTION[†]

An important technique for measuring the band gap energy of a semiconductor is the absorption of incident photons by the material. In this experiment photons of selected wavelength are directed at the sample, and the relative transmission of the various photons is observed. Since photons with energies greater than the band gap energy are absorbed while photons with energies less than

---

[†]In this context the word "optical" does not necessarily imply that the photons absorbed are in the visible part of the spectrum. Many semiconductors absorb photons in the infrared region, but this is included in the term "optical absorption."

the band gap are transmitted, this experiment gives an accurate measure of the band gap energy.

It is apparent that a photon with energy $h\nu \geq E_g$ can be absorbed in a semiconductor (Fig. 4-1). Since the valence band contains many electrons and the conduction band has many empty states into which the electrons may be excited, the probability of photon absorption is high. As Fig. 4-1 indicates, an electron excited to the conduction band by optical absorption may initially have more energy than is common for conduction band electrons (almost all electrons are near E_c unless the sample is very heavily doped). Thus the excited electron loses energy to the lattice in scattering events until its velocity reaches the thermal equilibrium velocity of other conduction band electrons. The electron and hole created by this absorption process are *excess carriers;* since they are out of balance with their environment, they must eventually recombine. While the excess carriers exist in their respective bands, however, they are free to contribute to the conductivity of the material.

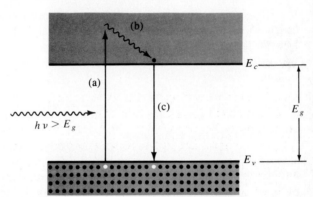

Figure 4-1
Optical absorption of a photon with $h\nu > E_g$: (a) an EHP is created during photon absorption; (b) the excited electron gives up energy to the lattice by scattering events; (c) the electron recombines with a hole in the valence band.

A photon with energy less than E_g is unable to excite an electron from the valence band to the conduction band. Thus in a pure semiconductor, there is negligible absorption of photons with $h\nu < E_g$. This explains why some materials are transparent in certain wavelength ranges. We are able to "see through" certain insulators, such as a good NaCl crystal, because a large energy gap containing no electron states exists in the material. If the band gap is about 2 eV wide, only long wavelengths (infrared) and the red part of the visible spectrum are transmitted; on the other hand, a band gap of about 3 eV allows infrared and the entire visible spectrum to be transmitted.

If a beam of photons with $h\nu > E_g$ falls on a semiconductor, there will be some predictable amount of absorption, determined by the properties of the material. We would expect the ratio of transmitted to incident light intensity to depend on the photon wavelength and the thickness of the sample. To calculate this dependence, let us assume that a photon beam of intensity $\mathbf{I}_0$ (photons/cm^2-s) is directed at a sample of thickness l (Fig. 4-2). The beam contains only photons of wavelength λ, selected by a monochromator. As the beam passes through the sample, its intensity at a distance x from the surface can be calcu-

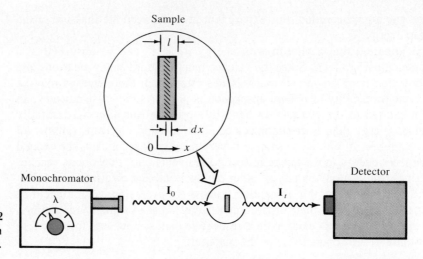

Figure 4-2
Optical absorption
experiment.

lated by considering the probability of absorption within any increment dx. Since a photon which has survived to x without absorption has no memory of how far it has traveled, its probability of absorption in any dx is constant. Thus the degradation of the intensity $-d\mathbf{I}(x)/dx$ is proportional to the intensity remaining at x:

$$-\frac{d\mathbf{I}(x)}{dx} = \boldsymbol{\alpha}\mathbf{I}(x) \tag{4-1}$$

The solution to this equation is

$$\mathbf{I}(x) = \mathbf{I}_0 e^{-\alpha x} \tag{4-2}$$

and the intensity of light transmitted through the sample thickness l is

$$\mathbf{I}_t = \mathbf{I}_0 e^{-\alpha l} \tag{4-3}$$

The coefficient $\boldsymbol{\alpha}$ is called the *absorption coefficient* and has units of cm^{-1}. This coefficient will of course vary with the photon wavelength and with the material. In a typical plot of $\boldsymbol{\alpha}$ vs. wavelength (Fig. 4-3), there is negli-

Figure 4-3
Dependence of
optical absorption
coefficient $\boldsymbol{\alpha}$ for a
semiconductor on
the wavelength of
incident light.

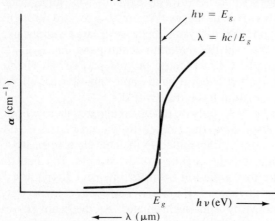

gible absorption at long wavelengths (*hν* small) and considerable absorption of photons with energies larger than E_g. According to Eq. (2-2), the relation between photon energy and wavelength is $E = hc/\lambda$. If E is given in electron volts and λ in micrometers, this becomes $E = 1.24/\lambda$.

Figure 4-4 indicates the band gap energies of some of the common semiconductors, relative to the visible, infrared, and ultraviolet portions of the spectrum. We observe that GaAs, Si, Ge, and InSb lie outside the visible region, in the infrared. Other semiconductors, such as GaP and CdS, have band gaps wide enough to pass photons in the visible range. It is important to note here that a semiconductor absorbs photons with energies equal to the band gap, *or larger*. Thus Si absorbs not only band gap light (~1 *μ*m) but also shorter wavelengths, including those in the visible part of the spectrum.

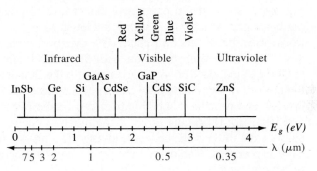

Figure 4-4
Band gaps of some common semiconductors relative to the optical spectrum.

When electron-hole pairs are generated in a semiconductor, or when carriers are excited into higher impurity levels from which they fall to their equilibrium states, light can be given off by the material. Many of the semiconductors are well suited for light emission, particularly the compound semiconductors with direct band gaps. The general property of light emission is called *luminescence*.[†] This overall category can be subdivided according to the excitation mechanism: If carriers are excited by photon absorption, the radiation resulting from the recombination of the excited carriers is called *photoluminescence;* if the excited carriers are created by high-energy electron bombardment of the material, the mechanism is called *cathodoluminescence;* if the excitation occurs by the introduction of current into the sample, the resulting luminescence is called *electroluminescence.* Other types of excitation are possible, but these three are the most important for device applications.

**4.2
LUMINESCENCE**

[†]The emission processes considered here should not be confused with radiation due to incandescence which occurs in heated materials. The various luminescent mechanisms can be considered "cold" processes as compared to the "hot" process of incandescence, which increases with temperature. In fact, most luminescent processes become more efficient as the temperature is lowered.

4.2.1 Photoluminescence

The simplest example of light emission from a semiconductor occurs for direct excitation and recombination of an EHP, as depicted in Fig. 3-5a. If the recombination occurs directly rather than via a defect level, band gap light is given off in the process. For steady state excitation, the recombination of EHPs occurs at the same rate as the generation, and one photon is emitted for each photon absorbed. Direct recombination is a fast process; the mean lifetime of the EHP is usually on the order of 10^{-8} s or less. Thus the emission of photons stops within approximately 10^{-8} s after the excitation is turned off. Such fast luminescent processes are often referred to as *fluorescence*. In some materials, however, emission continues for periods up to seconds or minutes after the excitation is removed. These slow processes are called *phosphorescence*, and the materials are called *phosphors*. An example of a slow process is shown in Fig. 4-5. This material contains a defect level (perhaps due to an impurity) in the band gap which has a strong tendency to temporarily capture (*trap*) electrons from the conduction band. The events depicted in the figure are as follows: (a) An incoming photon with $h\nu_1 > E_g$ is absorbed, creating an EHP; (b) the excited electron gives up energy to the lattice by scattering until it nears the bottom of the conduction band; (c) the electron is *trapped* by the impurity level E_t and remains trapped until it can be thermally reexcited to the conduction band (d); (e) finally direct recombination occurs as the electron falls to an empty state in the valence band, giving off a photon ($h\nu_2$) of approximately the band gap energy. The delay time between excitation and recombination can be relatively long if the probability of thermal reexcitation from the trap (d) is small. Even longer delay times result if the electron is retrapped several times before recombination. If the trapping probability is greater than the probability of recombination, an electron may make several trips between the trap and the conduction band before recombination finally occurs. In such a material the emission of phosphorescent light persists for a relatively long time after the excitation is removed.

The color of light emitted by a phosphor such as ZnS depends primarily on the impurities present, since many radiative transitions involve impurity levels

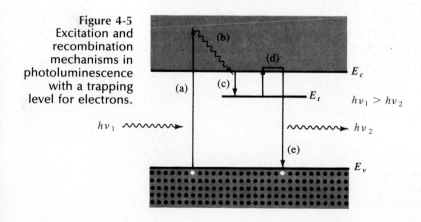

Figure 4-5
Excitation and recombination mechanisms in photoluminescence with a trapping level for electrons.

$h\nu_1 > h\nu_2$

within the band gap. This selection of colors is particularly useful in the fabrication of a color television screen, as discussed in the next section.

One of the most common examples of photoluminescence is the fluorescent lamp. Typically such a lamp is composed of a glass tube filled with gas (e.g., a mixture of argon and mercury), with a fluorescent coating on the inside of the tube. When an electric discharge is induced between electrodes in the tube, the excited atoms of the gas emit photons, largely in the visible and ultra-violet regions of the spectrum. This light is absorbed by the luminescent coating, and visible photons are emitted. The efficiency of such a lamp is considerably better than that of an incandescent bulb, and the wavelength mixture in light given off can be adjusted by proper selection of the fluorescent material.

A 0.46-μm-thick sample of GaAs is illuminated with monochromatic light of $h\nu = 2$ eV. The absorption coefficient α is 5×10^4 cm^{-1}. The power incident on the sample is 10 mW.

EXAMPLE 4-1

(a) Find the total energy absorbed by the sample per second (J/s).
(b) Find the rate of excess thermal energy given up by the electrons to the lattice before recombination (J/s).
(c) Find the number of photons per second given off from recombination events, assuming perfect quantum efficiency.

(a) From Eq. (4-3),

SOLUTION

$$\mathbf{I}_t = \mathbf{I}_0 e^{-\alpha l} = 10^{-2} \exp(-5 \times 10^4 \times 0.46 \times 10^{-4})$$
$$= 10^{-2} e^{-2.3} = 10^{-3} \text{ W}$$

Thus the absorbed power is

$$10 - 1 = 9 \text{ mW} = 9 \times 10^{-3} \text{ J/s} \quad \simeq \mathbf{I}_1$$

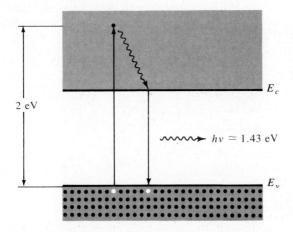

(b) The fraction of each photon energy unit which is converted to heat is

$$\frac{2 - 1.43}{2} = 0.285$$

Thus the amount of energy converted to heat per second is

$$0.285 \times 9 \times 10^{-3} = 2.57 \times 10^{-3} \text{ J/s}$$

(c) Assuming one emitted photon for each photon absorbed (perfect quantum efficiency), we have

$$\frac{9 \times 10^{-3} \text{ J/s}}{1.6 \times 10^{-19} \text{ J/eV} \times 2 \text{ eV/photon}} = 2.81 \times 10^{16} \text{ photons/s}$$

Alternative solution: Recombination radiation accounts for $9 - 2.57 = 6.43$ mW at 1.43 eV/photon.

$$\frac{6.43 \times 10^{-3}}{1.6 \times 10^{-19} \times 1.43} = 2.81 \times 10^{16} \text{ photons/s}$$

4.2.2 Cathodoluminescence

The most common example of the excitation of luminescent materials by energetic electrons is the cathode-ray tube (CRT). This light-emitting tube is the basis of the oscilloscope, television set, and other visual-display systems. The basic principle of the CRT is the selective excitation of a phosphorescent screen by a beam of energetic electrons within a vacuum tube (Fig. 4-6). Electrons are emitted from a heated cathode and are accelerated by an electric field to a positively charged anode. A beam of electrons passes through a hole in the anode and proceeds to the deflection system where the path of the electrons is bent by an electric or a magnetic field. By varying the deflection, the electron beam can be directed to any spot on the screen. As the energetic electrons collide with the phosphor coating, energy is given up to the phosphor as electrons are excited to higher states within the material. When these electrons recombine, the screen gives off light at the desired spot. In a television picture tube, the intensity and position of the beam is varied to give a visible replica of the image being received.

Figure 4-6
Simplified view of a
cathode-ray tube.

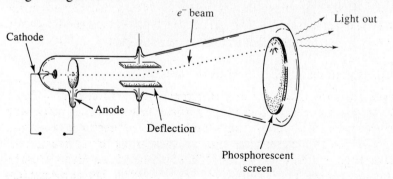

In a color television tube the screen is coated with a pattern of phosphor dots, which are doped to emit the primary additive colors—red, green, and blue. To excite the proper dots, three electron beams are swept together through the pattern. Each beam controls the excitation of one type of phosphor dot. By varying the relative intensity of light emitted by each of the dots, a broad range of colors can be synthesized.

4.2.3   Electroluminescence

There are many ways by which electrical energy can be used to generate photon emission in a solid. In LEDs an electric current causes the injection of minority carriers into regions of the crystal where they can recombine with majority carriers, resulting in the emission of recombination radiation. This important effect (*injection electroluminescence*) will be discussed in Chapters 6 and 10 in terms of p-n junction theory.

The first electroluminescent effect to be observed was the emission of photons by certain phosphors in an alternating electric field (the Destriau effect). In this device, a phosphor powder such as ZnS is held in a binder material (often a plastic) of a high dielectric constant. When an a-c electric field is applied, light is given off by the phosphor. Such cells can be useful as lighting panels, although their efficiency has thus far been too low for most applications and their reliability is poor.

When excess electrons and holes are created in a semiconductor, there is a corresponding increase in the conductivity of the sample as indicated by Eq. (3-43). If the excess carriers arise from optical excitation, the resulting increase in conductivity is called *photoconductivity*. This is an important effect, with useful applications in the analysis of semiconductor materials and in the operation of several types of devices. In this section we shall examine the mechanisms by which excess electrons and holes recombine and apply the recombination kinetics to the analysis of photoconductive devices. However, the importance of recombination is not limited to cases in which the excess carriers are created optically. In fact, virtually every semiconductor device depends in some way on the recombination of excess electrons and holes. Therefore, the concepts developed in this section will be used extensively in the analyses of diodes, transistors, lasers, and other devices in later chapters.

**4.3
CARRIER LIFETIME
AND PHOTO-
CONDUCTIVITY**

4.3.1   Direct Recombination of Electrons and Holes

It was pointed out in Section 3.1.4 that electrons in the conduction band of a semiconductor may make transitions to the valence band (i.e, recombine with holes in the valence band) either directly or indirectly. In direct recombination, an excess population of electrons and holes decays by electrons falling from the conduction band to empty states (holes) in the valence band. Energy lost by an electron in making the transition is given up as a photon. Direct recombination occurs *spontaneously;* that is, the probability that an electron and a hole will recombine is constant in time. As in the case of carrier scattering, this constant probability leads us to expect an exponential solution for the decay of the excess carriers. In this case the rate of decay of electrons at any time t is proportional to the number of electrons remaining at t and the number of holes, with some constant of proportionality for recombination, α_r. The *net* rate of change in the conduction band electron concentration is the thermal generation rate $\alpha_r n_i^2$ from Eq. (3-7) minus the recombination rate

$$\frac{dn(t)}{dt} = \alpha_r n_i^2 - \alpha_r n(t) p(t) \tag{4-4}$$

Let us assume the excess electron-hole population is created at $t = 0$, for example by a short flash of light, and the initial excess electron and hole concentrations Δn and Δp are equal.[†] Then as the electrons and holes recombine in pairs, the instantaneous concentrations of excess carriers $\delta n(t)$ and $\delta p(t)$ are also equal. Thus we can write the total concentrations of Eq. (4-4) in terms of the equilibrium values n_0 and p_0 and the excess carrier concentrations $\delta n(t) = \delta p(t)$. Using Eq. (3-24) we have

$$\frac{d\delta n(t)}{dt} = \alpha_r n_i^2 - \alpha_r [n_0 + \delta n(t)][p_0 + \delta p(t)]$$

$$= -\alpha_r [(n_0 + p_0)\delta n(t) + \delta n^2(t)] \tag{4-5}$$

This nonlinear equation would be difficult to solve in its present form. Fortunately, it can be simplified for the case of low-level injection. If the excess carrier concentrations are small, we can neglect the δn^2 term. Furthermore, if the material is extrinsic, we can usually neglect the term representing the equilibrium minority carriers. For example, if the material is p-type ($p_0 \gg n_0$), Eq. (4-5) becomes

$$\frac{d\delta n(t)}{dt} = -\alpha_r p_0 \,\delta n(t) \tag{4-6}$$

The solution to this equation is an exponential decay from the original excess carrier concentration Δn:

$$\delta n(t) = \Delta n e^{-\alpha_r p_0 t} = \Delta n e^{-t/\tau_n} \tag{4-7}$$

Excess electrons in a p-type semiconductor recombine with a decay constant $\tau_n = (\alpha_r p_0)^{-1}$, called the *recombination lifetime*. Since the calculation is made in terms of the minority carriers, τ_n is often called the *minority carrier lifetime*. The decay of excess holes in n-type material occurs with $\tau_p = (\alpha_r n_0)^{-1}$. In the case of direct recombination, the excess majority carriers decay at exactly the same rate as the minority carriers.

[†]We will use $\delta n(t)$ and $\delta p(t)$ to mean instantaneous excess carrier concentrations, and Δn, Δp for their values at $t = 0$. Later we will use similar symbolism for spatial distributions, such as $\delta n(x)$ and $\Delta n(x = 0)$.

A numerical example may be helpful in visualizing the approximations made in the analysis of direct recombination. Let us assume a sample of GaAs is doped with 10^{15} acceptors/cm^3. The intrinsic carrier concentration of GaAs is approximately 10^6 cm^{-3}; thus the minority electron concentration is $n_0 = n_i^2/p_0 = 10^{-3}$ cm^{-3}. Certainly the approximation $p_0 \gg n_0$ is valid in this case. Now if 10^{14} EHP/cm^3 are created at $t = 0$, we can calculate the decay of these carriers in time. The approximation $\delta n \ll p_0$ is reasonable, as Fig. 4-7 indicates. This figure shows the decay in time of the excess populations for a carrier recombination lifetime of $\tau_n = \tau_p = 10^{-8}$ s.

EXAMPLE 4-2

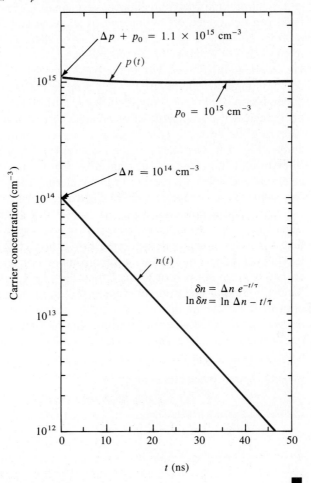

Figure 4-7
Decay of excess electrons and holes by recombination, for $\Delta n = \Delta p = 0.1 p_0$, with n_0 negligible, and $\tau = 10$ ns (Example 4-2). The exponential decay of $\delta n(t)$ is linear on this semilogarithmic graph.

There is a large percentage change in the minority carrier electron concentration in this example and a small percentage change in the majority hole concentration. Basically, the approximations of extrinsic material and low-level injection allow us to represent $n(t)$ in Eq. (4-4) by the excess concentration $\delta n(t)$ and $p(t)$ by the equilibrium value p_0. Figure 4-7 indicates that this is a good approximation for the example. A more general expression for the carrier lifetime is

$$\tau_n = \frac{1}{\alpha_r(n_0 + p_0)} \tag{4-8}$$

This expression is valid for n- or p-type material if the injection level is low.

4.3.2   Indirect Recombination; Trapping

In column IV semiconductors and in certain compounds, the probability of direct electron-hole recombination is very small (Appendix III). There is some band gap light given off by materials such as Si and Ge during recombination, but this radiation is very weak and may be detected only by sensitive equipment. The vast majority of the recombination events in indirect materials occur via *recombination levels* within the band gap, and the resulting energy loss by recombining electrons is usually given up to the lattice as heat rather than by the emission of photons. Any impurity or lattice defect can serve as a recombination center if it is capable of receiving a carrier of one type and subsequently capturing the opposite type of carrier, thereby annihilating the pair. For example, Fig. 4-8 illustrates a recombination level E_r which is below E_F at equilibrium and therefore is substantially filled with electrons. When excess electrons and holes are created in this material, each EHP recombines at E_r in two steps: (a) hole capture and (b) electron capture.

Since the recombination centers in Fig. 4-8 are filled at equilibrium, the first event in the recombination process is hole capture. It is important to note that this event is equivalent to an electron at E_r falling to the valence band, leaving behind an empty state in the recombination level. Thus in hole capture, energy is *given up* as heat to the lattice. Similarly, energy is given up when a conduction band electron subsequently falls to the empty state in E_r. When both of these events have occurred, the recombination center is back to its original state (filled with an electron), but an EHP is missing. Thus one EHP recombination has taken place, and the center is ready to participate in another recombination event by capturing a hole.

Figure 4-8
Capture processes at a recombination level: (a) hole capture at a filled recombination center; (b) electron capture at an empty center.

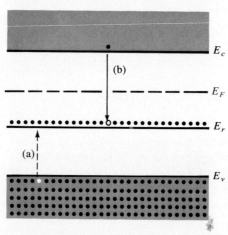

The carrier lifetime resulting from indirect recombination is somewhat more complicated than is the case for direct recombination, since it is necessary to account for unequal times required for capturing each type of carrier. In particular, recombination is often delayed by the tendency for a captured carrier to be thermally reexcited to its original band before capture of the opposite type of carrier can occur (Section 4.2.1). For example, if electron capture (b) does not follow immediately after hole capture (a) in Fig. 4-8, the hole may be thermally reexcited to the valence band. Energy is required for this process, which is equivalent to a valence band electron being raised to the empty state in the recombination level. This process delays the recombination, since the hole must be captured again before recombination can be completed.

When a carrier is trapped temporarily at a center and then is reexcited without recombination taking place, the process is often called *temporary trapping*. Although the nomenclature varies somewhat, it is common to refer to an impurity or defect center as a *trapping center* (or simply *trap*) if, after capture of one type of carrier, the most probable next event is reexcitation. If the most probable next event is capture of the opposite type of carrier, the center is predominately a recombination center. The recombination can be slow or fast, depending on the average time the first carrier is held before the second carrier is captured. In general, trapping levels located deep in the band gap are slower in releasing trapped carriers than are the levels located near one of the bands. This results from the fact that more energy is required, for example, to reexcite a trapped electron from a center near the middle of the gap to the conduction band than is required to reexcite an electron from a level closer to the conduction band.

As an example of impurity levels in semiconductors, Fig. 4-9 shows the energy level positions of various impurities in Si. In this diagram a superscript indicates whether the impurity is positive (donor) or negative (acceptor) when ionized. Some impurities introduce multiple levels in the band gap; for example, Zn introduces a level (Zn^-) located 0.31 eV above the valence band and a second level ($Zn^=$) near the middle of the gap. Each Zn impurity atom is capable of accepting two electrons from the semiconductor, one in the lower level and then one in the upper level.

The effects of recombination and trapping can be measured by a *photoconductive decay* experiment. As Fig. 4-7 shows, a population of excess electrons and holes disappears with a decay constant characteristic of the particular recombination process. The conductivity of the sample during the decay is

$$\sigma(t) = q[n(t)\mu_n + p(t)\mu_p] \qquad (4\text{-}9)$$

Therefore, the time dependence of the carrier concentrations can be monitored by recording the sample resistance as a function of time. A typical experimental arrangement is shown schematically in Fig. 4-10. A source of short pulses of light is required, along with an oscilloscope for displaying the sample voltage as the resistance varies. Microsecond light pulses can be obtained by periodically discharging a capacitor through a flash tube containing a gas such as

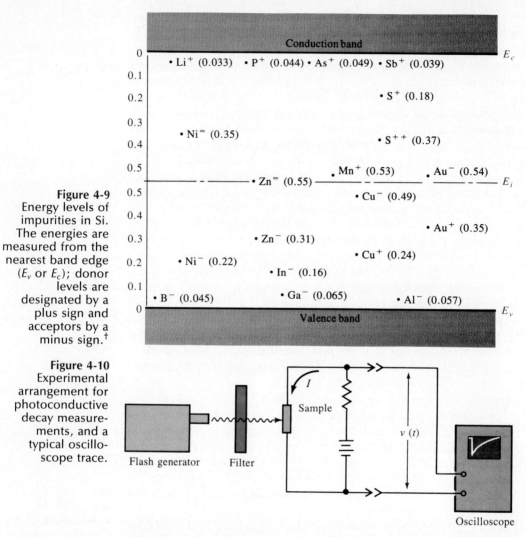

Figure 4-9
Energy levels of impurities in Si. The energies are measured from the nearest band edge (E_v or E_c); donor levels are designated by a plus sign and acceptors by a minus sign.[†]

Figure 4-10
Experimental arrangement for photoconductive decay measurements, and a typical oscilloscope trace.

xenon. For shorter pulses, special techniques such as the use of a pulsed laser (Chapter 10) must be used.

4.3.3 Steady State Carrier Generation; Quasi-Fermi Levels

In the discussion above we emphasized the transient decay of an excess EHP population. However, the various recombination mechanisms are also impor-

[†]References: S.M. Sze and J.C. Irvin, "Resistivity, Mobility, and Impurity Levels in GaAs, Ge, and Si at 300°K," *Solid State Electronics,* vol. 11, pp. 599–602 (June 1968); E. Schibli and A.G. Milnes, "Deep Impurities in Silicon," *Materials Science and Engineering,* vol. 2, pp. 173–180 (1967).

tant in a sample at thermal equilibrium or with a steady state EHP generation–recombination balance.[†] For example, a semiconductor at equilibrium experiences thermal generation of EHPs at a rate $g(T) = g_i$ described by Eq. (3-7). This generation is balanced by the recombination rate so that the equilibrium concentrations of carriers n_0 and p_0 are maintained:

$$g(T) = \alpha_r n_i^2 = \alpha_r n_0 p_0 \qquad (4\text{-}10)$$

This equilibrium rate balance can include generation from defect centers as well as band-to-band generation.

If a steady light is shone on the sample, an optical generation rate g_{op} will be added to the thermal generation, and the carrier concentrations n and p will increase to new steady state values. We can write the balance between generation and recombination in terms of the equilibrium carrier concentrations and the departures from equilibrium δn and δp:

$$g(T) + g_{op} = \alpha_r np = \alpha_r (n_0 + \delta n)(p_0 + \delta p) \qquad (4\text{-}11)$$

For steady state recombination and no trapping, $\delta n = \delta p$; thus Eq. (4-11) becomes

$$g(T) + g_{op} = \alpha_r n_0 p_0 + \alpha_r [(n_0 + p_0)\delta n + \delta n^2] \qquad (4\text{-}12)$$

The term $\alpha_r n_0 p_0$ is just equal to the thermal generation rate $g(T)$. Thus, neglecting the δn^2 term for low-level excitation, we can rewrite Eq. (4-12) as

$$g_{op} = \alpha_r (n_0 + p_0)\delta n = \frac{\delta n}{\tau_n} \qquad (4\text{-}13)$$

The excess carrier concentration can be written as

$$\delta n = \delta p = g_{op}\tau_n \qquad (4\text{-}14)$$

More general expressions are given in Eq. (4-16), which allow for the case $\tau_p \neq \tau_n$, when trapping is present.

As a numerical example, let us assume that 10^{13} EHP/cm^3 are created optically   **EXAMPLE 4-3**
every microsecond in a Si sample with $n_0 = 10^{14}$ cm^{-3} and $\tau_n = \tau_p = 2$ μsec. The steady state excess electron (or hole) concentration is then 2×10^{13} cm^{-3} from Eq. (4-14). While the percentage change in the majority electron concentration is small, the minority carrier concentration changes from

$$p_0 = n_i^2/n_0 = (2.25 \times 10^{20})/10^{14} = 2.25 \times 10^6 \text{ cm}^{-3} \qquad (equilibrium)$$

[†]The term *equilibrium* refers to a condition of no external excitation except for temperature, and no net motion of charge (e.g., a sample at a constant temperature, in the dark, with no fields applied). *Steady state* refers to a nonequilibrium condition in which all processes are constant and are balanced by opposing processes (e.g., a sample with a constant current or a constant optical generation of EHPs just balanced by recombination).

to

$$p = 2 \times 10^{13} \text{ cm}^{-3} \qquad (steady\ state)$$

Note that the equilibrium equation $n_0 p_0 = n_i^2$ cannot be used with the subscripts removed; that is, $np \neq n_i^2$ when excess carriers are present.

∎

It is often desirable to refer to the steady state electron and hole concentrations in terms of Fermi levels, which can be included in band diagrams for various devices. The Fermi level E_F used in Eq. (3-25) is meaningful only when no excess carriers are present. However, we can write expressions for the steady state concentrations in the same *form* as the equilibrium expressions by defining separate *quasi-Fermi levels* F_n and F_p for electrons and holes. The resulting carrier concentration equations

$$n = n_i e^{(F_n - E_i)/kT} \tag{4-15a}$$

$$p = n_i e^{(E_i - F_p)/kT} \tag{4-15b}$$

can be considered as defining relations for the quasi-Fermi levels.[†]

EXAMPLE 4-4   In Example 4-3, the steady state electron concentration is

$$n = n_0 + \delta n = 1.2 \times 10^{14} = (1.5 \times 10^{10}) e^{(F_n - E_i)/0.0259}$$

where $kT \simeq 0.0259$ eV at room temperature. Thus the electron quasi-Fermi level position $F_n - E_i$ is found from

$$F_n - E_i = 0.0259 \ln(8 \times 10^3) = 0.233 \text{ eV}$$

and F_n lies 0.233 eV above the intrinsic level. By a similar calculation, the hole quasi-Fermi level lies 0.186 eV below E_i (Fig. 4-11). In this example, the equilibrium Fermi level is $0.0259 \ln(6.67 \times 10^3) = 0.228$ eV above the intrinsic level.

Figure 4-11
Quasi-Fermi levels F_n and F_p for a Si sample with $n_0 = 10^{14}$ cm^{-3}, $\tau_p = 2 \ \mu$s, and $g_{op} = 10^{19}$ EHP/cm^3-s (Example 4-4).

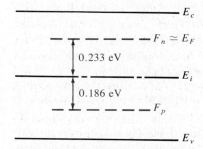

∎

[†]In some texts the quasi-Fermi level is called *IMREF*, which is Fermi spelled backward.

The quasi-Fermi levels of Fig. 4-11 illustrate dramatically the deviation from equilibrium caused by the optical excitation; the steady state F_n is only slightly above the equilibrium E_F, whereas F_p is greatly displaced below E_F. From the figure it is obvious that the excitation causes a large percentage change in minority carrier hole concentration and a relatively small change in the electron concentration.

In summary, the quasi-Fermi levels F_n and F_p are the steady state analogues of the equilibrium Fermi level E_F. When excess carriers are present, the deviations of F_n and F_p from E_F indicate how far the electron and hole populations are from the equilibrium values n_0 and p_0. A given concentration of excess EHPs causes a large shift in the minority carrier quasi-Fermi level compared with that for the majority carriers. The separation of the quasi-Fermi levels $F_n - F_p$ is a direct measure of the deviation from equilibrium (at equilibrium $F_n = F_p = E_F$). The concept of quasi-Fermi levels is very useful in visualizing minority and majority carrier concentrations in devices where these quantities vary with position.

4.3.4 Photoconductive Devices

There are a number of applications for devices which change their resistance when exposed to light. For example, such light detectors can be used in the home to control automatic night lights which turn on at dusk and turn off at dawn. They can also be used to measure illumination levels, as in exposure meters for cameras. Many systems include a light beam aimed at the photoconductor, which signals the presence of an object between the source and detector. Such systems are useful in moving-object counters, burglar alarms, and many other applications. Detectors are used in optical signaling systems in which information is transmitted by a light beam and is received at a photoconductive cell.

Considerations in choosing a photoconductor for a given application include the sensitive wavelength range, time response, and optical sensitivity of the material. In general, semiconductors are most sensitive to photons with energies equal to the band gap or slightly more energetic than band gap. Less energetic photons are not absorbed, and photons with $h\nu \gg E_g$ are absorbed at the surface and contribute little to the bulk conductivity. Therefore, the table of band gaps (Appendix III) indicates the photon energies to which most semiconductor photodetectors respond. For example, CdS ($E_g = 2.42$ eV) is commonly used as a photoconductor in the visible range, and narrow-gap materials such as Ge (0.67 eV) and InSb (0.18 eV) are useful in the infrared portion of the spectrum. Some photoconductors respond to excitations of carriers from impurity levels within the band gap and therefore are sensitive to photons of less than band gap energy.

The optical sensitivity of a photoconductor can be evaluated by examining the steady state excess carrier concentrations generated by an optical generation rate g_{op}. If the mean time each carrier spends in its respective band before capture is τ_n and τ_p, we have

$$\delta n = \tau_n g_{op} \quad \text{and} \quad \delta p = \tau_p g_{op} \tag{4-16}$$

and the photoconductivity change is

$$\Delta\sigma = qg_{op}(\tau_n\mu_n + \tau_p\mu_p) \tag{4-17}$$

For simple recombination, τ_n and τ_p will be equal. If trapping is present, however, one of the carriers may spend little time in its band before being trapped. From Eq. (4-17) it is obvious that for maximum photoconductive response, we want high mobilities and long lifetimes. Some semiconductors are especially good candidates for photoconductive devices because of their high mobility; for example, InSb has an electron mobility of about 10^5 cm^2/V-s and therefore is used as a sensitive infrared detector in many applications.

The time response of a photoconductive cell is limited by the recombination times, the degree of carrier trapping, and the time required for carriers to drift through the device in an electric field. Often these properties can be adjusted by proper choice of material and device geometry, but in some cases improvements in response time are made at the expense of sensitivity. For example, the drift time can be reduced by making the device short, but this substantially reduces the responsive area of the device. In addition, it is often desirable that the device have a large dark resistance, and for this reason, shortening the length may not be practical. There is usually a compromise between sensitivity, response time, dark resistance, and other requirements in choosing a device for a particular application.

**4.4
DIFFUSION
OF CARRIERS**

When excess carriers are created nonuniformly in a semiconductor, the electron and hole concentrations vary with position in the sample. Any such spatial variation (*gradient*) in n and p calls for a net motion of the carriers from regions of high carrier concentration to regions of low carrier concentration. This type of motion is called *diffusion* and represents an important charge transport process in semiconductors. The two basic processes of current conduction are diffusion due to a carrier gradient and drift in an electric field.

4.4.1 Diffusion Processes

When a bottle of perfume is opened in one corner of a closed room, the scent is soon detected throughout the room. If there is no convection or other net motion of the air, the scent spreads by diffusion. The diffusion is the natural result of the *random motion* of the individual molecules. Consider, for example, a volume of arbitrary shape with scented air molecules inside and unscented molecules outside the volume. All the molecules undergo random thermal motion and collisions with other molecules. Thus each molecule moves in an arbitrary direction until it collides with another air molecule, after which it moves in a new direction. If the motion is truly random, a molecule at the edge of the volume has equal probabilities of moving into or out of the volume on its next step (assuming the curvature of the surface is negligible on the molecular scale). Therefore, after a mean free time $\bar{t}$, half the molecules at the edge will have moved into the volume and half will have moved out of the volume. The net effect is that the volume containing scented molecules has increased. This process will continue until the molecules are uniformly distributed in the room. Only then will a given volume gain as many molecules as it loses in a given time. In other words, net diffusion will continue as long as gradients exist in the distribution of scented molecules.

Carriers in a semiconductor diffuse in a carrier gradient by random thermal motion and scattering from the lattice and impurities. For example, a pulse of excess electrons injected at $x = 0$ at time $t = 0$ will spread out in time as shown in Fig. 4-12. Initially, the excess electrons are concentrated at $x = 0$; as time passes, however, electrons diffuse to regions of low electron concentration until finally $n(x)$ is constant.

We can calculate the rate at which the electrons diffuse in a one-dimensional problem by considering an arbitrary distribution $n(x)$ such as Fig. 4-13a. Since the mean free path $\bar{l}$ between collisions is a small incremental distance, we can divide x into segments $\bar{l}$ wide, with $n(x)$ evaluated at the center of each segment (Fig. 4-13b).

The electrons in segment (1) to the left of x_0 in Fig. 4-13b have equal chances of moving left or right, and in a mean free time $\bar{t}$ one-half of them will move into segment (2). The same is true of electrons within one mean free path of x_0 to the right; one-half of these electrons will move through x_0 from right to left in a mean free time. Therefore, the *net* number of electrons passing x_0 from left to right in one mean free time is $\frac{1}{2}(n_1 \bar{l} A) - \frac{1}{2}(n_2 \bar{l} A)$, where the area

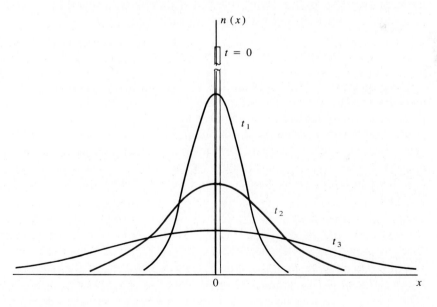

$n(x)$

$t = 0$

t_1

t_2

t_3

0

x

Figure 4-12
Spreading of a
pulse of electrons
by diffusion.

Figure 4-13
An arbitrary
electron
concentration
gradient in one
dimension:
(a) division of $n(x)$
into segments of
length equal to a
mean free path for
the electrons;
(b) expanded view
of two of the
segments centered
at x_0.

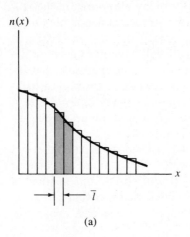

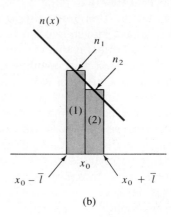

(a)   (b)

Figure 4-13 An arbitrary electron concentration gradient in one dimension: (a) division of $n(x)$ into segments of length equal to a mean free path for the electrons; (b) expanded view of two of the segments centered at x_0.

perpendicular to x is A. The rate of electron flow in the $+x$-direction per unit area (the electron flux density ϕ_n) is given by

$$\phi_n(x_0) = \frac{\bar{l}}{2\bar{t}}(n_1 - n_2) \tag{4-18}$$

Since the mean free path $\bar{l}$ is a small differential length, the difference in electron concentration $(n_1 - n_2)$ can be written as

$$n_1 - n_2 = \frac{n(x) - n(x + \Delta x)}{\Delta x}\bar{l} \tag{4-19}$$

where x is taken at the center of segment (1) and $\Delta x = \bar{l}$. In the limit of small Δx (i.e., small mean free path $\bar{l}$ between scattering collisions), Eq. (4-18) can be written in terms of the carrier gradient $dn(x)/dx$:

$$\phi_n(x) = \frac{\bar{l}^2}{2\bar{t}} \lim_{\Delta x \to 0} \frac{n(x) - n(x + \Delta x)}{\Delta x} = \frac{-\bar{l}^2}{2\bar{t}} \frac{dn(x)}{dx} \tag{4-20}$$

The quantity $\bar{l}^2/2\bar{t}$ [†] is called the *electron diffusion coefficient* D_n, with units cm^2/s. The minus sign in Eq. (4-20) arises from the definition of the derivative; it simply indicates that the net motion of electrons due to diffusion is in the direction of *decreasing* electron concentration. This is the result we expect, since net diffusion occurs from regions of high particle concentration to regions of low particle concentration. By identical arguments, we can show that holes in a hole concentration gradient move with a diffusion coefficient D_p. Thus

$$\phi_n(x) = -D_n \frac{dn(x)}{dx} \tag{4-21a}$$

[†]If motion in three dimensions were included, the diffusion would be smaller in the x-direction. Actually, the diffusion coefficient should be calculated from the true energy distributions and scattering mechanisms. Diffusion coefficients are usually determined experimentally for a particular material, as described in Section 4.4.5.

$$\phi_p(x) = -D_p \frac{dp(x)}{dx} \tag{4-21b}$$

The diffusion current crossing a unit area (the current density) is the particle flux density multiplied by the charge of the carrier:

$$J_n(\text{diff.}) = -(-q)D_n \frac{dn(x)}{dx} = +qD_n \frac{dn(x)}{dx} \tag{4-22a}$$

$$J_p(\text{diff.}) = -(+q)D_p \frac{dp(x)}{dx} = -qD_p \frac{dp(x)}{dx} \tag{4-22b}$$

It is important to note that electrons and holes move together in a carrier gradient [Eqs. (4-21)], but the resulting currents are in opposite directions [Eqs. (4-22)] because of the opposite charge of electrons and holes.

4.4.2 Diffusion and Drift of Carriers; Built-in Fields

If an electric field is present in addition to the carrier gradient, the current densities will each have a drift component and a diffusion component

$$J_n(x) = q\mu_n n(x)\mathscr{E}(x) + qD_n \frac{dn(x)}{dx} \tag{4-23a}$$

$$\text{drift} \qquad\qquad \text{diffusion}$$

$$J_p(x) = q\mu_p p(x)\mathscr{E}(x) - qD_p \frac{dp(x)}{dx} \tag{4-23b}$$

and the total current density is the sum of the contributions due to electrons and holes:

$$J(x) = J_n(x) + J_p(x) \tag{4-24}$$

We can best visualize the relation between the particle flow and the currents of Eqs. (4-23) by considering a diagram such as shown in Fig. 4-14. In this figure an electric field is assumed to be in the x-direction, along with carrier distributions $n(x)$ and $p(x)$ which decrease with increasing x. Thus the derivatives in Eqs. (4-21) are negative, and diffusion takes place in the $+x$-direction. The resulting electron and hole diffusion currents [J_n(diff.) and

Drift and diffusion directions for electrons and holes in a carrier gradient and an electric field. Particle flow directions are indicated by dashed arrows, and the resulting currents are indicated by solid arrows.   **Figure 4-14**

J_p(diff.)] are in opposite directions, according to Eqs. (4-22). Holes drift in the direction of the electric field [ϕ_p(drift)], whereas electrons drift in the opposite direction because of their negative charge. The resulting drift current is in the $+x$-direction in each case. Note that the drift and diffusion components of the current are additive for holes when the field is in the direction of decreasing hole concentration, whereas the two components are subtractive for electrons under similar conditions. The total current may be due primarily to the flow of electrons or holes, depending on the relative concentrations and the relative magnitudes and directions of electric field and carrier gradients.

An important result of Eqs. (4-23) is that minority carriers can contribute significantly to the current through diffusion. Since the drift terms are proportional to carrier concentration, minority carriers seldom provide much drift current. On the other hand, diffusion current is proportional to the *gradient* of concentration. For example, in n-type material the minority hole concentration p may be many orders of magnitude smaller than the electron concentration n, but the gradient dp/dx may be significant. As a result, minority carrier currents through diffusion can sometimes be as large as majority carrier currents.

In discussing the motion of carriers in an electric field, we should indicate the influence of the field on the energies of electrons in the band diagrams. Assuming an electric field $\mathcal{E}(x)$ in the x-direction, we can draw the energy bands as in Fig. 4-15, to include the change in potential energy of electrons in the field. Since electrons drift in a direction opposite to the field, we expect the potential energy for electrons to increase in the direction of the field, as in Fig. 4-15. The electrostatic potential $\mathcal{V}(x)$ varies in the opposite direction, since it is defined in terms of positive charges and is therefore related to the electron potential energy $E(x)$ displayed in the figure by $\mathcal{V}(x) = E(x)/(-q)$.

From the definition of electric field,

$$\mathcal{E}(x) = -\frac{d\mathcal{V}(x)}{dx} \tag{4-25}$$

Figure 4-15
Energy band diagram of a semiconductor in an electric field $\mathcal{E}(x)$.

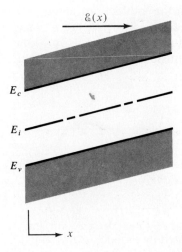

we can relate $\mathscr{E}(x)$ to the electron potential energy in the band diagram by choosing some reference in the band for the electrostatic potential. We are interested only in the spatial variation in $\mathscr{V}(x)$ for Eq. (4-25). Choosing E_i as a convenient reference, we can relate the electric field to this reference by

$$\mathscr{E}(x) = -\frac{d\mathscr{V}(x)}{dx} = -\frac{d}{dx}\left[\frac{E_i}{(-q)}\right] = \frac{1}{q}\frac{dE_i}{dx} \qquad (4\text{-}26)$$

Therefore, the variation of band energies with $\mathscr{E}(x)$ as drawn in Fig. 4-15 is correct. The direction of the slope in the bands relative to $\mathscr{E}$ is simple to remember: Since the diagram indicates electron energies, we know the slope in the bands must be such that electrons drift "downhill" in the field. Therefore, $\mathscr{E}$ points "uphill" in the band diagram.

At equilibrium, no net current flows in a semiconductor. Thus any fluctuation which would begin a diffusion current also sets up an electric field which redistributes carriers by drift. An examination of the requirements for equilibrium indicates that the diffusion coefficient and mobility must be related. Setting Eq. (4-23b) equal to zero for equilibrium, we have

$$\mathscr{E}(x) = \frac{D_p}{\mu_p}\frac{1}{p(x)}\frac{dp(x)}{dx} \qquad (4\text{-}27)$$

Using Eq. (3-25b) for $p(x)$,

$$\mathscr{E}(x) = \frac{D_p}{\mu_p}\frac{1}{kT}\left(\frac{dE_i}{dx} - \frac{dE_F}{dx}\right) \qquad (4\text{-}28)$$

The equilibrium Fermi level does not vary with x, and the derivative of E_i is given by Eq. (4-26). Thus Eq. (4-28) reduces to

$$\frac{D}{\mu} = \frac{kT}{q} \qquad (4\text{-}29)$$

This result is obtained for either carrier type. This important equation is called the *Einstein relation*. It allows us to calculate either D or μ from a measurement of the other. Table 4-1 lists typical values of D and μ for several semiconductors at room temperature. It is clear from these values that $D/\mu \simeq 0.026$ V.

Table 4-1. Diffusion coefficient and mobility of electrons and holes for several semiconductors at 300 K.

	D_n (cm^2/s)	D_p (cm^2/s)	μ_n (cm^2/V-s)	μ_p (cm^2/V-s)
Ge	100	50	3900	1900
Si	35	12.5	1350	480
GaAs	220	10	8500	400

An important result of the balance of drift and diffusion at equilibrium is that *built-in* fields accompany gradients in E_i [see Eq. (4-26)]. Such gradients in the bands at equilibrium (E_F constant) can arise when the band gap varies due to changes in alloy composition. More commonly, built-in fields result from doping gradients. For example, a donor distribution $N_d(x)$ causes a gradient in $n_0(x)$, which must be balanced by a built-in electric field $\mathscr{E}(x)$.

EXAMPLE 4-5   An intrinsic Si sample is doped with donors from one side such that $N_d = N_0 \exp(-ax)$. (a) Find an expression for $\mathscr{E}(x)$ at equilibrium over the range for which $N_d \gg n_i$. (b) Evaluate $\mathscr{E}(x)$ when $a = 1(\mu m)^{-1}$. (c) Sketch a band diagram such as in Fig. 4-15 and indicate the direction of $\mathscr{E}$.

SOLUTION   (a) From Eq. (4-23a),

$$\mathscr{E}(x) = -\frac{D_n}{\mu_n}\frac{dn/dx}{n} = -\frac{kT}{q}\frac{N_0(-a)e^{-ax}}{N_0 e^{-ax}} = +\frac{kT}{q}a$$

We notice for this exponential impurity distribution, $\mathscr{E}(x)$ depends on a but not on N_0 or x.

(b) $\mathscr{E}(x) = 0.0259(10^4) = 259$ V/cm

(c)

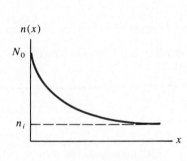
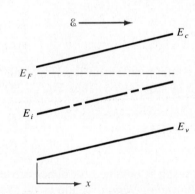

4.4.3   Diffusion and Recombination; The Continuity Equation

In the discussion of diffusion of excess carriers, we have thus far neglected the important effects of recombination. These effects must be included in a description of conduction processes, however, since recombination can cause a variation in the carrier distribution. For example, consider a differential length Δx of a semiconductor sample with area A in the yz-plane (Fig. 4-16). The hole current density leaving the volume, $J_p(x + \Delta x)$, can be larger or smaller than the current density entering, $J_p(x)$, depending on the generation and recombination of carriers taking place within the volume. The net increase in hole concentration per unit time, $\partial p/\partial t$, is the difference between the hole flux per unit

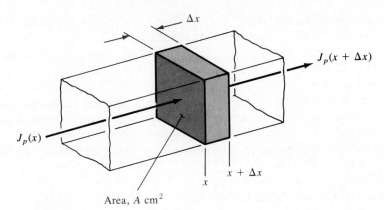

Figure 4-16
Current entering
and leaving a
volume $\Delta x A$.

volume entering and leaving, minus the recombination rate. We can convert hole current density to hole particle flux density by dividing J_p by q. The current densities are already expressed per unit area; thus dividing $J_p(x)/q$ by Δx gives the number of carriers per unit volume entering $\Delta x A$ per unit time, and $(1/q)J_p(x + \Delta x)/\Delta x$ is the number leaving per unit volume and time:

$$\frac{\partial p}{\partial t}\bigg|_{x \to x + \Delta x} = \frac{1}{q}\frac{J_p(x) - J_p(x + \Delta x)}{\Delta x} - \frac{\delta p}{\tau_p} \qquad (4\text{-}30)$$

$$\begin{array}{ccc} \text{Rate of} & = & \text{increase of hole concentra-} & - & \text{recombination} \\ \text{hole buildup} & & \text{tion in } \Delta x A \text{ per unit time} & & \text{rate} \end{array}$$

As Δx approaches zero, we can write the current change in derivative form:

$$\frac{\partial p(x, t)}{\partial t} = \frac{\partial \delta p}{\partial t} = -\frac{1}{q}\frac{\partial J_p}{\partial x} - \frac{\delta p}{\tau_p} \qquad (4\text{-}31a)$$

The expression (4-31a) is called the *continuity equation* for holes. For electrons we can write

$$\frac{\partial \delta n}{\partial t} = \frac{1}{q}\frac{\partial J_n}{\partial x} - \frac{\delta n}{\tau_n} \qquad (4\text{-}31b)$$

since the electronic charge is negative.

When the current is carried strictly by diffusion (negligible drift), we can replace the currents in Eqs. (4-31) by the expressions for diffusion current; for example, for electron diffusion we have

$$J_n(\text{diff.}) = qD_n\frac{\partial \delta n}{\partial x} \qquad (4\text{-}32)$$

Substituting this into Eq. (4-31b) we obtain the *diffusion equation* for electrons,

$$\boxed{\frac{\partial \delta n}{\partial t} = D_n\frac{\partial^2 \delta n}{\partial x^2} - \frac{\delta n}{\tau_n}} \qquad (4\text{-}33a)$$

and similarly for holes,

$$\frac{\partial \delta p}{\partial t} = D_p \frac{\partial^2 \delta p}{\partial x^2} - \frac{\delta p}{\tau_p}$$ (4-33b)

These equations are useful in solving transient problems of diffusion with recombination. For example, a pulse of electrons in a semiconductor (Fig. 4-12) spreads out by diffusion and disappears by recombination. To solve for the electron distribution in time, $n(x, t)$, we would begin with the diffusion equation, Eq. (4-33a).

4.4.4 Steady State Carrier Injection; Diffusion Length

In many problems a steady state distribution of excess carriers is maintained, such that the time derivatives in Eqs. (4-33) are zero. In the steady state case the diffusion equations become

$$\frac{d^2 \delta n}{dx^2} = \frac{\delta n}{D_n \tau_n} \equiv \frac{\delta n}{L_n^2}$$ (4-34a)

$$\frac{d^2 \delta p}{dx^2} = \frac{\delta p}{D_p \tau_p} \equiv \frac{\delta p}{L_p^2} \qquad (steady\ state)$$ (4-34b)

where $L_n \equiv \sqrt{D_n \tau_n}$ is called the electron *diffusion length* and L_p is the diffusion length for holes. We no longer need partial derivatives, since the time variation is zero for steady state.

The physical significance of the diffusion length can be understood best by an example. Let us assume that excess holes are somehow injected into a semi-infinite semiconductor bar at $x = 0$, and the steady state hole injection maintains a constant excess hole concentration at the injection point $\delta p(x = 0) = \Delta p$. The injected holes diffuse along the bar, recombining with a characteristic life-time τ_p. In steady state we expect the distribution of excess holes to decay to zero for large values of x, because of the recombination (Fig. 4-17). For this

Figure 4-17 Injection of holes at $x = 0$, giving a steady state hole distribution $p(x)$ and a resulting diffusion current density $J_p(x)$.

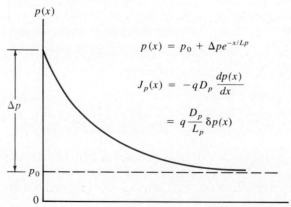

$$p(x) = p_0 + \Delta p e^{-x/L_p}$$

$$J_p(x) = -q D_p \frac{dp(x)}{dx}$$

$$= q \frac{D_p}{L_p} \delta p(x)$$

problem we use the steady state diffusion equation for holes, Eq. (4-34b). The solution to this equation has the form

$$\delta p(x) = C_1 e^{x/L_p} + C_2 e^{-x/L_p} \qquad (4\text{-}35)$$

We can evaluate C_1 and C_2 from the boundary conditions. Since recombination must reduce $\delta p(x)$ to zero for large values of x, $\delta p = 0$ at $x = \infty$, and therefore $C_1 = 0$. Similarly, the condition $\delta p = \Delta p$ at $x = 0$ gives $C_2 = \Delta p$, and the solution is

$$\boxed{\delta p(x) = \Delta p e^{-x/L_p}} \qquad (4\text{-}36)$$

The injected excess hole concentration dies out exponentially in x due to recombination, and the diffusion length L_p represents the distance at which the excess hole distribution is reduced to $1/e$ of its value at the point of injection. We can show that L_p *is the average distance a hole diffuses before recombining*. To calculate an average diffusion length, we must obtain an expression for the probability that an injected hole recombines in a particular interval dx. The probability that a hole injected at $x = 0$ *survives* to x without recombination is $\delta p(x)/\Delta p = \exp(-x/L_p)$, the ratio of the steady state concentrations at x and 0. On the other hand, the probability that a hole at x will *recombine* in the subsequent interval dx is

$$\frac{\delta p(x) - \delta p(x + dx)}{\delta p(x)} = \frac{-(d\delta p(x)/dx)\,dx}{\delta p(x)} = \frac{1}{L_p}dx \qquad (4\text{-}37)$$

Thus the total probability that a hole injected at $x = 0$ will recombine in a given dx is the product of the two probabilities:

$$(e^{-x/L_p})\left(\frac{1}{L_p}dx\right) = \frac{1}{L_p}e^{-x/L_p}\,dx \qquad (4\text{-}38)$$

Then, using the usual averaging techniques described by Eq. (2-21), the average distance a hole diffuses before recombining is

$$\langle x \rangle = \int_0^\infty x \frac{e^{-x/L_p}}{L_p}\,dx = L_p \qquad (4\text{-}39)$$

The steady state distribution of excess holes causes diffusion, and therefore a hole current, in the direction of decreasing concentration. From Eqs. (4-22b) and (4-36) we have

$$J_p(x) = -qD_p\frac{dp}{dx} = -qD_p\frac{d\delta p}{dx} = q\frac{D_p}{L_p}\Delta p e^{-x/L_p} = q\frac{D_p}{L_p}\delta p(x) \quad (4\text{-}40)$$

Since $p(x) = p_0 + \delta p(x)$, the space derivative involves only the excess concentration. We notice that since $\delta p(x)$ is proportional to its derivative for an exponential distribution, the diffusion current at any x is just proportional to the excess concentration δp at that position.

Although this example seems rather restricted, its usefulness will become apparent in Chapter 5 in the discussion of p-n junctions. The injection of minority carriers across a junction often leads to exponential distributions as in Eq. (4-36), with the resulting diffusion current of Eq. (4-40).

4.4.5 The Haynes–Shockley Experiment

One of the classic semiconductor experiments is the demonstration of drift and diffusion of minority carriers, first performed by J. R. Haynes and W. Shockley in 1951 at the Bell Telephone Laboratories. The experiment allows independent measurement of the minority carrier mobility μ and diffusion coefficient D. The basic principles of the Haynes–Shockley experiment are as follows: A pulse of holes is created in an n-type bar (for example) that contains an electric field (Fig. 4-18); as the pulse drifts in the field and spreads out by diffusion, the excess hole concentration is monitored at some point down the bar; the time required for the holes to drift a given distance in the field gives a measure of the mobility; and the spreading of the pulse during a given time is used to calculate the diffusion coefficient.

In Fig. 4-18 a pulse of excess carriers is created by a light flash at some point $x = 0$ in an n-type semiconductor ($n_0 \gg p_0$). We assume that the excess carriers have a negligible effect on the electron concentration but change the hole concentration significantly. The excess holes drift in the direction of the electric field and eventually reach the point $x = L$, where they are monitored. By measuring the drift time t_d, we can calculate the drift velocity v_d and, therefore, the hole mobility:

$$\mathsf{v}_d = \frac{L}{t_d} \tag{4-41}$$

$$\mu_p = \frac{\mathsf{v}_d}{\mathscr{E}} \tag{4-42}$$

Figure 4-18
Drift and diffusion of a hole pulse in an n-type bar: (a) sample geometry; (b) position and shape of the pulse for several times during its drift down the bar.

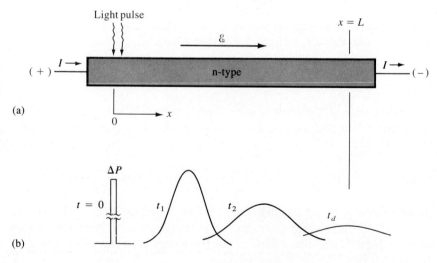

Thus the hole mobility can be calculated directly from a measurement of the drift time for the pulse as it moves down the bar. In contrast with the Hall effect (Section 3.4.5), which can be used with resistivity to obtain the *majority* carrier mobility, the Haynes–Shockley experiment is used to measure the *minority* carrier mobility.

As the pulse drifts in the $\mathscr{E}$ field it also spreads out by diffusion. By measuring the spread in the pulse, we can calculate D_p. To predict the distribution of holes in the pulse as a function of time, let us first reexamine the case of diffusion of a pulse *without drift, neglecting recombination* (Fig. 4-12). The equation which the hole distribution must satisfy is the time-dependent diffusion equation, Eq. (4-33b). For the case of negligible recombination (τ_p long compared with the times involved in the diffusion), we can write the diffusion equation as

$$\frac{\partial \delta p(x,t)}{\partial t} = D_p \frac{\partial^2 \delta p(x,t)}{\partial x^2} \tag{4-43}$$

The function which satisfies this equation is called a *gaussian distribution*,

$$\delta p(x,t) = \left[\frac{\Delta P}{2\sqrt{\pi D_p t}}\right] e^{-x^2/4D_p t} \tag{4-44}$$

where ΔP is the number of holes per unit area created over a negligibly small distance at $t = 0$. The factor in brackets indicates that the peak value of the pulse (at $x = 0$) decreases with time, and the exponential factor predicts the spread of the pulse in the positive and negative x-directions (Fig. 4-19). If we designate the peak value of the pulse as $\hat{\delta p}$ at any time (say t_d), we can use Eq. (4-44) to calculate D_p from the value of δp at some point x. The most convenient choice is the point $\Delta x/2$, at which δp is down by $1/e$ of its peak value $\hat{\delta p}$. At this point we can write

$$e^{-1}\hat{\delta p} = \hat{\delta p}\, e^{-(\Delta x/2)^2/4 D_p t_d} \tag{4-45}$$

$$D_p = \frac{(\Delta x)^2}{16 t_d} \tag{4-46}$$

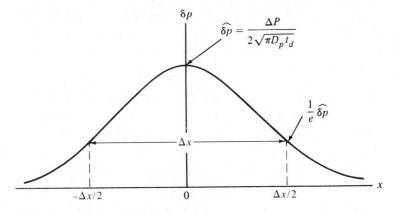

Figure 4-19
Calculation of D_p from the shape of the δp distribution after time t_d. No drift or recombination is included.

$$\hat{\delta p} = \frac{\Delta P}{2\sqrt{\pi D_p t_d}}$$

Since Δx cannot be measured directly, we use an experimental setup such as Fig. 4-20, which allows us to display the pulse on an oscilloscope as the carriers pass under a detector. As we shall see in Chapter 5, a forward-biased p-n junction serves as an excellent injector of minority carriers, and a reverse-biased junction serves as a detector. The measured quantity in Fig. 4-20 is the pulse width Δt displayed on the oscilloscope in time. It is related to Δx by the drift velocity, as the pulse drifts past the detector point (2).

$$\Delta x = \Delta t \mathsf{v}_d = \Delta t \frac{L}{t_d} \qquad (4\text{-}47)$$

EXAMPLE 4-6 An n-type Ge sample is used in the Haynes–Shockley experiment shown in Fig. 4-20. The length of the sample is 1 cm, and the probes (1) and (2) are separated by 0.95 cm. The battery voltage E_0 is 2 V. A pulse arrives at point (2) 0.25 ms after injection at (1); the width of the pulse Δt is 117 μs. Calculate the hole mobility and diffusion coefficient, and check the results against the Einstein relation.

SOLUTION

$$\mu_p = \frac{\mathsf{v}_d}{\mathscr{E}} = \frac{0.95/(0.25 \times 10^{-3})}{2/1} = 1900 \text{ cm}^2/(\text{V-s})$$

$$D_p = \frac{(\Delta x)^2}{16 t_d} = \frac{(\Delta t L)^2}{16 t_d^3}$$

$$= \frac{(117 \times 0.95)^2 \times 10^{-12}}{16(0.25)^3 \times 10^{-9}} = 49.4 \text{ cm}^2/\text{s}$$

$$\frac{D_p}{\mu_p} = \frac{49.4}{1900} = 0.026 = \frac{kT}{q}$$

∎

4.4.6   Gradients in the Quasi-Fermi Levels

In Section 3.5 we saw that equilibrium implies no gradient in the Fermi level E_F. In contrast, any combination of drift and diffusion implies a gradient in the steady state quasi-Fermi level.

We can use the results of Eqs. (4-23), (4-26), and (4-29) to demonstrate the power of the concept of quasi-Fermi levels in semiconductors [see Eq. (4-15)]. If we take the general case of nonequilibrium electron concentration with drift and diffusion, we must write the total electron current as

$$J_n(x) = q\mu_n n(x)\mathscr{E}(x) + qD_n \frac{dn(x)}{dx} \qquad (4\text{-}48)$$

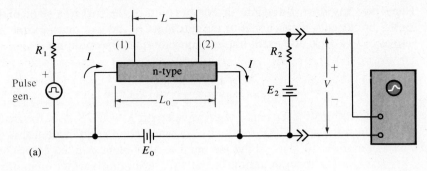

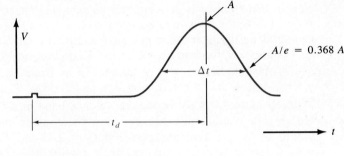

(b)

Figure 4-20
The Haynes–
Shockley
experiment:
(a) circuit
schematic;
(b) typical trace on
the oscilloscope
screen.

where the gradient in electron concentration is

$$\frac{dn(x)}{dx} = \frac{d}{dx}\left[n_i e^{(F_n - E_i)/kT}\right] = \frac{n(x)}{kT}\left(\frac{dF_n}{dx} - \frac{dE_i}{dx}\right) \qquad (4\text{-}49)$$

Using the Einstein relation, the total electron current becomes

$$J_n(x) = q\mu_n n(x)\mathscr{E}(x) + \mu_n n(x)\left[\frac{dF_n}{dx} - \frac{dE_i}{dx}\right] \qquad (4\text{-}50)$$

But Eq. (4-26) indicates that the subtractive term in the brackets is just $q\mathscr{E}(x)$, giving a direct cancellation of $q\mu_n n(x)\mathscr{E}(x)$ and leaving

$$J_n(x) = \mu_n n(x)\frac{dF_n}{dx} \qquad (4\text{-}51)$$

Thus, the processes of electron drift and diffusion are summed up by the spatial variation of the quasi-Fermi level. The same derivation can be made for holes, and we can write the current due to drift and diffusion in the form of a *modified Ohm's law*

$$J_n(x) = q\mu_n n(x)\frac{d(F_n/q)}{dx} = \sigma_n(x)\frac{d(F_n/q)}{dx} \qquad (4\text{-}52a)$$

$$J_p(x) = q\mu_p p(x)\frac{d(F_p/q)}{dx} = \sigma_p(x)\frac{d(F_p/q)}{dx} \qquad (4\text{-}52b)$$

Therefore, any drift, diffusion, or combination of the two in a semiconductor results in currents proportional to the gradients of the two quasi-Fermi levels. Conversely, a lack of current implies constant quasi-Fermi levels.

PROBLEMS

4.1.   A 100-mW laser beam with wavelength $\lambda = 6328$ Å is focused onto an InP sample 100 μm thick. The absorption coefficient at this wavelength is 10^5 cm^{-1}. Find the number of photons emitted per second by radiative recombination in the InP, assuming perfect quantum efficiency. What power is delivered to the sample as heat?

4.2.   Construct a semilogarithmic plot such as Fig. 4-7 for GaAs doped with 10^{15} donors/cm^3 and having 2×10^{14} EHP/cm^3 created uniformly at $t = 0$. Assume that $\tau_n = \tau_p = 20$ ns.

4.3.   With E_F located 0.4 eV above the valence band in a Si sample, what charge state would you expect for most Ga atoms in the sample? What would be the predominant charge state of Zn? Au? *Note:* By charge state we mean neutral, singly positive, doubly negative, etc.

4.4.   A Si sample is doped with 10^{16} cm^{-3} Sb. How many Zn atoms/cm^3 must be added to exactly compensate this material ($n_0 = p_0 = n_i$)?

4.5.   A sample is doped with donors such that $n_0 = Gx$ for $n_0 \gg n_i$, where G is a constant. Find the built-in electric field $\mathscr{E}(x)$.

4.6.   A Ge sample with $n_0 = 10^{17}$ cm^{-3} is optically excited at 300 K such that $g_{op} = 10^{20}$ EHP/cm^3-s and $\tau_n = \tau_p = 10$ μs. What is the separation of the quasi-Fermi levels ($F_n - F_p$)? Draw a band diagram such as Fig. 4-11.

4.7.   For the steady state excess hole distribution shown in Fig. 4-17, find the expression for the hole quasi-Fermi level position F_p as a function of x while $p(x) \gg p_0$.

4.8.   Calculate the recombination coefficient α_r for the low-level excitation described in Prob. 4.2. Assume that this value of α_r applies when the GaAs sample is uniformly exposed to a steady state optical generation rate $g_{op} = 10^{20}$ EHP/cm^3-s. Find the steady state excess carrier concentration $\Delta n = \Delta p$.

4.9.   Design and sketch a photoconductor using a 5-μm-thick film of CdS, assuming that $\tau_n = \tau_p = 10^{-6}$ s and $N_d = 10^{14}$ cm^{-3}. The dark resistance (with $g_{op} = 0$) should be 10 MΩ, and the device must fit in a square 0.5 cm on a side; therefore, some sort of folded or zigzag pattern is in order. With an excitation of $g_{op} = 10^{21}$ EHP/cm^3-s, what is the resistance change?

4.10. The current required to feed the hole injection at $x = 0$ in Fig. 4-17 is obtained by evaluating Eq. (4-40) at $x = 0$. The result is $I_p(x = 0)$ $= qAD_p\Delta p/L_p$. Show that this current can be calculated by integrating the charge stored in the steady state hole distribution $\delta p(x)$ and then dividing by the average hole lifetime τ_p. Explain why this approach gives $I_p(x = 0)$.

4.11. Assume that a photoconductor in the shape of a bar of length L and area A has a constant voltage V applied, and it is illuminated such that g_{op} EHP/cm^3-s are generated uniformly throughout. If $\mu_n \gg \mu_p$, we can assume the optically induced change in current ΔI is dominated by the mobility μ_n and lifetime τ_n for electrons. Show that $\Delta I = qALg_{op}\tau_n/\tau_t$ for this photoconductor, where τ_t is the transit time of electrons drifting down the length of the bar.

4.12. In Fig. 4-17 the steady state excess hole concentration at $x = 0$ is $\Delta p = 10^{15}$ cm^{-3}. The semi-infinite Ge bar has a cross section $A = 10^{-2}$ cm^2. The hole diffusion length L_p is 10^{-3} cm, and the hole lifetime is 10^{-4} s.
   (a) What is the steady state stored charge Q_p in the exponential excess hole distribution?
   (b) What is the hole current $I_p(x = 0)$ feeding this steady state distribution?
   (c) What is the slope of the distribution (in cm^{-4}) at $x = 0$?

4.13. Boron is diffused into an intrinsic Si sample, giving the acceptor distribution shown in Figure P4-13. Sketch the equilibrium band diagram and show the direction of the resulting electric field, for $N_a(x) \gg n_i$. Repeat for phosphorus, with $N_d(x) \gg n_i$.

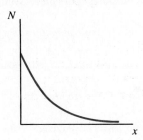

N

x

Figure P4-13

4.14. Show that $np = n_i^2 e^{(F_n - F_p)/kT}$ in steady state.

4.15. We wish to use the Haynes–Shockley experiment to calculate the hole lifetime τ_p in an n-type sample. Assume the peak voltage of the pulse displayed on the oscilloscope screen is proportional to the hole concentration under the collector terminal at time t_d, and that the displayed pulse can be approximated as a gaussian, as in Eq. (4-44), which decays due to recombination by e^{-t/τ_p}. The electric field is varied and the following data taken: For $t_d = 200\ \mu s$, the peak is 20 mV; for $t_d = 50\ \mu s$, the peak is 80 mV. What is τ_p?

*4.16. Consider a sample of GaAs ($n_i = 10^6\ cm^{-3}$ at 300 K) doped with 10^{15} donors per cm^3 illuminated with the 5145 Å line of an argon ion laser. For GaAs at 5145 Å, $\alpha = 10^4\ cm^{-1}$. Calculate and plot the steady state excess electron profile $\delta n(x)$ in the region within 5 μm of the surface for photon fluxes of 10^{15}, 10^{17}, and 10^{19} photons $cm^{-2}\ s^{-1}$ using low-level injection assumptions and directly solving Eq. (4-12). For this problem, assume that $\tau_n = \tau_p = 10^{-6}$ s. Neglect diffusion.

*4.17. For the sample of Prob. 4.16, calculate and plot the steady state excess electron profile $\delta n(x)$ in the region within 5 μm of the surface for a photon flux of 10^{19} photons $cm^{-2}\ s^{-1}$ using low-level injection assumptions and directly solving Eq. (4-12) for values of α_r of 10^{-9}, 10^{-7}, and $10^{-5}\ cm^3\ s^{-1}$.

*4.18. Using the results of problem 4.16 obtained for a photon flux of 10^{15} photons $cm^{-2}\ s^{-1}$, calculate and plot the transient excess carrier profile 1, 2, and 5 ns after the laser flux is interrupted, by integrating Eq. (4-5) within each depth interval, using $10^{-6}\ cm^3\ s^{-1}$ for α_r. In this case, ignore carrier diffusion.

READING LIST

Blakemore, J. S. *Semiconductor Statistics.* New York: Dover Publications, 1987.

Hodgson, J. N. *Optical Absorption and Dispersion in Solids.* London: Chapman and Hall, 1970.

Pankove, J. I. *Optical Processes in Semiconductors.* Englewood Cliffs, N.J.: Prentice Hall, 1971.

Pierret, R. F. *Semiconductor Fundamentals,* 2d ed. Reading, Mass.: Addison-Wesley, 1988.

Ridley, B. K. *Quantum Processes in Semiconductors.* 2d ed. Oxford: Clarendon Press, 1988.

Shockley, W. *Electrons and Holes in Semiconductors.* New York: Van Nostrand Reinhold, 1950.

Shur, M. *GaAs Devices and Circuits.* New York: Plenum Press, 1987.

Thomas, G. A. "An Electron-Hole Liquid." *Scientific American,* 234, no. 6 (June 1976): 28–37.

Van Der Ziel, A. *Solid State Physical Electronics*. 3d ed. Englewood Cliffs, N.J.: Prentice Hall, 1976.

Wolfe, C. M., N. Holonyak, and G. E. Stillman. *Physical Properties of Semiconductors*. Englewood Cliffs, N.J.: Prentice Hall, 1989.

chapter 5

JUNCTIONS

Most semiconductor devices contain at least one junction between p-type and n-type material. These p-n junctions are fundamental to the performance of functions such as rectification, amplification, switching, and other operations in electronic circuits. In this chapter we shall discuss the equilibrium state of the junction and the flow of electrons and holes across a junction under steady state and transient conditions. This is followed by a discussion of metal-semiconductor junctions and heterojunctions between semiconductors having different band gaps. With the background provided in this chapter on junction properties, we can then discuss specific devices in later chapters.

5.1
FABRICATION
OF p-n JUNCTIONS

Before investigating their electrical properties, we shall discuss how p-n junctions are made. The technology of junction fabrication is, of course, a broad subject which includes the accumulated knowledge and experience of many research and manufacturing groups. Without attempting a thorough description of these manufacturing techniques, we can investigate some of the more basic methods of forming junctions and making contacts to them in mountings suitable for devices. The type of junction we seek is a change from n-type to p-type material within a *single crystal*. Therefore, we may change the dopant from donors to acceptors during the growth of the crystal. Alternatively, we may introduce impurities of one type into regions of a crystal which was grown with lighter doping of the opposite type. In either case, only the doping is changed at the junction; there are no gross changes in the lattice structure itself.

5.1.1   Grown Junctions

A method of junction fabrication that was popular in the early days of semi-conductor device development is the *grown junction* technique. In this method the dopant is abruptly changed in the melt during the process of crystal growth. As an example, let us assume an n-type Si crystal containing 10^{14} phosphorus atoms/cm^3 is grown from a melt. The growth process is stopped temporarily while sufficient boron atoms are added to the melt to result in 6×10^{14} boron atoms/cm^3 in the crystal. This is a sufficient concentration of acceptors to compensate the original donors and leave a net acceptor concentration of $N_a - N_d = 5 \times 10^{14}$ cm^{-3}. This process (called *counterdoping*) can be continued as more donors and acceptors are alternately added.

The original grown junction technique has been supplanted by the more flexible methods of introducing junctions into crystals after growth. An important exception, however, is the *epitaxial* growth of p-n junctions (Section 1.4), which is used widely in integrated circuits and for other applications. An epitaxial layer of one type can be grown on a substrate of the opposite type to form a sharp, single-crystal junction. We shall discuss the use of epitaxial junctions in more detail in Chapter 9, when integrated circuit fabrication is presented.

5.1.2   Alloyed Junctions

A convenient technique for making p-n junctions is the alloying of a metal containing doping atoms on a semiconductor with the opposite type of dopant. This process was used in the 1950s to produce diodes and transistors. As an example of alloying, a sample of n-type Ge can be heated with a pellet of In on it to form a small local melt from which a p-n junction grows on the parent crystal (Fig. 5-1). As the In and Ge are heated to about 160°C, a molten solution of the two materials forms (Fig. 5-1b). The proportion of Ge to In in this alloy melt is well above the solubility at lower temperature. As the temperature is lowered, Ge grows out of the alloy mixture because of the reduced solubility. At the interface between the cooling mixture and the Ge crystal, a *regrown region* of crystalline Ge is formed. This regrown region has the same lattice structure as the Ge parent crystal, but it now has a high concentration of In

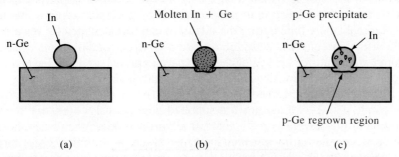

(a)   (b)   (c)

The alloying process: (a) pellet of In in contact with n-type Ge; (b) molten phase of In and Ge mixture during heating; (c) cross-sectional view of an alloyed junction, showing p-type regrown region and precipitation of Ge in the In.

Figure 5-1

atoms. If $N_a > N_d$ in this region, it is p-type. The regrowing of the p-type Ge crystal onto the underlying Ge substrate is similar to the process of liquid-phase epitaxy discussed in Section 1.4.2. In the present case, however, the Ge substrate serves not only as the seed crystal for the growing layer but also provides the Ge in the melt. The resulting p-n junction consists of the interface between the regrown p-type Ge region and the n-type Ge sample.

5.1.3 Diffused Junctions

In the 1960s alloying was supplanted by impurity diffusion as the most common technique for forming p-n junctions. This method, along with the use of selective masking to control junction geometry, made possible the development of integrated circuits. Selective diffusion is an impressive technique in its controllability, accuracy, and versatility, as we shall see in Chapter 9. As a preliminary illustration of this point, however, we note that selective diffusion and other IC techniques are sufficiently accurate to allow the fabrication of thousands of p-n junction devices, properly interconnected, on a Si chip.

The diffusion of impurities into a solid is basically the same type of process as discussed in Section 4.4 in terms of excess carriers. In each case, diffusion is a result of random motion, and particles diffuse in the direction of decreasing concentration gradient. The random motion of impurity atoms in a solid is, of course, rather limited unless the temperature is high. Thus diffusion of doping impurities into a semiconductor such as Si is accomplished at temperatures in the neighborhood of 1000°C. At these temperatures, many atoms in the semiconductor move out of their lattice sites, leaving vacancies into which impurity atoms can move. Most of the common doping impurities diffuse by this type of vacancy motion and occupy lattice positions in the crystal after it is cooled.

If a wafer of n-type Si is placed in a furnace at 1000°C, in a gaseous atmosphere containing a high concentration of boron atoms, a sharp impurity concentration gradient occurs at the surface of the sample. Since the temperature is high enough to allow slow random motion of atoms, boron establishes a certain solubility concentration at the surface and begins to diffuse into the sample (Fig. 5-2). The diffusion coefficient, describing the ability of an impurity to wander into the sample, is known for most of the common impurities in many semiconductors as a function of temperature; thus the distribution of impurities in the sample at any time during the diffusion can be calculated from a solution of the diffusion equation with appropriate boundary conditions. If the source of boron atoms at the surface of the sample mentioned above is limited (e.g., a given number of atoms deposited on the Si surface before diffusion), a gaussian distribution as described by Eq. (4-44) (for $x > 0$) is obtained. On the other hand, if the boron atoms are supplied continuously, such that the concentration at the surface is maintained at a constant value, the distribution follows what is called a *complementary error function.* In either case, there is some point in the sample at which the introduced acceptor concentration just equals the background donor concentration in the originally n-type sample. This point is the location of the p-n junction; to the left of this point in the sample of Fig. 5-2, acceptor atoms predominate and the material is p-type,

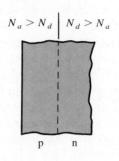

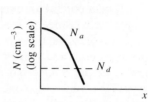

Figure 5-2
Impurity
concentration
profile for
diffusion.

whereas to the right of the junction, the background donor atoms predominate and the material is n-type. The depth of the junction beneath the surface of the sample can be controlled accurately by the time and temperature of the diffusion (Prob. 5.22).

A typical diffusion furnace for Si device production is shown in Fig. 5-3. The furnace core surrounds a silica tube through which an inert gas flows. The

Figure 5-3
Silicon wafers
being loaded into a
diffusion furnace.
(Photograph
courtesy of
American
Microsystems, Inc.)

Si wafers are placed in the tube during diffusion, and the impurity atoms are introduced into the gas flow over the samples. Common impurity source materials for diffusions in Si are B_2O_3, BBr_3, and BCl_3 for boron; phosphorus sources include PH_3, P_2O_5, and $POCl_3$. Solid sources are placed in the silica tube upstream from the sample or in a separate heating zone of the furnace; gaseous sources can be metered directly into the gas flow system; and with liquid sources the inert carrier gas is bubbled through the liquid before being introduced into the furnace tube. The Si wafers are held in a silica "boat" (Fig. 5-3) which can be pushed into position in the furnace and removed by a silica rod.

It is important to remember the degree of cleanliness required in these processing steps. Since typical doping concentrations represent one part per million or less, cleanliness and purity of materials is critically important. Thus the impurity source and carrier gas in Fig. 5-3 must be extremely pure; the silica tube, sample holder and pushrod must be cleaned and etched in hydrofluoric acid (HF) before use (once in use, the tube cleanliness can be maintained if no unwanted impurities are introduced); finally, the Si wafers themselves must undergo an elaborate cleaning procedure before diffusion, including a final etch containing HF to remove any SiO_2 from the surface (this may be done selectively, as discussed below).

Selective diffusions can be made in Si by opening "windows" in a layer of SiO_2 grown on the surface of the wafer. Hydrofluoric acid removes any unmasked SiO_2, but does not disturb the Si itself. It is convenient that Si oxidizes in the form of SiO_2, since this makes possible the basic oxide masking process needed for Si device fabrication. Wafers of Si can be placed in a furnace containing O_2 gas to grow an SiO_2 layer (less than 0.1 μm thick) on the surface. The oxidized wafer is in effect encapsulated by silica glass, which blocks penetration by most impurity atoms. By removing the oxide in desired regions of the wafer, diffusion can be carried out at these controlled locations (Fig. 5-4). The process used for selectively removing the oxide involves covering the surface of the oxide with an acid-resistant coating except where the diffusion windows are needed; then an HF etch or a plasma-enhanced etch removes the SiO_2 at the uncoated locations. The coating is a photosensitive organic material called *photoresist (PR)*, which can be polymerized by ultraviolet light, X-rays, or an electron or ion beam. If the exposing radiation is passed through a *mask* containing the desired pattern, the coating can be polymerized everywhere except where the pattern is to appear. Then the unpolymerized areas can be removed prior to etching, which removes the SiO_2 in the uncoated areas. This technique is known as *photolithography*. The masks for this process are usually generated by a computer, then reduced photographically to the proper size. The close tolerances available with this technique make possible the intricate diffusion patterns necessary for Si ICs. We shall discuss mask fabrication in more detail in Section 9.2, including more recent methods than those discussed here.

The example of Fig. 5-4 illustrates the selective diffusion of many isolated p-n junctions on a single wafer. In most cases the oxide layer is left on the

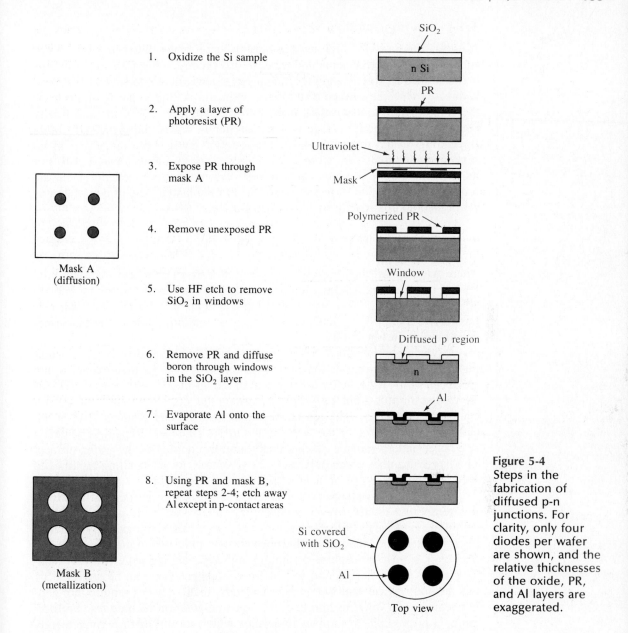

1. Oxidize the Si sample

2. Apply a layer of photoresist (PR)

3. Expose PR through mask A

4. Remove unexposed PR

Mask A
(diffusion)

5. Use HF etch to remove SiO_2 in windows

6. Remove PR and diffuse boron through windows in the SiO_2 layer

7. Evaporate Al onto the surface

8. Using PR and mask B, repeat steps 2-4; etch away Al except in p-contact areas

Mask B
(metallization)

SiO₂ labels: SiO$_2$; n Si; PR; Ultraviolet; Mask; Polymerized PR; Window; Diffused p region; n; Al; Si covered with SiO$_2$; Al; Top view

Figure 5-4
Steps in the fabrication of diffused p-n junctions. For clarity, only four diodes per wafer are shown, and the relative thicknesses of the oxide, PR, and Al layers are exaggerated.

wafer, and a metal such as Al is evaporated onto the wafer and patterns are defined photolithographically to contact the diffused layers. Then the individual devices can be separated by sawing or by scribing and breaking the wafer. The final steps of the process are mounting of individual devices in appropriate packages and connecting leads to the Al contact regions. Very accurate lead bonders are available for bonding Au or Al wire (about one thousandth of an inch in diameter) to the device and then to the package leads.

5.1.4 Ion Implantation

A useful alternative to high-temperature diffusion is the direct implantation of energetic ions into the semiconductor. In this process a beam of impurity ions is accelerated to kinetic energies ranging from several keV to several MeV and is directed onto the surface of the semiconductor. As the impurity atoms enter the crystal, they give up their energy to the lattice in collisions and finally come to rest at some average penetration depth, called the *projected range.* Depending on the impurity and its implantation energy, the range in a given semiconductor may vary from a few hundred angstroms to about 1 μm. For most implantations the ions come to rest distributed almost evenly about the projected range R_p, as shown in Fig. 5-5a. An implanted dose of ϕ ions/cm^2 is distributed approximately by a gaussian formula

$$N(x) = \frac{\phi}{\sqrt{2\pi}\,\Delta R_p} \exp\left[-\frac{1}{2}\left(\frac{x - R_p}{\Delta R_p}\right)^2\right] \tag{5-1}$$

where ΔR_p, called the *straggle*, measures the half-width of the distribution at $e^{-1/2}$ of the peak Fig. (5-5a). Both R_p and ΔR_p increase with increasing implantation energy. By performing several implantations at different energies, it is possible to synthesize a desired impurity distribution, such as the uniformly doped region in Fig. 5-5b.

A typical ion implanter is shown schematically in Fig. 5-6. A gas containing the desired impurity is ionized within the *source* and is then extracted into the *acceleration tube*. After acceleration to the desired kinetic energy, the ions are passed through a *mass separator* to ensure that only the desired ion species enters the *drift tube*.[†] The ion beam is then focused and scanned electrostatically over the surface of the wafer in the *target chamber*. Repetitive scanning in a raster pattern provides exceptionally uniform doping of the wafer surface. The target chamber commonly includes automatic wafer-handling facilities to speed up the process of implanting many wafers per hour.

An obvious advantage of implantation is that it can be done at relatively low temperatures; this means that doping layers can be implanted without disturbing previously diffused regions. The ions can be blocked by metal or photoresist layers; therefore, the photolithographic techniques described in Section 5.1.3 can be used to define ion implanted doping patterns. Very shallow (tenths of a micron) and well-defined doping layers can be achieved by this method. As we shall see in later chapters, many devices require thin doping regions and may be improved by ion implantation techniques. Furthermore, it is possible to implant impurities which do not diffuse conveniently into semiconductors.

One of the major advantages of implantation is the precise control of doping concentration it provides. Since the ion beam current can be measured accurately during implantation, a precise quantity of impurity can be introduced.

[†]In many ion implanters the mass separation occurs before the ions are accelerated to high energy.

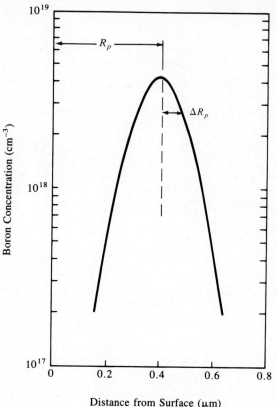

(a)

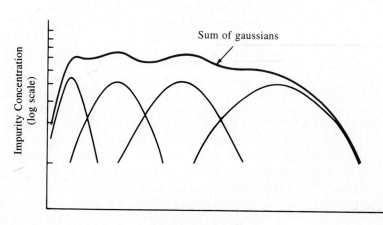

Distance from Surface

(b)

Figure 5-5
Distributions of implanted impurities: (a) gaussian distribution of boron atoms about a projected range R_p (in this example, a dose of 10^{14} B atoms/cm^2 implanted at 100 keV); (b) a relatively flat distribution obtained by summing four gaussians implanted at selected energies and doses.

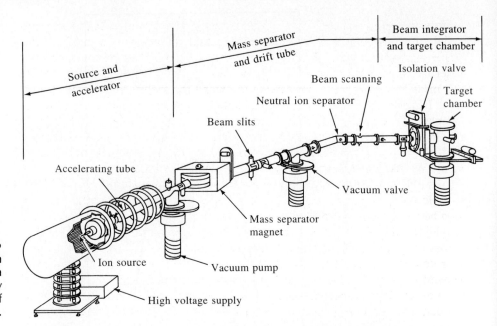

Figure 5-6
Schematic diagram of the 400-keV ion implantation facility at the University of Texas.

This control over doping level, along with the uniformity of the implant over the wafer surface, make ion implantation particularly attractive for the fabrication of Si integrated circuits (Chapter 9).

One problem with this doping method is the lattice damage which results from collisions between the ions and the lattice atoms. However, most of this damage can be removed in Si by heating the crystal after the implantation. This process is called *annealing*. Although Si can be heated to temperatures in excess of 1000°C without difficulty, GaAs and some other compounds tend to dissociate at high temperatures. For example, As evaporation from the surface of GaAs during annealing damages the sample. Therefore, it is common to encapsulate the GaAs with a thin layer of silicon nitride during the anneal. Another approach to annealing either Si or compounds is to heat the sample only briefly (e.g., 10 s) using a flash lamp, rather than a conventional furnace.

**5.2
EQUILIBRIUM
CONDITIONS**

In this chapter we wish to develop both a useful mathematical description of the p-n junction and a strong qualitative understanding of its properties. There must be some compromise in these two goals, since a complete mathematical treatment would obscure the essentially simple physical features of junction operation, while a completely qualitative description would not be useful in making calculations. The approach, therefore, will be to describe the junction mathematically while neglecting small effects which add little to the basic solution. In Section 5.6 we shall include several deviations from the simple theory.

The mathematics of p-n junctions is greatly simplified for the case of the *step junction,* which has uniform p doping on one side of a sharp junction and uniform n doping on the other side. This model represents alloyed and epi-taxial junctions quite well; the diffused junction, however, is actually *graded* ($N_d - N_a$ varies over a significant distance on either side of the junction). After the basic ideas of junction theory are explored for the step junction, we can make the appropriate corrections to extend the theory to the graded junction. In these discussions we shall assume one-dimensional current flow in samples of uniform cross-sectional area.

In this section we investigate the properties of the step junction at equilibrium (i.e., with no external excitation and no net currents flowing in the device). We shall find that the difference in doping on each side of the junction causes a potential difference between the two types of material. This is a reasonable result, since we would expect some charge transfer because of diffusion between the p material (many holes) and the n material (many electrons). In addition, we shall find that there are four components of current which flow across the junction due to the drift and diffusion of electrons and holes. These four components combine to give zero net current for the equilibrium case. However, the application of bias to the junction increases some of these current components with respect to others, giving net current flow. If we understand the nature of these four current components, a sound view of p-n junction operation, with or without bias, will follow.

5.2.1 The Contact Potential

Let us consider separate regions of p- and n-type semiconductor material, brought together to form a junction (Fig. 5-7). This is not a practical way of forming a device, but it does allow us to discover the requirements of equilibrium at a junction. Before they are joined, the n material has a large concentration of electrons and few holes, whereas the converse is true for the p material. Upon joining the two regions (Fig. 5-7), we expect diffusion of carriers to take place because of the large carrier concentration gradients at the junction. Thus holes diffuse from the p side into the n side, and electrons diffuse from n to p. The resulting diffusion current cannot build up indefinitely, however, because an opposing electric field is created at the junction (Fig. 5-7b). If the two regions were boxes of red air molecules and green molecules (perhaps due to appropriate types of pollution), eventually there would be a homogeneous mixture of the two after the boxes were joined. This cannot occur in the case of the charged particles in a p-n junction because of the development of space charge and the electric field $\mathscr{E}$. If we consider that electrons diffusing from n to p leave behind uncompensated[†] donor ions (N_d^+)

[†]We recall that neutrality is maintained in the bulk materials of Fig. 5-7a by the presence of one electron for each ionized donor ($n = N_d^+$) in the n material and one hole for each ionized acceptor ($p = N_a^-$) in the p material (neglecting minority carriers). Thus, if electrons leave n, some of the positive donor ions near the junction are left uncompensated, as in Fig. 5-7b. The donors and acceptors are fixed in the lattice, in contrast to the mobile electrons and holes.

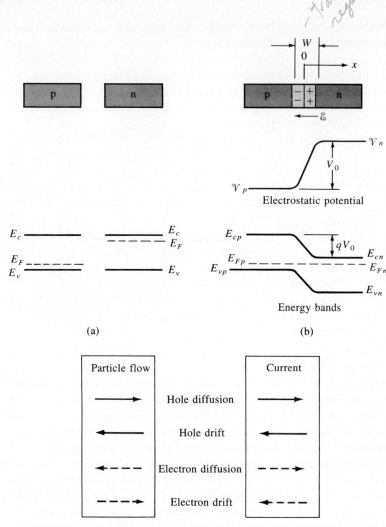

Figure 5-7
Properties of an equilibrium p-n junction: (a) isolated, neutral regions of p-type and n-type material and energy bands for the isolated regions; (b) junction, showing space charge in the transition region W, the resulting electric field $\mathscr{E}$ and contact potential V_0, and the separation of the energy bands; (c) directions of the four components of particle flow within the transition region, and the resulting current directions.

in the n material, and holes leaving the p region leave behind uncompensated acceptors (N_a^-), it is easy to visualize the development of a region of positive space charge near the n side of the junction and negative charge near the p side. The resulting electric field is directed from the positive charge toward the negative charge. Thus $\mathscr{E}$ is in the direction opposite to that of diffusion current for each type of carrier (recall electron current is opposite to the direction of electron flow). Therefore, the field creates a drift component of current from n to p, opposing the diffusion current (Fig. 5-7c).

Since we know that no *net* current can flow across the junction at equilibrium, the current due to the drift of carriers in the $\mathscr{E}$ field must exactly cancel the diffusion current. Furthermore, since there can be no net buildup of electrons or holes on either side as a function of time, the drift and diffusion currents must cancel for *each* type of carrier.

$$J_p(\text{drift}) + J_p(\text{diff.}) = 0 \qquad (5\text{-}2a)$$

$$J_n(\text{drift}) + J_n(\text{diff.}) = 0 \qquad (5\text{-}2b)$$

Therefore, the electric field $\mathscr{E}$ builds up to the point where the net current is zero at equilibrium. The electric field appears in some region W about the junction, and there is an equilibrium potential difference V_0 across W. In the electrostatic potential diagram of Fig. 5-7b, there is a gradient in potential in the direction opposite to $\mathscr{E}$, in accordance with the fundamental relation $\mathscr{E}(x) = -d\mathcal{V}(x)/dx.$[†] We assume the electric field is zero in the neutral regions outside W. Thus there is a constant potential $\mathcal{V}_n$ in the neutral n material, a constant $\mathcal{V}_p$ in the neutral p material, and a potential difference $V_0 = \mathcal{V}_n - \mathcal{V}_p$ between the two. The region W is called the *transition region*,[‡] and the potential difference V_0 is called the *contact potential*. The contact potential appearing across W is a *built-in* potential barrier, in that it is necessary to the maintenance of equilibrium at the junction; it does not imply any external potential. Indeed, the contact potential cannot be measured by placing a voltmeter across the devices, because new contact potentials are formed at each probe, just canceling V_0. By definition V_0 is an equilibrium quantity, and no net current can result from it.

The contact potential separates the bands as in Fig. 5-7b; the valence and conduction energy bands are higher on the p side of the junction than on the n side by the amount $qV_0.$[§] The separation of the bands at equilibrium is just that required to make the Fermi level constant throughout the device. We discussed the lack of spatial variation of the Fermi level at equilibrium in Section 3.5. Thus if we know the band diagram, including E_F, for each separate material (Fig. 5-7a), we can find the band separation for the junction at equilibrium simply by drawing a diagram such as Fig. 5-7b with the Fermi levels aligned.

To obtain a quantitative relationship between V_0 and the doping concentrations on each side of the junction, we must use the requirements for equi-

---

[†]When we write $\mathscr{E}(x)$, we refer to the value of $\mathscr{E}$ as computed in the x-direction. This value will of course be negative, since it is directed opposite to the true direction of $\mathscr{E}$ as shown in Fig. 5-7b.

[‡]Other names for this region are the *space charge region*, since space charge exists within W while neutrality is maintained outside this region, and the *depletion region*, since W is almost depleted of carriers compared with the rest of the crystal. The contact potential V_0 is also called the *diffusion potential*, since it represents a potential barrier which diffusing carriers must surmount in going from one side of the junction to the other.

[§]The electron energy diagram of Fig. 5-7b is related to the electrostatic potential diagram by $-q$, the negative charge on the electron. Since $\mathcal{V}_n$ is a higher potential than $\mathcal{V}_p$ by the amount V_0, the electron energies on the n side are *lower* than those on the p side by qV_0.

librium in the drift and diffusion current equations. For example, the drift and diffusion components of the hole current just cancel at equilibrium:

$$J_p(x) = q\left[\underbrace{\mu_p p(x)\mathscr{E}(x)}_{\text{drift}} - \underbrace{D_p \frac{dp(x)}{dx}}_{\text{diffusion}}\right] = 0 \qquad (5\text{-}3)$$

This equation can be rearranged to obtain

$$\frac{\mu_p}{D_p}\mathscr{E}(x) = \frac{1}{p(x)}\frac{dp(x)}{dx} \qquad (5\text{-}4)$$

where the x-direction is arbitrarily taken from p to n. The electric field can be written in terms of the gradient in the potential, $\mathscr{E}(x) = -d\mathscr{V}(x)/dx$, so that Eq. (5-4) becomes

$$-\frac{q}{kT}\frac{d\mathscr{V}(x)}{dx} = \frac{1}{p(x)}\frac{dp(x)}{dx} \qquad (5\text{-}5)$$

with the use of the Einstein relation for μ_p/D_p. This equation can be solved by integration over the appropriate limits. In this case we are interested in the potential on either side of the junction, $\mathscr{V}_p$ and $\mathscr{V}_n$, and the hole concentration just at the edge of the transition region on either side, p_p and p_n. For a step junction it is reasonable to take the electron and hole concentration in the neutral regions outside the transition region as their equilibrium values. Since we have assumed a one-dimensional geometry, p and $\mathscr{V}$ can be taken reasonably as functions of x only. Integration of Eq. (5-5) gives

$$-\frac{q}{kT}\int_{\mathscr{V}_p}^{\mathscr{V}_n} d\mathscr{V} = \int_{p_p}^{p_n}\frac{1}{p}\,dp$$

$$-\frac{q}{kT}(\mathscr{V}_n - \mathscr{V}_p) = \ln p_n - \ln p_p = \ln\frac{p_n}{p_p} \qquad (5\text{-}6)$$

The potential difference $\mathscr{V}_n - \mathscr{V}_p$ is the contact potential V_0 (Fig. 5-7b). Thus we can write V_0 in terms of the equilibrium hole concentrations on either side of the junction:

$$V_0 = \frac{kT}{q}\ln\frac{p_p}{p_n} \qquad (5\text{-}7)$$

If we consider the step junction to be made up of material with N_a acceptors/cm^3 on the p side and a concentration of N_d donors on the n side, we can write Eq. (5-7) as

$$V_0 = \frac{kT}{q}\ln\frac{N_a}{n_i^2/N_d} = \frac{kT}{q}\ln\frac{N_a N_d}{n_i^2} \qquad (5\text{-}8)$$

by considering the majority carrier concentration to be the doping concentration on each side.

Another useful form of Eq. (5-7) is

$$\frac{p_p}{p_n} = e^{qV_0/kT} \qquad (5\text{-}9)$$

By using the equilibrium condition $p_p n_p = n_i^2 = p_n n_n$, we can extend Eq. (5-9) to include the electron concentrations on either side of the junction:

$$\boxed{\frac{p_p}{p_n} = \frac{n_n}{n_p} = e^{qV_0/kT}} \qquad (5\text{-}10)$$

This relation will be very valuable in calculation of the I–V characteristics of the junction.

EXAMPLE 5-1

An n-type sample of Ge contains $N_d = 10^{16}$ cm^{-3}. A junction is formed by alloying with In at 160°C. Assume that the acceptor concentration in the regrown region equals the solid solubility at the alloying temperature.
(a) Calculate the Fermi level positions at 300 K in the p and n regions.
(b) Draw an equilibrium band diagram for the junction and determine the contact potential V_0 from the diagram.
(c) Compare the results of part (b) with V_0 as calculated from Eq. (5-8).

From Appendix V, $N_a \simeq 3 \times 10^{18}$ cm^{-3}.
(a) Assume that $p_p = N_a$ and $n_n = N_d$. On the p side:

$$p_p = n_i e^{(E_i - E_{Fp})/kT}$$

Thus

$$E_{ip} - E_{Fp} = kT \ln \frac{p_p}{n_i} = 0.0259 \ln \frac{3 \times 10^{18}}{2.5 \times 10^{13}} = 0.303 \text{ eV}$$

~See page 79 for table

Similarly,

$$E_{Fn} - E_{in} = 0.0259 \ln \frac{10^{16}}{2.5 \times 10^{13}} = 0.155 \text{ eV}$$

The resulting band diagram is shown in Fig. 5-8.
(b)
$$qV_0 = E_{cp} - E_{cn} = E_{ip} - E_{in}$$
$$V_0 = 0.303 + 0.155 = 0.458 \text{ V}$$

(c)
$$V_0 = 0.0259 \ln \frac{3 \times 10^{18} \times 10^{16}}{6.25 \times 10^{26}}$$
$$= 0.0259 \ln(4.8 \times 10^7) = 0.458 \text{ V}$$

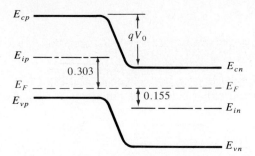

Figure 5-8
Equilibrium band diagram for the p-n junction of Example 5-1.

5.2.2 Equilibrium Fermi Levels

We have observed that the Fermi level must be constant throughout the device at equilibrium. This observation can be easily related to the results of the previous section. Since we have assumed that p_n and p_p are given by their equilibrium values outside the transition region, we can write Eq. (5-9) in terms of the basic definitions of these quantities using Eq. (3-19):

$$\frac{p_p}{p_n} = e^{qV_0/kT} = \frac{N_v e^{-(E_{Fp}-E_{vp})/kT}}{N_v e^{-(E_{Fn}-E_{vn})/kT}} \tag{5-11a}$$

$$e^{qV_0/kT} = e^{(E_{Fn}-E_{Fp})/kT} e^{(E_{vp}-E_{vn})/kT} \tag{5-11b}$$

$$qV_0 = E_{vp} - E_{vn} \tag{5-12}$$

The Fermi level and valence band energies are written with subscripts to indicate the p side and the n side of the junction.

From Fig. 5-7b the energy bands on either side of the junction are separated by the contact potential V_0 times the electronic charge q; thus the energy difference $E_{vp} - E_{vn}$ is just qV_0. Equation (5-12) results from the fact that the Fermi levels on either side of the junction are equal at equilibrium ($E_{Fn} - E_{Fp} = 0$). When bias is applied to the junction, the potential barrier is raised or lowered from the value of the contact potential, and the Fermi levels on either side of the junction are shifted with respect to each other by an energy in electron volts numerically equal to the applied voltage in volts.

5.2.3 Space Charge at a Junction

Within the transition region, electrons and holes are in transit from one side of the junction to the other. Some electrons diffuse from n to p, and some are swept by the electric field from p to n (and conversely for holes); there are, however, very few carriers within the transition region at any given time, since the electric field serves to sweep out carriers which have wandered into W. To a good approximation, we can consider the space charge within the transition region as due only to the uncompensated donor and acceptor ions. The charge density within W is plotted in Fig. 5-9b. Neglecting carriers within the space charge region, the charge density on the n side is just q times the concentration of donor ions N_d, and the negative charge density on the p side is $-q$ times the

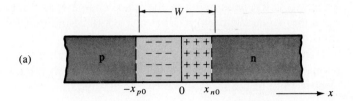

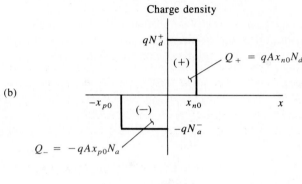

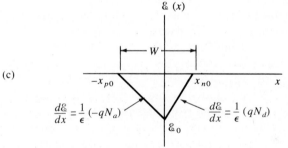

Figure 5-9
Space charge and electric field distribution within the transition region of a p-n junction with $N_d > N_a$: (a) the transition region, with $x = 0$ defined at the metallurgical junction; (b) charge density within the transition region, neglecting the free carriers; (c) the electric field distribution, where the reference direction for $\mathscr{E}$ is arbitrarily taken as the $+x$-direction.

concentration of acceptors N_a. The assumption of carrier depletion within W and neutrality outside W is known as the *depletion approximation*.

Since the dipole about the junction must have an equal number of charges on either side $(Q_+ = |Q_-|)$,[†] the transition region may extend into the p and n regions unequally, depending on the relative doping of the two sides. For example, if the p side is more lightly doped than the n side $(N_a < N_d)$, the space charge region must extend farther into the p material than into the n, to "uncover" an equivalent amount of charge. For a sample of cross-sectional area A, the total uncompensated charge on either side of the junction is

$$qAx_{p0}N_a = qAx_{n0}N_d \tag{5-13}$$

[†]A simple way of remembering this equal charge requirement is to note that electric flux lines must begin and end on charges of opposite sign. Therefore, if Q_+ and Q_- were not of equal magnitude, the electric field would not be contained within W but would extend farther into the p or n regions until the enclosed charges became equal.

where x_{p0} is the penetration of the space charge region into the p material, and x_{n0} is the penetration into n. The total width of the transition region (W) is the sum of x_{p0} and x_{n0}.

$W = x_{p0} + x_{n0}$

To calculate the electric field distribution within the transition region, we begin with the point form of *Gauss's law*, which relates the gradient of the electric field to the local space charge at any point x:

$$\frac{d\mathscr{E}(x)}{dx} = \frac{q}{\epsilon}(p - n + N_d^+ - N_a^-) \tag{5-14}$$

This equation is greatly simplified within the transition region if we neglect the contribution of the carriers ($p - n$) to the space charge. With this approximation we have two regions of constant space charge:

$$\frac{d\mathscr{E}}{dx} = \frac{q}{\epsilon}N_d, \qquad 0 < x < x_{n0} \tag{5-15a}$$

$$\frac{d\mathscr{E}}{dx} = -\frac{q}{\epsilon}N_a, \qquad -x_{p0} < x < 0 \tag{5-15b}$$

assuming complete ionization of the impurities ($N_d^+ = N_d$, and $N_a^- = N_a$). We can see from these two equations that a plot of $\mathscr{E}(x)$ vs. x within the transition region has two slopes, positive ($\mathscr{E}$ increasing with x) on the n side and negative ($\mathscr{E}$ becoming more negative as x increases) on the p side. There is some maximum value of the field $\mathscr{E}_0$ at $x = 0$ (the metallurgical junction between the p and n materials), and $\mathscr{E}(x)$ is everywhere negative within the transition region (Fig. 5-9c). These conclusions come from Gauss's law, but we could predict the qualitative features of Fig. 5-9 without equations. We expect the electric field $\mathscr{E}(x)$ to be negative throughout W, since we know that the $\mathscr{E}$ field actually points in the $-x$-direction, from n to p (i.e., from the positive charges of the transition region dipole toward the negative charges). The electric field is assumed to go to zero at the edges of the transition region, since we are neglecting any small $\mathscr{E}$ field in the neutral n or p regions. Finally, there must be a maximum $\mathscr{E}_0$ at the junction, since this point is between the charges Q_+ and Q_- on either side of the transition region. All the electric flux lines pass through the $x = 0$ plane, so this is the obvious point of maximum electric field.

The value of $\mathscr{E}_0$ can be found by integrating either part of Eq. (5-15) with appropriate limits (see Fig. 5-9c in choosing the limits of integration).

$$\int_{\mathscr{E}_0}^{0} d\mathscr{E} = \frac{q}{\epsilon}N_d\int_{0}^{x_{n0}} dx, \qquad 0 < x < x_{n0} \tag{5-16a}$$

$$\int_{0}^{\mathscr{E}_0} d\mathscr{E} = -\frac{q}{\epsilon}N_a\int_{-x_{p0}}^{0} dx, \qquad -x_{p0} < x < 0 \tag{5-16b}$$

Therefore, the maximum value of the electric field is

$$\mathscr{E}_0 = -\frac{q}{\epsilon}N_d x_{n0} = -\frac{q}{\epsilon}N_a x_{p0} \tag{5-17}$$

It is simple to relate the electric field to the contact potential V_0, since the $\mathcal{E}$ field at any x is the negative of the potential gradient at that point:

$$\mathcal{E}(x) = -\frac{d\mathcal{V}(x)}{dx} \quad \text{or} \quad -V_0 = \int_{-x_{p0}}^{x_{n0}} \mathcal{E}(x)\,dx \qquad (5\text{-}18)$$

Thus the negative of the contact potential is simply the area under the $\mathcal{E}(x)$ vs. x triangle. This relates the contact potential to the width of the depletion region:

$$V_0 = -\frac{1}{2}\mathcal{E}_0 W = \frac{1}{2}\frac{q}{\epsilon}N_d x_{n0} W \qquad (5\text{-}19)$$

Since the balance of charge requirement is $x_{n0}N_d = x_{p0}N_a$, and W is simply $x_{p0} + x_{n0}$, we can write $x_{n0} = WN_a/(N_a + N_d)$ in Eq. (5-19):

$$V_0 = \frac{1}{2}\frac{q}{\epsilon}\frac{N_a N_d}{N_a + N_d}W^2 \qquad (5\text{-}20)$$

By solving for W, we have an expression for the width of the transition region in terms of the contact potential, the doping concentrations, and known constants q and ϵ.

$$W = \left[\frac{2\epsilon V_0}{q}\left(\frac{N_a + N_d}{N_a N_d}\right)\right]^{1/2} = \left[\frac{2\epsilon V_0}{q}\left(\frac{1}{N_a} + \frac{1}{N_d}\right)\right]^{1/2} \qquad (5\text{-}21)$$

There are several useful variations of Eq. (5-21); for example, V_0 can be written in terms of the doping concentrations with the aid of Eq. (5-8):

$$W = \left[\frac{2\epsilon kT}{q^2}\left(\ln\frac{N_a N_d}{n_i^2}\right)\left(\frac{1}{N_a} + \frac{1}{N_d}\right)\right]^{1/2} \qquad (5\text{-}22)$$

We can also calculate the <u>penetration of the transition region</u> into the n and p materials:

$$x_{p0} = \frac{WN_d}{N_a + N_d} = \frac{W}{1 + N_a/N_d} = \left\{\frac{2\epsilon V_0}{q}\left[\frac{N_d}{N_a(N_a + N_d)}\right]\right\}^{1/2} \qquad (5\text{-}23a)$$

$$x_{n0} = \frac{WN_a}{N_a + N_d} = \frac{W}{1 + N_d/N_a} = \left\{\frac{2\epsilon V_0}{q}\left[\frac{N_a}{N_d(N_a + N_d)}\right]\right\}^{1/2} \qquad (5\text{-}23b)$$

As expected, Eqs. (5-23) predict that the transition region extends farther into the side with the lighter doping. For example, if $N_a \ll N_d$, x_{p0} is large compared with x_{n0}. This agrees with our qualitative argument that a deep penetration is necessary in lightly doped material to "uncover" the same amount of space charge as for a short penetration into heavily doped material.

Another important result of Eq. (5-21) is that the transition width W varies as the square root of the potential across the region. In the derivation to this point, we have considered only the equilibrium contact potential V_0. In Section 5.3 we shall see that an applied voltage can increase or decrease the potential across the transition region by aiding or opposing the equilibrium electric field. Therefore, Eq. (5-21) predicts that an applied voltage will increase or decrease the width of the transition region as well.

EXAMPLE 5-2   Aluminum is alloyed into an n-type Si sample ($N_d = 10^{16}$ cm^{-3}), forming an abrupt junction of circular cross section, with a diameter of 20 mils. Assume that the acceptor concentration in the alloyed regrown region is $N_a = 4 \times 10^{18}$ cm^{-3}. Calculate V_0, x_{n0}, x_{p0}, Q_+, and $\mathscr{E}_0$ for this junction at equilibrium (300 K). Sketch $\mathscr{E}(x)$ and charge density to scale, as in Fig. 5-9.

SOLUTION   From Eq. (5-8),

$$V_0 = \frac{kT}{q} \ln \frac{N_a N_d}{n_i^2} = 0.0259 \ln \frac{4 \times 10^{34}}{2.25 \times 10^{20}}$$

$$= 0.0259 \ln(1.78 \times 10^{14}) = 0.85 \text{ V}$$

From Eq. (5-21),

$$W = \left[\frac{2\epsilon V_0}{q}\left(\frac{1}{N_a} + \frac{1}{N_d}\right)\right]^{1/2} \qquad \mathscr{E} \neq \mathscr{E}_0 \, \mathscr{E}R. \overset{8.85 \times 10^{-14}}{\underset{\text{see App 3}}{}}$$

$$= \left[\frac{2(11.8 \times 8.85 \times 10^{-14})(0.85)}{1.6 \times 10^{-19}}(0.25 \times 10^{-18} + 10^{-16})\right]^{1/2}$$

$$= 3.34 \times 10^{-5} \text{ cm} = 0.334 \text{ }\mu\text{m}$$

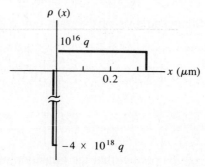

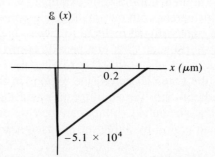

From Eq. (5-23),

$$x_{n0} = \frac{3.34 \times 10^{-5}}{1 + 0.0025} \simeq 0.333 \ \mu m$$

$$x_{p0} = \frac{3.34 \times 10^{-5}}{1 + 400} \simeq 8.3 \times 10^{-8} \ cm = 8.3 \ \text{Å}$$

Note that $x_{n0} \simeq W$. $A = \pi r^2 = \pi (2.54 \times 10^{-2})^2 = 2.03 \times 10^{-3} \ cm^2$

$$Q_+ = -Q_- = qAx_{n0}N_d = (1.6 \times 10^{-19})(2.03 \times 10^{-3})(3.33 \times 10^{-5})(10^{16})$$

$$= 1.08 \times 10^{-10} \ C$$

$$\mathscr{E}_0 = \frac{-qN_d x_{n0}}{\epsilon} = \frac{-(1.6 \times 10^{-19})(10^{16})(3.33 \times 10^{-5})}{(11.8)(8.85 \times 10^{-14})}$$

$$= -5.1 \times 10^4 \ V/cm$$

■

One useful feature of a p-n junction is that current flows quite freely in the p to n direction when the p region has a positive external voltage bias relative to n (forward bias and forward current), whereas virtually no current flows when p is made negative relative to n (reverse bias and reverse current). This asymmetry of the current flow makes the p-n junction diode very useful as a rectifier. While rectification is an important application, it is only the beginning of a host of uses for the biased junction. As we shall see in Chapter 6, biased p-n junctions can be used as voltage-variable capacitors, photocells, light emitters, and many more devices which are basic to modern electronics. Two or more junctions can be used to form transistors and controlled switches.

In this section we begin with a qualitative description of current flow in a biased junction. With the background of the previous section, the basic features of current flow are relatively simple to understand, and these qualitative concepts form the basis for the analytical description of forward and reverse currents in a junction.

5.3.1   Qualitative Description of Current Flow at a Junction

We assume that an applied voltage bias V appears across the transition region of the junction rather than in the neutral n and p regions. Of course, there will be some voltage drop in the neutral material, if a current flows through it. But in most p-n junction devices, the length of each region is small compared with its area, and the doping is usually moderate to heavy; thus the resistance is small in each neutral region, and only a small voltage drop can be maintained outside the space charge (transition) region. For almost all calculations it is valid to assume that an applied voltage appears entirely across the transition region. We shall take V to be positive when the external bias is positive on the p side relative to the n side.

Since an applied voltage changes the electrostatic potential barrier and thus the electric field within the transition region, we would expect changes in the various components of current at the junction (Fig. 5-10). In addition, the separation of the energy bands is affected by the applied bias, along with the width of the depletion region. Let us begin by examining qualitatively the effects of bias on the important features of the junction.

The *electrostatic potential barrier* at the junction is lowered by a forward bias V_f from the equilibrium contact potential V_0 to the smaller value $V_0 - V_f$.

Figure 5-10
Effects of a bias at a
p-n junction;
transition region
width and electric
field, electrostatic
potential, energy
band diagram, and
particle flow and
current directions
within *W* for
(a) equilibrium,
(b) forward bias,
and (c) reverse
bias.

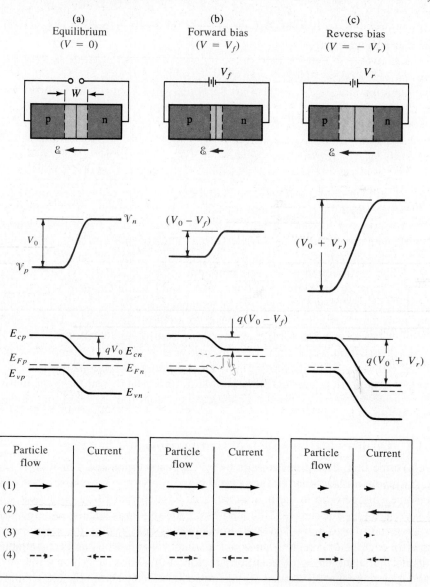

(1) Hole diffusion   (3) Electron diffusion
(2) Hole drift   (4) Electron drift

This lowering of the potential barrier occurs because a forward bias (p positive with respect to n) raises the electrostatic potential on the p side relative to the n side. For a reverse bias ($V = -V_r$) the opposite occurs; the electrostatic potential of the p side is depressed relative to the n side, and the potential barrier at the junction becomes larger ($V_0 + V_r$).

The *electric field* within the transition region can be deduced from the potential barrier. We notice that the field decreases with forward bias, since the applied electric field opposes the built-in field. With reverse bias the field at the junction is increased by the applied field, which is in the same direction as the equilibrium field.

The change in electric field at the junction calls for a change in the *transition region width W*, since it is still necessary that a proper number of positive and negative charges (in the form of uncompensated donor and acceptor ions) be exposed for a given value of the $\mathscr{E}$ field. Thus we would expect the width W to decrease under forward bias (smaller $\mathscr{E}$, fewer uncompensated charges) and to increase under reverse bias. Equations (5-21) and (5-23) can be used to calculate W, x_{p0}, and x_{n0} if V_0 is replaced by the new barrier height $V_0 - V$.[†]

The *separation of the energy bands* is a direct function of the electrostatic potential barrier at the junction. The height of the electron energy barrier is simply the electronic charge q times the height of the electrostatic potential barrier. Thus the bands are separated less [$q(V_0 - V_f)$] under forward bias than at equilibrium, and more [$q(V_0 + V_r)$] under reverse bias. We assume the Fermi level deep inside each neutral region is essentially the equilibrium value (we shall return to this assumption later); therefore, the shifting of the energy bands under bias implies a separation of the Fermi levels on either side of the junction, as mentioned in Section 5.2.2. Under forward bias, the Fermi level on the n side E_{Fn} is above E_{Fp} by the energy qV_f; for reverse bias, E_{Fp} is qV_r joules higher than E_{Fn}. *In energy units of electron volts, the Fermi levels in the two neutral regions are separated by an energy (eV) numerically equal to the applied voltage (V).*

The *diffusion current* is composed of majority carrier electrons on the n side surmounting the potential energy barrier to diffuse to the p side, and holes surmounting their barrier from p to n.[‡] There is a distribution of energies for

---

[†]With bias applied to the junction, the 0 in the subscripts of x_{n0} and x_{p0} does not imply equilibrium. Instead, it signifies the origin of a new set of coordinates $x_n = 0$ and $x_p = 0$, as defined later in Fig. 5-12.

[‡]Remember that the potential energy barriers for electrons and holes are directed oppositely. The barrier for electrons is apparent from the energy band diagram, which is always drawn for electron energies. For holes, the potential energy barrier at the junction has the same shape as the electrostatic potential barrier (the conversion factor between electrostatic potential and hole energy is $+q$). A simple check of these two barrier directions can be made by asking the directions in which carriers are swept by the $\mathscr{E}$ field within the transition region—a hole is swept in the direction of $\mathscr{E}$, from n to p (swept down the potential "hill" for holes); an electron is swept opposite to $\mathscr{E}$, from p to n (swept down the potential energy "hill" for electrons).

electrons in the n-side conduction band (Fig. 3-16), and some electrons in the high-energy "tail" of the distribution have enough energy to diffuse from n to p at equilibrium in spite of the barrier. With forward bias, however, the barrier is lowered (to $V_0 - V_f$), and many more electrons in the n-side conduction band have sufficient energy to diffuse from n to p over the smaller barrier. Therefore, the electron diffusion current can be quite large with forward bias. Similarly, more holes can diffuse from p to n under forward bias because of the lowered barrier. For reverse bias the barrier becomes so large ($V_0 + V_r$) that virtually no electrons in the n-side conduction band or holes in the p-side valence band have enough energy to surmount it. Therefore, the diffusion current is usually negligible for reverse bias.

The *drift current* is relatively insensitive to the height of the potential barrier. This sounds strange at first, since we normally think in terms of material with ample carriers, and therefore we expect drift current to be simply proportional to the applied field. The reason for this apparent anomaly is the fact that the drift current is limited *not* by *how fast* carriers are swept down the barrier, *but* rather *how often*. For example, minority carrier electrons on the p side which wander into the transition region will be swept down the barrier by the $\mathscr{E}$ field, giving rise to the electron component of drift current. However, this current is small not because of the size of the barrier, but because there are very few minority electrons in the p side to participate. Every electron on the p side which diffuses to the transition region will be swept down the potential energy hill, whether the hill is large or small. The electron drift current does not depend on how fast an individual electron is swept from p to n, but rather on how many electrons are swept down the barrier per second. Similar comments apply regarding the drift of minority holes from the n side to the p side of the junction. To a good approximation, therefore, the electron and hole drift currents at the junction are independent of the applied voltage.

The supply of minority carriers on each side of the junction required to participate in the drift component of current is generated by thermal excitation of electron–hole pairs. For example, an EHP created near the junction on the p side provides a minority electron in the p material. If the EHP is generated within a diffusion length L_n of the transition region, this electron can diffuse to the junction and be swept down the barrier to the n side. The resulting current due to drift of generated carriers across the junction is commonly called the *generation current* since its magnitude depends entirely on the rate of generation of EHPs. As we shall discuss later, this generation current can be increased greatly by optical excitation of EHPs near the junction (the p-n junction *photodiode*).

The *total current* crossing the junction is composed of the sum of the diffusion and drift components. As Fig. 5-10 indicates, the electron and hole diffusion currents are both directed from p to n (although the particle flow directions are opposite to each other), and the drift currents are from n to p. The *net* current crossing the junction is zero at equilibrium, since the drift and diffusion components cancel for each type of carrier (the equilibrium electron and hole components need not be equal, as in Fig. 5-10, as long as the net hole current and the net electron current are each zero). Under reverse bias, both diffusion components are negligible because of the large barrier at the junction, and the only current is the relatively small (and essentially voltage-independent) generation current from n to p. This generation current is shown in Fig. 5-11, in a sketch of a typical $I–V$ plot for a p-n junction. In this figure the positive direction for the current I is taken from p to n, and the applied voltage V is positive when the positive battery terminal is connected to p and the negative terminal to n. The only current flowing in this p-n junction diode for negative V is the small current I(gen.) due to carriers generated in the transition region or minority carriers which diffuse to the junction and are collected. The current at $V = 0$ (equilibrium) is zero since the generation and diffusion currents cancel:[†]

$$I = I(\text{diff.}) - |I(\text{gen.})| = 0 \quad \text{for } V = 0 \qquad (5\text{-}24)$$

As we shall see in the next section, an applied forward bias $V = V_f$ increases the probability that a carrier can diffuse across the junction, by the factor $\exp (qV_f/kT)$. Thus the diffusion current under forward bias is given by its equilibrium value multiplied by $\exp (qV/kT)$; similarly, for reverse bias the diffusion current is the equilibrium value reduced by the same factor, with $V = -V_r$. Since the equilibrium diffusion current is equal in magnitude to $|I(\text{gen.})|$, the diffusion current with applied bias is simply $|I(\text{gen.})| \exp(qV/kT)$. The total current I is then the diffusion current minus the absolute value of the generation current, which we will now refer to as I_0:

$$I = I_0(e^{qV/kT} - 1) \qquad (5\text{-}25)$$

In Eq. (5-25) the applied voltage V can be positive or negative, $V = V_f$ or $V = -V_r$. When V is positive and greater than a few kT/q ($kT/q = 0.0259$ V

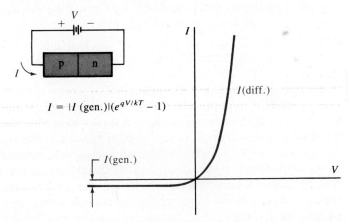

$$I = |I(\text{gen.})|(e^{qV/kT} - 1)$$

Figure 5-11
$I–V$ characteristic of a p-n junction.

[†]The total current I is the sum of the generation and diffusion components. However, these components are oppositely directed, I(diff.) being positive and I(gen.) being negative for the chosen reference direction. To avoid confusion of signs, we use here the magnitude of the drift current $|I(\text{gen.})|$ and include its negative sign in Eq. (5-24). Thus when we write the term $-|I(\text{gen.})|$, there is no doubt that the generation current is in the negative current direction. This approach emphasizes the fact that the two components of current add with opposite signs to give the total current.

at room temperature), the exponential term is much greater than unity. The current thus increases exponentially with forward bias. When V is negative (reverse bias), the exponential term approaches zero and the current is $-I_0$, which is in the n to p (negative) direction. This negative generation current is also called the _reverse saturation current._ The striking feature of Fig. 5-11 is the nonlinearity of the I–V characteristic. Current flows relatively freely in the forward direction of the diode, but almost no current flows in the reverse direction.

5.3.2   Carrier Injection

From the discussion in the previous section, we expect the minority carrier concentration on each side of a p-n junction to vary with the applied bias because of variations in the diffusion of carriers across the junction. The equilibrium ratio of hole concentrations on each side

$$\frac{p_p}{p_n} = e^{qV_0/kT} \tag{5-26}$$

becomes with bias (Fig. 5-10)

$$\frac{p(-x_{p0})}{p(x_{n0})} = e^{q(V_0-V)/kT} \tag{5-27}$$

This equation uses the altered barrier $V_0 - V$ to relate the steady state hole concentrations on the two sides of the transition region with either forward or reverse bias (V positive or negative). For low-level injection we can neglect changes in the majority carrier concentrations, which vary only slightly with bias compared with their equilibrium values. With this simplification we can write the ratio of Eq. (5-26) to (5-27) as

$$\frac{p(x_{n0})}{p_n} = e^{qV/kT} \qquad \text{taking } p(-x_{p0}) = p_p \tag{5-28}$$

With forward bias, Eq. (5-28) suggests a greatly increased minority carrier hole concentration at the edge of the transition region on the n side $p(x_{n0})$ than was the case at equilibrium. Conversely, the hole concentration $p(x_{n0})$ under reverse bias (V negative) is reduced below the equilibrium value p_n. The exponential increase of the hole concentration at x_{n0} with forward bias is an example of _minority carrier injection_. As Fig. 5-12 suggests, a forward bias V results in a steady state injection of excess holes into the n region and electrons into the p region. We can easily calculate the excess hole concentration Δp_n at the edge of the transition region x_{n0} by subtracting the equilibrium hole concentration from Eq. (5-28),

$$\Delta p_n = p(x_{n0}) - p_n = p_n(e^{qV/kT} - 1) \tag{5-29}$$

and similarly for excess electrons on the p side,

$$\Delta n_p = n(-x_{p0}) - n_p = n_p(e^{qV/kT} - 1) \tag{5-30}$$

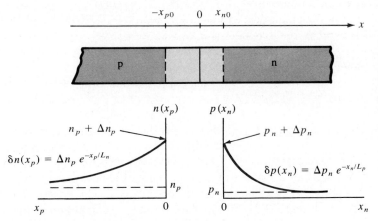

Figure 5-12
Minority carrier distributions on the two sides of the transition region for a forward-biased p-n junction. This figure provides definitions of distances x_n and x_p measured from the transition region edges.

From our study of diffusion of excess carriers in Section 4.4.4, we expect that injection leading to a steady concentration of Δp_n excess holes at x_{n0} will produce a *distribution* of excess holes in the n material. As the holes diffuse deeper into the n region, they recombine with electrons in the n material, and the resulting excess hole distribution is obtained as a solution of the diffusion equation, Eq. (4-34b). If the n region is long compared with the hole diffusion length L_p, the solution is exponential, as in Eq. (4-36). Similarly, the injected electrons in the p material diffuse and recombine, giving an exponential distribution of excess electrons. For convenience, let us define two new coordinates (Fig. 5-12): Distances measured in the *x*-direction in the n material from x_{n0} will be designated x_n; distances in the p material measured in the $-x$-direction with $-x_{p0}$ as the origin will be called x_p. This convention will simplify the mathematics considerably. We can write the diffusion equation as in Eq. (4-34) for each side of the junction and solve for the distributions of excess carriers (δn and δp) assuming long p and n regions:

$$\delta n(x_p) = \Delta n_p e^{-x_p/L_n} = n_p(e^{qV/kT} - 1)e^{-x_p/L_n} \qquad (5\text{-}31a)$$

$$\delta p(x_n) = \Delta p_n e^{-x_n/L_p} = p_n(e^{qV/kT} - 1)e^{-x_n/L_p} \qquad (5\text{-}31b)$$

The hole diffusion current at any point x_n in the n material can be calculated from Eq. (4-40):

$$I_p(x_n) = -qAD_p \frac{d\,\delta p(x_n)}{dx_n} = qA\frac{D_p}{L_p}\Delta p_n e^{-x_n/L_p} = qA\frac{D_p}{L_p}\delta p(x_n) \quad (5\text{-}32)$$

where A is the cross-sectional area of the junction. Thus the hole diffusion current at each position x_n is proportional to the excess hole concentration at that point.[†] The total hole current injected into the n material at the junction can be obtained simply by evaluating Eq. (5-32) at x_{n0}:

[†]With carrier injection due to bias, it is clear that the equilibrium Fermi levels cannot be used to describe carrier concentrations in the device. It is necessary to use the concept of quasi-Fermi levels, taking into account the spatial variations of the carrier concentrations.

$$I_p(x_n = 0) = \frac{qAD_p}{L_p} \Delta p_n = \frac{qAD_p}{L_p} p_n(e^{qV/kT} - 1) \tag{5-33}$$

By a similar analysis, the injection of electrons into the p material leads to an electron current at the junction of

$$I_n(x_p = 0) = -\frac{qAD_n}{L_n} \Delta n_p = -\frac{qAD_n}{L_n} n_p(e^{qV/kT} - 1) \tag{5-34}$$

The minus sign in Eq. (5-34) means that the electron current is opposite to the x_p-direction; that is, the true direction of I_n is in the $+x$-direction, adding to I_p in the total current (Fig. 5-13). If we neglect recombination in the transition region, we can consider that each injected electron reaching $-x_{p0}$ must pass through x_{n0}. Thus the total diode current I at x_{n0} can be calculated as the sum of $I_p(x_n = 0)$ and $-I_n(x_p = 0)$. If we take the $+x$-direction as the reference direction for the total current I, we must use a minus sign with $I_n(x_p)$ to account for the fact that x_p is defined in the $-x$-direction:

$$I = I_p(x_n = 0) - I_n(x_p = 0) = \frac{qAD_p}{L_p} \Delta p_n + \frac{qAD_n}{L_n} \Delta n_p \tag{5-35}$$

$$\boxed{I = qA\left(\frac{D_p}{L_p}p_n + \frac{D_n}{L_n}n_p\right)(e^{qV/kT} - 1) = I_0(e^{qV/kT} - 1)} \tag{5-36}$$

Equation (5-36) is the *diode equation*, having the same form as the qualitative relation Eq. (5-25). Nothing in the derivation excludes the possibility that the bias voltage V can be negative; thus the diode equation describes the total current through the diode for either forward or reverse bias. We can calculate the current for reverse bias by letting $V = -V_r$:

$$I = qA\left(\frac{D_p}{L_p}p_n + \frac{D_n}{L_n}n_p\right)(e^{-qV_r/kT} - 1) \tag{5-37}$$

If V_r is larger than a few kT/q, the total current is just the reverse saturation current

$$I = -qA\left(\frac{D_p}{L_p}p_n + \frac{D_n}{L_n}n_p\right) = -I_0 \tag{5-38}$$

Another simple and instructive way of calculating the total current is to consider the injected current as supplying the carriers for the excess distributions (Fig. 5-13b). For example, $I_p(x_n = 0)$ must supply enough holes per second to maintain the steady state exponential distribution $\delta p(x_n)$ as the holes recombine. The total positive charge stored in the excess carrier distribution at any instant of time is

$$Q_p = qA \int_0^\infty \delta p(x_n)dx_n = qA \Delta p_n \int_0^\infty e^{-x_n/L_p} dx_n = qAL_p\Delta p_n \tag{5-39}$$

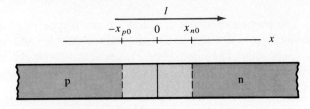

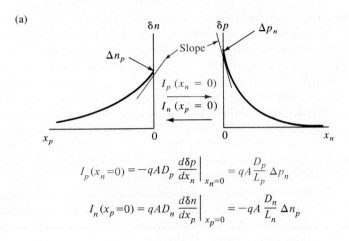

(a)

$$I_p(x_n=0) = -qAD_p \frac{d\delta p}{dx_n}\bigg|_{x_n=0} = qA\frac{D_p}{L_p}\Delta p_n$$

$$I_n(x_p=0) = qAD_n \frac{d\delta n}{dx_p}\bigg|_{x_p=0} = -qA\frac{D_n}{L_n}\Delta n_p$$

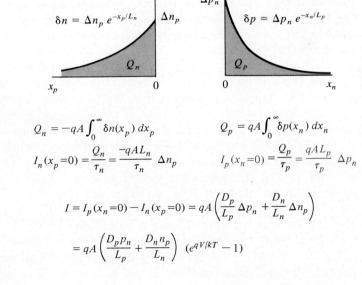

(b)

$$Q_n = -qA\int_0^\infty \delta n(x_p)\, dx_p \qquad Q_p = qA\int_0^\infty \delta p(x_n)\, dx_n$$

$$I_n(x_p=0) = \frac{Q_n}{\tau_n} = \frac{-qAL_n}{\tau_n}\Delta n_p \qquad I_p(x_n=0) = \frac{Q_p}{\tau_p} = \frac{qAL_p}{\tau_p}\Delta p_n$$

$$I = I_p(x_n=0) - I_n(x_p=0) = qA\left(\frac{D_p}{L_p}\Delta p_n + \frac{D_n}{L_n}\Delta n_p\right)$$

$$= qA\left(\frac{D_p p_n}{L_p} + \frac{D_n n_p}{L_n}\right)(e^{qV/kT}-1)$$

Figure 5-13
Two methods for calculating junction current from the excess minority carrier distributions: (a) diffusion currents at the edges of the transition region; (b) charge in the distributions divided by the minority carrier lifetimes.

The average lifetime of a hole in the n-type material is τ_p. Thus, on the average, this entire charge distribution recombines and must be replenished every τ_p seconds. The injected hole current at $x_n = 0$ needed to maintain the distribution is simply the total charge divided by the average time of replacement:

$$I_p(x_n = 0) = \frac{Q_p}{\tau_p} = qA\frac{L_p}{\tau_p}\Delta p_n = qA\frac{D_p}{L_p}\Delta p_n \qquad (5\text{-}40)$$

using $D_p/L_p = L_p/\tau_p$.

This is the same result as Eq. (5-33), which was calculated from the diffusion currents. Similarly, we can calculate the negative charge stored in the distribution $\delta n(x_p)$ and divide by τ_n to obtain the injected electron current in the p material. This method, called the *charge control approximation*, illustrates the important fact that the minority carriers injected into either side of a p-n junction diffuse into the neutral material and recombine with the majority carriers. The minority carrier current [for example, $I_p(x_n)$] decreases exponentially with distance into the neutral region. Thus several diffusion lengths away from the junction, most of the total current is carried by the majority carriers. We shall discuss this point in more detail in the following section.

In summary, we can calculate the current at a p-n junction in two ways (Fig. 5-13): (a) from the slopes of the excess minority carrier distributions at the two edges of the transition regions and (b) from the steady state charge stored in each distribution. We add the hole current injected into the n material $I_p(x_n = 0)$ to the electron current injected into the p material $I_n(x_p = 0)$, after including a minus sign with $I_n(x_p)$ to conform with the conventional definition of positive current in the $+x$-direction. We are able to add these two currents because of the assumption that no recombination takes place within the transition region.[†] Thus we effectively have the total electron and hole current at one point in the device (x_{n0}). Since the total current must be constant throughout the device (despite variations in the current components), I as described by Eq. (5-36) is the total current at every position x in the diode.

One implication of Eq. (5-36) is that the total current at the junction is dominated by injection of carriers from the more heavily doped side into the side with lesser doping. For example, if the p material is very heavily doped and the n region is lightly doped, the minority carrier concentration on the p side (n_p) is negligible compared with the minority carrier concentration on the n side (p_n). Thus the diode equation can be approximated by injection of holes only, as in Eq. (5-33). This means that the charge stored in the minority carrier distributions is due mostly to holes on the n side (Fig. 5-14). This structure is called a p^+-n junction, where the + superscript simply means heavy doping. Another characteristic of the p^+-n or n^+-p structure is that the transition region extends primarily into the lightly doped region, as we found in the discussion of Eq. (5-23). Having one side heavily doped is a useful arrangement for many practical devices, as we shall see in our discussions of switching diodes and transistors. This type of junction is common in devices which are fabricated by

---

[†]This assumption will be discussed further in Section 5.6.2.

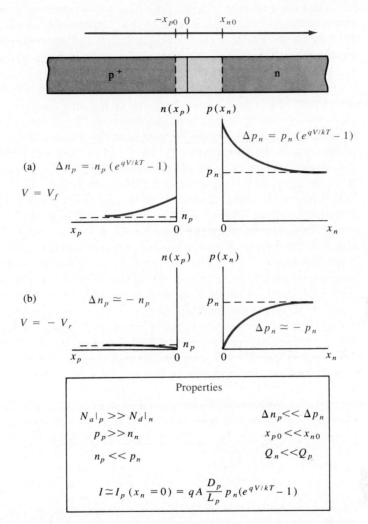

Figure 5-14
Properties of a p^+-n junction with (a) forward bias and (b) reverse bias.

counterdoping. For example, an n-type Si sample with $N_d = 10^{14}$ cm^{-3} can be used as the substrate for an implanted or diffused junction. If the doping of the p region is greater than 10^{19} cm^{-3} (typical of diffused junctions), the structure is definitely p^+-n, with n_p more than five orders of magnitude smaller than p_n. Since this configuration is common in device technology, we shall return to it in much of the following discussion.

In this discussion of carrier injection and minority carrier distributions, we have primarily assumed forward bias. The distributions for reverse bias can be obtained from the same equations (Fig. 5-14b), if a negative value of V is introduced. For example, if $V = -V_r$ (p negatively biased with respect to n), we can approximate Eq. (5-29) as

$$\Delta p_n = p_n(e^{q(-V_r)/kT} - 1) \simeq -p_n \qquad \text{for } V_r \gg kT/q \qquad (5\text{-}41)$$

and similarly $\Delta n_p \simeq -n_p$.

Chm 104 10AM

Thus for a reverse bias of more than a few tenths of a volt, the minority carrier concentration at each edge of the transition region becomes essentially zero as the excess concentration approaches the negative of the equilibrium concentration. The excess minority carrier concentrations in the neutral regions are still given by Eq. (5-31), so that depletion of carriers below the equilibrium values extends approximately a diffusion length beyond each side of the transition region. This reverse-bias depletion of minority carriers can be thought of as *minority carrier extraction,* analogous to the injection of forward bias. Physically, extraction occurs because minority carriers at the edges of the depletion region are swept down the barrier at the junction to the other side and are not replaced by an opposing diffusion of carriers. For example, when holes at x_{n0} are swept across the junction to the p side by the $\mathscr{E}$ field, a gradient in the hole distribution in the n material exists, and holes in the n region diffuse toward the junction. The steady state hole distribution in the n region has the inverted exponential shape of Fig. 5-14b. It is important to remember that although the reverse saturation current occurs at the junction by drift of carriers down the barrier, this current is fed from each side by diffusion toward the junction of minority carriers in the neutral regions. The rate of carrier drift across the junction (reverse saturation current) depends on the rate at which holes arrive at x_{n0} (and electrons at x_{p0}) by diffusion from the neutral material. These minority carriers are supplied by thermal generation, and we can show that the expression for the reverse saturation current, Eq.(5-38), represents the rate at which carriers are generated thermally within a diffusion length of each side of the transition region.

EXAMPLE 5-3   Consider a volume of n-type material of area A, with a length of one hole diffusion length L_p. The rate of thermal generation of holes within the volume is

$$AL_p\frac{p_n}{\tau_p} \qquad \text{since } g_{\text{th}} = \alpha_r n_i^2 = \alpha_r n_n p_n = \frac{p_n}{\tau_p}$$

Assume that each thermally generated hole diffuses out of the volume before it can recombine. The resulting hole current is $I = qAL_p p_n/\tau_p$, which is the same as the saturation current for a p^+-n junction. We conclude that saturation current is due to the collection of minority carriers thermally generated within a diffusion length of the junction.

■

5.3.3 Minority and Majority Carrier Currents

We have calculated the total diode current by considering only the minority carrier diffusion currents, evaluated at the edges of the transition region. This was possible because minority carriers are injected into each side under forward bias and extracted from each side for reverse bias. *The drift of minority carriers can be neglected* in the neutral regions *outside W,* because the minority carrier concentration is small compared with that of the majority carriers. If the minority carriers contribute to the total current at all, their contribution must be through diffusion (dependent on the *gradient* of the carrier concentration). Even a very small concentration of minority carriers can have an appreciable effect on the current if the spatial variation is large.

Calculation of the majority carrier currents in the two neutral regions is simple, once we have found the minority carrier current. Since the total current I must be constant throughout the device, the majority carrier component of current is just the difference between I and the minority component (Fig. 5-15). In the p^+-n junction, the total current is equal to the hole diffusion current at the edge of the transition region x_{n0}. Since $I_p(x_n)$ is proportional to the excess hole concentration at each position in the n material [Eq. (5-32)], it decreases exponentially in x_n with the decreasing $\delta p(x_n)$. Thus the electron component of current must increase appropriately with x_n to maintain the total current I. Far from the junction, the current in the n material is carried almost entirely by electrons. The physical explanation of this is that electrons must flow in from the n material (and ultimately from the negative terminal of the battery), to resupply electrons lost by recombination in the excess hole distribution near the junction. If the junction were p-n instead of p^+-n, so that electrons as well as holes were injected across the transition region, the electron current $I_n(x_n)$ would include sufficient electron flow to supply not only recombination near x_{n0}, but also injection of electrons into the p region. Of course, the flow of electrons in the n material toward the junction constitutes a current in the $+x$-direction, contributing to the total current I.

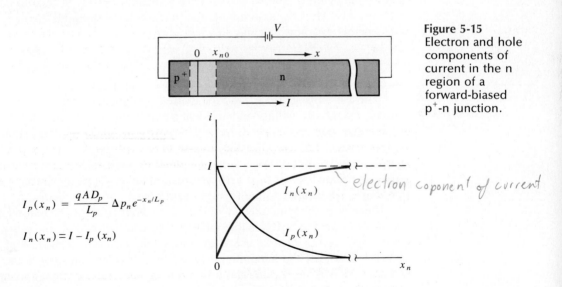

$$I_p(x_n) = \frac{qAD_p}{L_p} \Delta p_n e^{-x_n/L_p}$$

$$I_n(x_n) = I - I_p(x_n)$$

Figure 5-15
Electron and hole components of current in the n region of a forward-biased p^+-n junction.

Find an expression for the electron current in the n-type material of a forward-biased p-n junction.   **EXAMPLE 5-4**

SOLUTION   The total current is

$$I = qA\left(\frac{D_p}{L_p}p_n + \frac{D_n}{L_n}n_p\right)(e^{qV/kT} - 1)$$

The hole current on the n side is

$$I_p(x_n) = qA\frac{D_p}{L_p}p_n e^{-x_n/L_p}(e^{qV/kT} - 1)$$

Thus the electron current in the n material is

$$I_n(x_n) = I - I_p(x_n) = qA\left[\frac{D_p}{L_p}(1 - e^{-x_n/L_p})p_n + \frac{D_n}{L_n}n_p\right](e^{qV/kT} - 1)$$

This expression includes the supplying of electrons for recombination with the injected holes, and the injection of electrons across the junction into the p side.

∎

Before leaving the discussion of majority carrier currents, we should note that the electric field in the neutral regions cannot be zero as we previously assumed; otherwise, there would be no drift currents. Thus our assumption that all of the applied voltage appears across the transition region is not completely accurate. On the other hand, the majority carrier concentrations are usually large in the neutral regions, so that only a small $\mathscr{E}$ field is needed to drive the drift currents. Thus the assumption that junction voltage equals applied voltage is acceptable for most calculations.

5.4
REVERSE-BIAS
BREAKDOWN

We have found that a p-n junction biased in the reverse direction exhibits a small, essentially voltage-independent saturation current. This is true until a critical reverse bias is reached, for which _reverse breakdown_ occurs (Fig. 5-16). At this critical voltage (V_{br}) the reverse current through the diode increases sharply, and relatively large currents can flow with little further increase in voltage. The existence of a critical breakdown voltage introduces almost a right-angle appearance to the reverse characteristic of most diodes.

There is nothing inherently destructive about reverse breakdown. If the current is limited to a reasonable value by the external circuit, the p-n junction can be operated in reverse breakdown as safely as in the forward-bias condition. For example, the maximum reverse current which can flow in the device of Fig. 5-16 is $(E - V_{br})/R$; the series resistance R can be chosen to limit the current to a safe level for the particular diode used. If the current is not limited externally, the junction can be damaged by excessive reverse current, which overheats the device as the maximum power rating is exceeded. It is important to remember, however, that such destruction of the device is not necessarily due to mechanisms unique to reverse breakdown; similar results occur if the

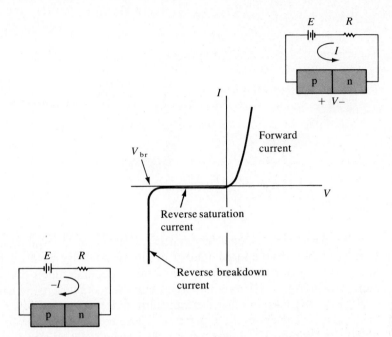

Figure 5-16
Reverse breakdown
in a p-n junction.

device passes excessive current in the forward direction.[†] As we shall see in the next chapter, useful devices called *breakdown diodes* are designed to operate in the reverse breakdown regime of their characteristics.

Reverse breakdown can occur by two mechanisms, each of which requires a critical electric field in the junction transition region. The first mechanism, called the *Zener effect,* is operative at low voltages (up to a few volts reverse bias). If the breakdown occurs at higher voltages (from a few volts to thousands of volts), the mechanism is *avalanche breakdown.* We shall discuss these two mechanisms in this section.

5.4.1 Zener Breakdown

When a heavily doped junction is reverse biased, the energy bands become crossed at relatively low voltages (i.e., the n-side conduction band appears opposite the p-side valence band). As Fig. 5-17 indicates, the crossing of the bands aligns the large number of empty states in the n-side conduction band opposite the many filled states of the p-side valence band. If the barrier separating these two bands is narrow, tunneling of electrons can occur, as discussed in Section 2.4.4. Tunneling of electrons from the p-side valence band to the n-side conduction band constitutes a reverse current from n to p; this is the *Zener effect*.

[†]The dissipated power (*IV*) in the junction is of course greater for a given current in the breakdown regime than would be the case for forward bias, simply because *V* is greater.

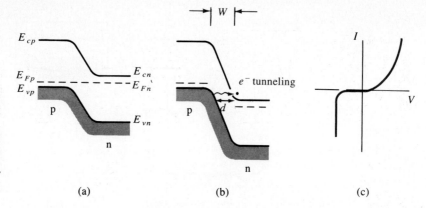

Figure 5-17
The Zener effect:
(a) heavily doped
junction at
equilibrium;
(b) reverse bias
with electron
tunneling from
p to n; (c) I–V
characteristic.

The basic requirements for tunneling current are a large number of electrons separated from a large number of empty states by a narrow barrier of finite height. Since the tunneling probability depends upon the width of the barrier (d in Fig. 5-17), it is important that the metallurgical junction be sharp and the doping high, so that the transition region W extends only a very short distance from each side of the junction. These requirements can be met, for example, by forming an alloyed p region in a heavily doped n-type sample. If the junction is not abrupt, or if either side of the junction is lightly doped, the transition region W will be too wide for tunneling.

As the bands are crossed (at a few tenths of a volt for a heavily doped junction), the tunneling distance d may be too large for appreciable tunneling. However, d becomes smaller as the reverse bias is increased, as we can demonstrate geometrically (Prob. 5.11). This assumes that the transition region width W does not increase appreciably with reverse bias. For low voltages and heavy doping on each side of the junction, this is a good assumption. However, if Zener breakdown does not occur with reverse bias of a few volts, avalanche breakdown will become dominant.

In the simple covalent bonding model, the Zener effect can be thought of as *field ionization* of the host atoms at the junction. That is, the reverse bias of a heavily doped junction causes a large electric field within W; at a critical field strength, electrons participating in covalent bonds may be torn from the bonds by the field and accelerated to the n side of the junction. The electric field required for this type of ionization is on the order of 10^6 V/cm.

5.4.2 Avalanche Breakdown

For lightly doped junctions electron tunneling is negligible, and instead, the breakdown mechanism involves the *impact ionization* of host atoms by energetic carriers. Normal lattice-scattering events can result in the creation of EHPs if the carrier being scattered has sufficient energy. For example, if the electric field $\mathscr{E}$ in the transition region is large, an electron entering from the p side may be accelerated to high enough kinetic energy to cause an ionizing

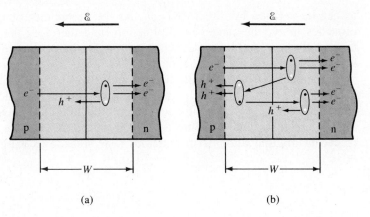

(a)   (b)

Electron–hole pairs created by impact ionization: (a) a single ionizing collision by an incoming electron; (b) primary, secondary, and tertiary collisions.

collision with the lattice (Fig. 5-18a). A single such interaction results in _car-rier multiplication;_ the original electron and the generated electron are both swept to the n side of the junction, and the generated hole is swept to the p side. The degree of multiplication can become very high if carriers generated within the transition region also have ionizing collisions with the lattice. For example, an incoming electron may have a collision with the lattice and create an EHP; each of these carriers has a chance of creating a new EHP, and each of those can also create an EHP, and so forth (Fig. 5-18b). This is an _avalanche_ proc-ess, since each incoming carrier can initiate the creation of a large number of new carriers.

We can make an approximate analysis of avalanche multiplication by as-suming that a carrier of either type has a probability P of having an ionizing collision with the lattice while being accelerated a distance W through the tran-sition region. Thus for n_{in} electrons entering from the p side, there will be Pn_{in} ionizing collisions and an EHP (secondary carriers) for each collision. After the Pn_{in} collisions by the primary electrons, we have the primary plus the sec-ondary electrons, $n_{in}(1 + P)$. After a collision, each EHP moves effectively a distance W within the transition region. For example, if an EHP is created at the center of the region, the electron drifts a distance $W/2$ to n and the hole $W/2$ to p. Thus the probability that an ionizing collision will occur due to the motion of the secondary carriers is still P in this simplified model. For $n_{in}P$ secondary pairs there will be $(n_{in}P)P$ ionizing collisions and $n_{in}P^2$ tertiary pairs. Summing up the total number of electrons out of the region at n after many collisions, we have

$$n_{out} = n_{in}(1 + P + P^2 + P^3 + \cdots) \tag{5-42}$$

assuming no recombination. In a more comprehensive theory we would in-clude recombination as well as different probabilities for ionizing collisions by electrons and holes. In our simple theory, the electron multiplication M_n is

$$M_n = \frac{n_{out}}{n_{in}} = 1 + P + P^2 + P^3 + \cdots = \frac{1}{1 - P} \tag{5-43}$$

as can be verified by direct division. As the probability of ionization P approaches unity, the carrier multiplication (and therefore the reverse current through the junction) increases without limit. Actually, the limit on the current will be dictated by the external circuit.

The relation between multiplication and P was easy to write in Eq. (5-43); however, the relation of P to parameters of the junction is much more complicated. Physically, we expect the ionization probability to increase with increasing electric field, and therefore to depend on the reverse bias. Measurements of carrier multiplication M in junctions near breakdown lead to an empirical relation

$$M = \frac{1}{1 - (V/V_{br})^n} \quad {\scriptstyle n=3 \text{-} 76}$$

(5-44)

where the exponent **n** varies from about 3 to 6, depending on the type of material used for the junction.

In general, the critical reverse voltage for breakdown increases with the band gap of the material, since more energy is required for an ionizing collision. Also, the peak electric field within W increases with increased doping on the more lightly doped side of the junction (Prob. 5.12). Therefore, V_{br} decreases as the doping increases, as Fig. 5-19 indicates.

Figure 5-19
Variation of avalanche breakdown voltage in abrupt p⁺-n junctions, as a function of donor concentration on the n side, for several semiconductors. [After S. M. Sze and G. Gibbons, *Applied Physics Letters*, vol. 8, p. 111 (1966).]

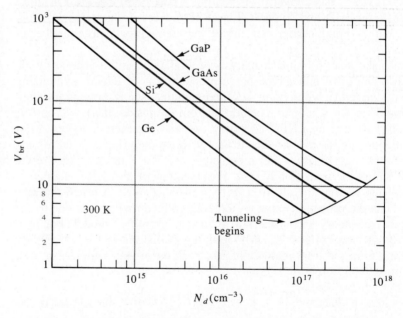

5.5
TRANSIENT AND A-C CONDITIONS

We have considered the properties of p-n junctions under equilibrium conditions and with steady state current flow. Most of the basic concepts of junction devices can be obtained from these properties, except for the important behavior of junctions under transient or a-c conditions. Since most solid state devices are used for switching or for processing a-c signals, we cannot claim to understand p-n junctions without knowing at least the basics of time-dependent processes. Unfortunately, a complete analysis of these effects involves more mathematical manipulation than is appropriate for an introductory discussion. Basically, the problem involves solving the various current flow equations in two simultaneous variables, space and time. We can, however, obtain the basic results for several special cases which represent typical time-dependent applications of junction devices.

In this section we investigate the important influence of excess carriers in transient and a-c problems. The switching of a diode from its forward state to its reverse state is analyzed to illustrate a typical transient problem. Finally, these concepts are applied to the case of small a-c signals to determine the equivalent capacitance of a p-n junction.

5.5.1   Time Variation of Stored Charge

Another look at the excess carrier distributions of a p-n junction under bias (e.g., Fig. 5-12) tells us that any change in current must lead to a change of charge stored in the carrier distributions. Since time is required in building up or depleting a charge distribution, however, the stored charge must inevitably lag behind the current in a time-dependent problem. This is inherently a capacitive effect, as we shall see in Section 5.5.3.

For a proper solution of a transient problem, we must use the time-dependent continuity equations, Eqs. (4-31). We can obtain each component of the current at position x and time t from these equations; for example, from Eq. (4-31a) we can write

$$-\frac{\partial J_p(x,t)}{\partial x} = q\frac{\delta p(x,t)}{\tau_p} + q\frac{\partial p(x,t)}{\partial t} \tag{5-45}$$

To obtain the instantaneous current density, we can integrate both sides at time t to obtain

$$J_p(0) - J_p(x) = q\int_0^x \left[\frac{\delta p(x,t)}{\tau_p} + \frac{\partial p(x,t)}{\partial t}\right]dx \tag{5-46}$$

For injection into a long n region from a p^+ region, we can take the current at $x_n = 0$ to be all hole current, and J_p at $x_n = \infty$ to be zero (Fig. 5-15). Then the total injected current, including time variations, is

$$i(t) = i_p(x_n = 0, t) = \frac{qA}{\tau_p}\int_0^\infty \delta p(x_n, t)\,dx_n + qA\frac{\partial}{\partial t}\int_0^\infty \delta p(x_n, t)\,dx_n$$

$$\boxed{i(t) = \frac{Q_p(t)}{\tau_p} + \frac{dQ_p(t)}{dt}} \tag{5-47}$$

This result indicates that the hole current injected across the p^+-n junction (and therefore approximately the total diode current) is determined by two charge storage effects: (1) the usual recombination term Q_p/τ_p in which the excess carrier distribution is replaced every τ_p seconds, and (2) a charge buildup (or depletion) term dQ_p/dt, which allows for the fact that the distribution of excess carriers can be increasing or decreasing in a time-dependent problem. For steady state the dQ_p/dt term is zero, and Eq. (5-47) reduces to Eq. (5-40), as expected. In fact, we could have written Eq. (5-47) intuitively rather than having obtained it from the continuity equation, since it is reasonable that the hole current injected at any given time must supply minority carriers for recombination and for whatever variations occur in the total stored charge.

We can solve for the stored charge as a function of time for a given current transient. For example, the step turn-off transient (Fig. 5-20a), in which a current I is suddenly removed at $t = 0$, leaves the diode with stored charge. Since the excess holes in the n region must die out by recombination with the matching excess electron population, some time is required for $Q_p(t)$ to reach zero. Solving Eq. (5-47) with Laplace transforms, with $i(t > 0) = 0$ and $Q_p(0) = I\tau_p$, we obtain

$$0 = \frac{1}{\tau_p}Q_p(s) + sQ_p(s) - I\tau_p$$

$$Q_p(s) = \frac{I\tau_p}{s + 1/\tau_p}$$

$$Q_p(t) = I\tau_p e^{-t/\tau_p} \tag{5-48}$$

As expected, the stored charge dies out exponentially from its initial value $I\tau_p$ with a time constant equal to the hole lifetime in the n material.

An important implication of Fig. 5-20 is that even though the current is suddenly terminated, the voltage across the junction persists until Q_p disappears. Since the excess hole concentration can be related to junction voltage by formulas derived in Section 5.3.2, we can presumably solve for $v(t)$. We already know that at any time during the transient, the excess hole concentration at $x_n = 0$ is

$$\Delta p_n(t) = p_n(e^{qv(t)/kT} - 1) \tag{5-49}$$

Figure 5-20
Effects of a step turn-off transient in a p$^+$-n diode: (a) current through the diode; (b) decay of stored charge in the n-region; (c) excess hole distribution in the n-region as a function of time during the transient.

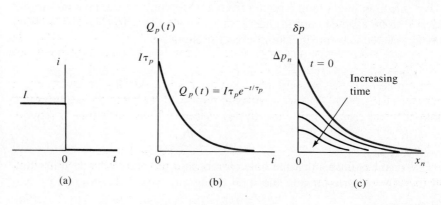

(a) (b) (c)

so that finding $\Delta p_n(t)$ will easily give us the transient voltage. Unfortunately, it is not simple to obtain $\Delta p_n(t)$ exactly from our expression for $Q_p(t)$. The problem is that the hole distribution does not remain in the convenient exponential form it has in steady state. As Fig. 5-20c suggests, the quantity $\delta p(x_n, t)$ becomes markedly nonexponential as the transient proceeds. For example, since the injected hole current is proportional to the gradient of the hole distribution at $x_n = 0$ (Fig. 5-13a), zero current implies zero gradient. Thus the slope of the distribution must be exactly zero at $x_n = 0$ throughout the transient.[†] This zero slope at the point of injection distorts the exponential distribution, particularly in the region near the junction. As time progresses in Fig. 5-20c, δp (and therefore δn) decreases as the excess electrons and holes recombine. To find the exact expression for $\delta p(x_n, t)$ during the transient would require a rather difficult solution of the time-dependent continuity equation.

An approximate solution for $v(t)$ can be obtained by assuming an exponential distribution for δp at every instant during the decay. This type of *quasi-steady state* approximation neglects distortion due to the slope requirement at $x_n = 0$ and the effects of diffusion during the transient. Thus we would expect the calculation to give rather crude results. On the other hand, such a solution can give us a feeling for the variation of junction voltage during the transient. If we take

$$\delta p(x_n, t) = \Delta p_n(t) e^{-x_n/L_p} \tag{5-50}$$

we have for the stored charge at any instant

$$Q_p(t) = qA \int_0^\infty \Delta p_n(t) e^{-x_n/L_p} \, dx_n = qAL_p \Delta p_n(t) \tag{5-51}$$

Relating $\Delta p_n(t)$ to $v(t)$ by Eq. (5-49) we have

$$\Delta p_n(t) = p_n(e^{qv(t)/kT} - 1) = \frac{Q_p(t)}{qAL_p} \tag{5-52}$$

Thus in the quasi-steady state approximation, the junction voltage varies according to

$$v(t) = \frac{kT}{q} \ln \left(\frac{I\tau_p}{qAL_p p_n} e^{-t/\tau_p} + 1 \right) \tag{5-53}$$

during the turn-off transient of Fig. 5-20. This analysis, while not accurate in its details, does indicate clearly that the voltage across a p-n junction cannot be changed instantaneously, and that stored charge can present a problem in a diode intended for switching applications.

Many of the problems of stored charge can be reduced by designing a p^+-n diode (for example) with a very narrow n region. If the n region is shorter than a hole diffusion length, very little charge is stored. Thus, little time is required

---

[†]We notice that, while the *magnitude* of δp cannot change instantaneously, the *slope* must go to zero immediately. This can occur in a small region near the junction with negligible redistribution of charge at $t = 0$.

to switch the diode on and off. This type of structure, called the _narrow base diode,_ is considered in Section 6.1.2 and in Prob. 6.5. The switching process can be made still faster by purposely adding recombination centers, such as Au atoms in Si, to increase the recombination rate.

5.5.2 Reverse Recovery Transient

In most switching applications a diode is switched from forward conduction to a reverse-biased state, and vice versa. The resulting stored charge transient is somewhat more complicated than for a simple turn-off transient, and therefore it requires slightly more analysis. An important result of this example is that a reverse current much larger than the normal reverse saturation current can flow in a junction during the time required for readjustment of the stored charge.

Let us assume a p^+-n junction is driven by a square wave generator that periodically switches from $+E$ to $-E$ volts (Fig. 5-21a). While E is positive the diode is forward biased, and in steady state the current I_f flows through the junction. If E is much larger than the small forward voltage of the junction, the source voltage appears almost entirely across the resistor, and the current is approximately $i = I_f \simeq E/R$. After the generator voltage is reversed ($t > 0$), the current must initially reverse to $i = I_r \simeq -E/R$. The reason for this unusually large reverse current through the diode is that the stored charge (and hence the junction voltage) cannot be changed instantaneously. Therefore, just as the current is reversed, the junction voltage remains at the small forward-bias value it had before $t = 0$. A voltage loop equation then tells us that the large reverse current $-E/R$ must flow temporarily. While the current is negative through the junction, the slope of the $\delta p(x_n)$ distribution must be positive at $x_n = 0$.

As the stored charge is depleted from the neighborhood of the junction (Fig. 5-21b), we can find the junction voltage again from Eq. (5-49). As long as Δp_n is positive, the junction voltage $v(t)$ is positive and small; thus $i \simeq -E/R$ until Δp_n goes to zero. When the stored charge is depleted and Δp_n becomes negative, the junction exhibits a negative voltage. Since the reverse-bias voltage of a junction can be large, the source voltage begins to divide between R and the junction. As time proceeds, the magnitude of the reverse current becomes smaller as more of $-E$ appears across the reverse-biased junction, until finally the only current is the small reverse saturation current which is characteristic of the diode. The time t_{sd} required for the stored charge (and therefore the junction voltage) to become zero is called the _storage delay time_. This delay time is an important figure of merit in evaluating diodes for switching applications. It is usually desirable that t_{sd} be small compared with the switching times required (Fig. 5-22). The critical parameter determining t_{sd} is the carrier lifetime (τ_p for the example of the p^+-n junction). Since the recombination rate determines the speed with which excess holes can disappear from the n region, we would expect t_{sd} to be proportional to τ_p. In fact, an exact analysis of the problem of Fig. 5-21 leads to the result

$$t_{sd} = \tau_p \left[\text{erf}^{-1}\left(\frac{I_f}{I_f + I_r}\right) \right]^2 \tag{5-54}$$

Storage
delay time

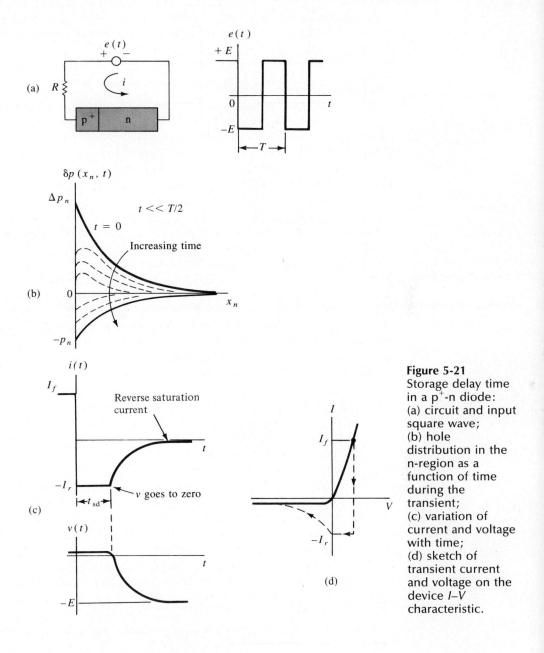

Figure 5-21
Storage delay time in a p$^+$-n diode:
(a) circuit and input square wave;
(b) hole distribution in the n-region as a function of time during the transient;
(c) variation of current and voltage with time;
(d) sketch of transient current and voltage on the device I–V characteristic.

where the error function (erf) is a tabulated function. Although the exact solution leading to Eq. (5-54) is too lengthy for us to consider here, an approximate result can be obtained from the quasi-steady state assumption.

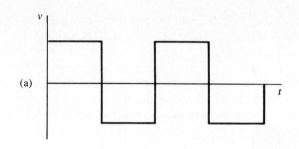

(a)

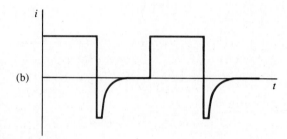

(b)

Figure 5-22
Effects of storage
delay time on
switching signal:
(a) switching
voltage; (b) diode
current.

EXAMPLE 5-5 Assume a p^+-n diode is biased in the forward direction, with a current I_f. At time $t = 0$ the current is switched to $-I_r$. Use the appropriate boundary conditions to solve Eq. (5-47) for $Q_p(t)$. Apply the quasi-steady state approximation to find the storage delay time t_{sd}.

SOLUTION From Eq. (5-47),

$$i(t) = \frac{Q_p(t)}{\tau_p} + \frac{dQ_p(t)}{dt} \qquad \text{for } t < 0, \quad Q_p = I_f \tau_p$$

Using Laplace transforms,

$$-\frac{I_r}{s} = \frac{Q_p(s)}{\tau_p} + sQ_p(s) - I_f\tau_p$$

$$Q_p(s) = \frac{I_f\tau_p}{s + 1/\tau_p} - \frac{I_r}{s(s + 1/\tau_p)}$$

$$Q_p(t) = I_f\tau_p e^{-t/\tau_p} + I_r\tau_p(e^{-t/\tau_p} - 1) = \tau_p[-I_r + (I_f + I_r)e^{-t/\tau_p}]$$

Assuming that $Q_p(t) = qAL_p \Delta p_n(t)$ as in Eq. (5-52),

$$\Delta p_n(t) = \frac{\tau_p}{qAL_p}[-I_r + (I_f + I_r)e^{-t/\tau_p}]$$

This is set to equal zero when $t = t_{sd}$, and we obtain:

$$t_{sd} = -\tau_p \ln\left[\frac{I_r}{I_f + I_r}\right] = \tau_p \ln\left(1 + \frac{I_f}{I_r}\right)$$

An important result of Eq. (5-54) is that τ_p can be calculated in a straightforward way from a measurement of storage delay time. In fact, measurement of t_{sd} from an experimental arrangement such as Fig. 5-21a is a common method of measuring lifetimes. In some cases this is a more convenient technique than the photoconductive decay measurement discussed in Section 4.3.2.

As in the case of the turn-off transient of the previous section, the storage delay time can be reduced by introducing recombination centers into the diode material, thus reducing the carrier lifetimes, or by utilizing the narrow base diode configuration.

5.5.3 Capacitance of p-n Junctions

There are basically two types of capacitance associated with a junction: (1) the *junction capacitance* due to the dipole in the transition region and (2) the *charge storage capacitance* arising from the lagging behind of voltage as current changes, due to charge storage effects.[†] Both of these capacitances are important, and they must be considered in designing p-n junction devices for use with time-varying signals. The junction capacitance (1) is dominant under reverse-bias conditions, and the charge storage capacitance (2) is dominant when the junction is forward biased. In many applications of p-n junctions, the capacitance is a limiting factor in the usefulness of the device; on the other hand, there are important applications in which the capacitance discussed here can be useful in circuit applications and in providing important information about the structure of the p-n junction.

The junction capacitance of a diode is easy to visualize from the charge distribution in the transition region (Fig. 5-9). The uncompensated acceptor ions on the p side provide a negative charge, and an equal positive charge results from the ionized donors on the n side of the transition region. The capacitance of the resulting dipole is slightly more difficult to calculate than is the usual parallel plate capacitance, but we can obtain it in a few steps.

Instead of the common expression $C = |Q/V|$, which applies to capacitors in which charge is a linear function of voltage, we must use the more general definition

$$C = \left| \frac{dQ}{dV} \right| \tag{5-55}$$

since the charge Q on each side of the transition region varies nonlinearly with the applied voltage. We can demonstrate this nonlinear dependence by reviewing the equations for the width of the transition region (W) and the resulting charge. The equilibrium value of W was found in Eq. (5-21) to be

$$W = \left[\frac{2\epsilon V_0}{q} \left(\frac{N_a + N_d}{N_a N_d} \right) \right]^{1/2} \quad (equilibrium) \tag{5-56}$$

[†]The capacitance (1) above is also referred to as *transition region capacitance* or *depletion layer capacitance;* (2) is often called the *diffusion capacitance*.

Since we are dealing with the nonequilibrium case with voltage V applied, we must use the altered value of the electrostatic potential barrier $(V_0 - V)$, as discussed in relation to Fig. 5-10. The proper expression for the width of the transition region is then *(Nonequilibrium)*

$$W = \left[\frac{2\epsilon(V_0 - V)}{q} \left(\frac{N_a + N_d}{N_a N_d} \right) \right]^{1/2} \qquad (\textit{with bias}) \qquad (5\text{-}57)$$

In this expression the applied voltage V can be either positive or negative to account for forward or reverse bias. As expected, the width of the transition region is increased for reverse bias and is decreased under forward bias. Since the uncompensated charge Q on each side of the junction varies with the transition region width, variations in the applied voltage result in corresponding variations in the charge, as required for a capacitor. The value of Q can be written in terms of the doping concentration and transition region width on each side of the junction (Fig. 5-9):

$$|Q| = qAx_{n0}N_d = qAx_{p0}N_a \qquad (5\text{-}58)$$

Relating the total width of the transition region W to the individual widths x_{n0} and x_{p0} from Eqs. (5-23) we have

$$x_{n0} = \frac{N_a}{N_a + N_d}W, \qquad x_{p0} = \frac{N_d}{N_a + N_d}W \qquad (5\text{-}59)$$

and therefore the charge on each side of the dipole is

$$|Q| = qA\frac{N_d N_a}{N_d + N_a}W = A\left[2q\epsilon(V_0 - V)\frac{N_d N_a}{N_d + N_a} \right]^{1/2} \qquad (5\text{-}60)$$

Thus the charge is indeed a nonlinear function of applied voltage. From this expression and the definition of capacitance in Eq. (5-55), we can calculate the junction capacitance C_j. Since the voltage that varies the charge in the transition region is the barrier height $(V_0 - V)$, we must take the derivative with respect to this potential difference:

$$C_j = \left| \frac{dQ}{d(V_0 - V)} \right| = \frac{A}{2}\left[\frac{2q\epsilon}{(V_0 - V)}\frac{N_d N_a}{N_d + N_a} \right]^{1/2} \qquad (5\text{-}61)$$

The quantity C_j is a *voltage-variable capacitance*, since C_j is proportional to $(V_0 - V)^{-1/2}$. There are several important applications for variable capacitors, including use in tuned circuits. The p-n junction device which makes use of the voltage-variable properties of C_j is called a *varactor*. We shall discuss this device further in Section 6.1.4.

Although the dipole charge is distributed in the transition region of the junction, the form of the parallel plate capacitor formula is obtained from the expressions for C_j and W:

$$C_j = \epsilon A\left[\frac{q}{2\epsilon(V_0 - V)}\frac{N_d N_a}{N_d + N_a} \right]^{1/2} = \frac{\epsilon A}{W} \qquad (5\text{-}62)$$

In analogy with the parallel plate capacitor, the transition region width W corresponds with the plate separation of the conventional capacitor.

In the case of an asymmetrically doped junction, the transition region extends primarily into the less heavily doped side, and the capacitance is determined by only one of the doping concentrations. For a p^+-n junction, $N_a \gg N_d$ and $x_{n0} \simeq W$, while x_{p0} is negligible. The capacitance is then

$$C_j = \frac{A}{2}\left[\frac{2q\epsilon}{V_0 - V}N_d\right]^{1/2} \qquad \text{for } p^+\text{-n} \qquad (5\text{-}63)$$

It is therefore possible to obtain the doping concentration of the lightly doped n region from a measurement of junction capacitance. For example, in a reverse-biased junction the applied voltage $V = -V_r$ can be made much larger than the contact potential V_0, so that the latter becomes negligible. If the area of the junction can be measured, a reliable value of N_d results from a measurement of C_j. However, these equations were obtained by assuming a sharp step junction. Certain modifications must be made in the case of a graded junction (Section 5.6.4 and Prob. 5.21).

The junction capacitance dominates the reactance of a p-n junction under reverse bias; for forward bias, however, the charge storage capacitance C_s becomes dominant. To calculate the capacitance due to charge storage effects, let us assume that a p^+-n junction is forward biased with a steady current I. The stored charge in the injected hole distribution is

$$Q_p = I\tau_p = qA\,\Delta p_n L_p = qAL_p p_n e^{qV/kT} \qquad \text{for } V \gg 0.0259 \text{ V} \quad (5\text{-}64)$$

The capacitance due to small changes in this stored charge is

$$C_s = \frac{dQ_p}{dV} = \frac{q^2}{kT}AL_p p_n e^{qV/kT} = \frac{q}{kT}I\tau_p \qquad (5\text{-}65)$$

Similarly, we can determine the *a-c conductance* by allowing small changes in the current:

$$G_s = \frac{dI}{dV} = \frac{qAL_p p_n}{\tau_p}\frac{d}{dV}(e^{qV/kT}) = \frac{q}{kT}I \qquad (5\text{-}66)$$

Thus the a-c component of current is

$$i(\text{a-c}) = G_s v(\text{a-c}) + C_s\frac{dv(\text{a-c})}{dt} \qquad (5\text{-}67)$$

where

$$G_s = \frac{q}{kT}I(\text{d-c}) \quad \text{and} \quad C_s = G_s\tau_p$$

The charge storage capacitance can be a serious limitation for forward-biased p-n junctions in high-frequency circuits. As in the case of the switching performance discussed in the two preceding sections, the high-frequency a-c response of a junction can be improved by reducing the carrier lifetime. Since C_s is proportional to τ_p, a short hole lifetime can make the forward-biased capacitance of a p^+-n junction acceptably small for many applications.

5.6
DEVIATIONS
FROM THE SIMPLE
THEORY

The approach we have taken in studying p-n junctions has focused on the basic principles of operation, neglecting secondary effects. This allows for a relatively uncluttered view of carrier injection and other junction properties, and illuminates the essential features of diode operation. To complete the description, however, we must now fill in a few details which can affect the operation of junction devices under special circumstances.

Most of the deviations from the simple theory can be treated by fairly straightforward modifications of the basic equations. In this section we shall investigate the most important deviations and alter the theory wherever possible. In a few cases, we shall simply indicate the approach to be taken and the result. The most important alterations to the simple diode theory are the effects of contact potential and changes in majority carrier concentration on carrier injection, recombination and generation within the transition region, ohmic effects, and the effects of graded junctions.

5.6.1 Effects of Contact Potential on Carrier Injection

If the forward-bias $I–V$ characteristics of various semiconductor diodes are compared, it becomes clear that the band gap has an important influence on carrier injection. For example, Fig. 5-23 compares the low-temperature characteristics of heavily doped diodes having various band gaps. One obvious feature of this figure is that the $I–V$ characteristics appear "square"; that is, the current is very small until a critical forward bias is reached, and then the current increases rapidly. This is typical of exponentials plotted on such a scale. However, it is significant that the limiting voltage is slightly less than the value of the band gap in electron volts.

Figure 5-23
$I–V$ characteristics of heavily doped p-n junction diodes at 77 K, illustrating the effects of contact potential on the forward current:
(a) Ge, $E_g \simeq 0.7$ eV;
(b) Si, $E_g \simeq 1.1$ eV;
(c) GaAs, $E_g \simeq 1.5$ eV;
(d) GaAsP, $E_g \simeq 2.0$ eV.

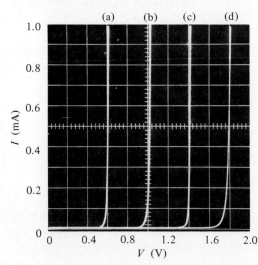

The reason for the small current at low voltages for these devices can be understood from a simple rearrangement of the diode equation. If we rewrite Eq. (5-36) for a forward-biased p^+-n diode (with $V \gg kT/q$) and include the exponential form for the minority carrier concentration p_n, we obtain

$$I = \frac{qAD_p}{L_p} p_n e^{qV/kT} = \frac{qAD_p}{L_p} N_v e^{[qV - (E_{Fn} - E_{vn})]/kT} \tag{5-68}$$

Hole injection into the n material is small if the forward bias V is much less than $(E_{Fn} - E_{vn})/q$. For a p^+-n diode, this quantity is essentially the contact potential, since the Fermi level is near the valence band on the p side. If the n region is also heavily doped, the contact potential is almost equal to the band gap (Fig. 5-24). This accounts for the dramatic increase in diode current near the band gap voltage in Fig. 5-23. Contributing to the small current at lower voltages is the fact that the minority carrier concentration $p_n = n_i^2/N_d$ is very small at low temperature (n_i small) and with heavy doping (N_d large).

The limiting forward bias across a p-n junction is equal to the contact potential, as in Fig. 5-24(b). This effect is not predicted by the simple diode equation, for which the current increases exponentially with applied voltage. The reason this important result is excluded in the simple theory is that in Eq. (5-28) we neglect changes in the majority carrier concentrations on either side of the junction. This assumption is valid only for low injection levels; for large injected carrier concentrations, the excess majority carriers become important compared with the majority doping. For example, at low injection $\Delta n_p = \Delta p_p$ is important compared with the equilibrium minority electron concentration n_p, but is negligible compared with the majority hole concentration p_p; this was the basis for neglecting Δp_p in Eq. (5-28). For high injection levels, however, Δp_p can be comparable to p_p and we must write Eq. (5-27) in the form

$$\frac{p(-x_{p0})}{p(x_{n0})} = \frac{p_p + \Delta p_p}{p_n + \Delta p_n} = e^{q(V_0 - V)/kT} = \frac{n_n + \Delta n_n}{n_p + \Delta n_p} \tag{5-69}$$

To find the more complete diode equation from this expression, we can follow essentially the same steps used in obtaining the low-level relation, Eq. (5-36). The algebra is more complicated, but the basic method is straight-

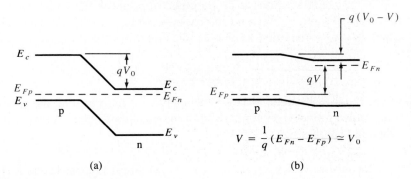

$$V = \frac{1}{q}(E_{Fn} - E_{Fp}) \simeq V_0$$

Figure 5-24
Examples of contact potential for a heavily doped p-n junction: (a) at equilibrium; (b) approaching the maximum forward bias $V = V_0$.

forward (Prob. 5.20). The resulting equation for the forward current through the diode is

$$I = qA\left[\frac{e^{qV/kT} - 1}{1 - e^{-2q(V_0 - V)/kT}}\right]\left[\frac{D_p p_n}{L_p}\left(1 + \frac{n_i^2}{p_p^2}e^{qV/kT}\right) + \frac{D_n n_p}{L_n}\left(1 + \frac{n_i^2}{n_n^2}e^{qV/kT}\right)\right]$$

(5-70)

For $V \ll V_0$ (low-level injection) and p_p, $n_n \gg n_i$, this equation reduces to the simple diode equation, Eq. (5-36). However, as the applied voltage V approaches the contact potential V_0, the denominator of the first bracketed term decreases toward zero. Thus the diode forward current increases rapidly near $V = V_0$, and *the contact potential is the limiting forward voltage for the junction.*[†]

5.6.2   Recombination and Generation in the Transition Region

In analyzing the p-n junction, we have assumed that recombination and thermal generation of carriers occur primarily in the neutral p and n regions, outside the transition region. In this model, <u>forward current in the diode is carried</u> by recombination of excess minority carriers injected into each neutral region by the junction. Similarly, the <u>reverse saturation current is</u> due to the thermal generation of EHPs in the neutral regions and the subsequent diffusion of the generated minority carriers to the transition region, where they are swept to the other side by the field. In many devices this model is adequate; however, a more complete description of junction operation should include recombination and generation within the transition region itself.

When a junction is forward biased, the transition region contains excess carriers of both types, which are in transit from one side of the junction to the other. Unless the width of the transition region W is very small compared with the carrier diffusion lengths L_n and L_p, significant recombination can take place within W. An accurate calculation of this recombination current is complicated by the fact that the recombination rate, which depends on the carrier concentrations [Eq. (4-5)], varies with position within the transition region. Analysis of the recombination kinetics shows that the current due to recombination within W is proportional to n_i and increases with forward bias according to approximately $\exp(qV/2kT)$. On the other hand, current due to recombination in the neutral regions is proportional to p_n and n_p [Eq. (5-36)] and therefore to n_i^2/N_d and n_i^2/N_a, and increases according to $\exp(qV/kT)$. The diode equation can be modified to include this effect by including the parameter **n**:

$$\boxed{I = I_0(e^{qV/nkT} - 1)}$$

(5-71)

where **n** varies between 1 and 2, depending on the material and temperature. Since **n** determines the departure from the ideal diode characteristic, it is often called the *ideality factor*.

[†]The forward voltage applied to the diode can exceed V_0 if ohmic effects are significant (Section 5.6.3), but the junction voltage itself will always be less than V_0.

The ratio of the two currents

$$\frac{I(\text{recombination in neutral regions})}{I(\text{recombination in transition region})} \propto \frac{n_i^2 e^{qV/kT}}{n_i e^{qV/2kT}} \propto n_i e^{qV/2kT} \qquad (5\text{-}72)$$

becomes small for wide band gap materials, low temperatures (small n_i), and for low voltage. Thus the forward current for low injection in a Si diode is likely to be dominated by recombination in the transition region, while a Ge diode may follow the usual diode equation. In either case, injection through W into the neutral regions becomes more important with increased voltage. Therefore, **n** in Eq. (5-71) may vary from ~2 at low voltage to ~1 at higher voltage.

Just as recombination within W can affect the forward characteristics, the reverse current through a junction can be influenced by carrier *generation* in the transition region. We found in Section 5.3.2 that the reverse saturation current can be accounted for by the thermal generation of EHPs within a diffusion length of either side of the transition region. The generated minority carriers diffuse to the transition region, where they are swept to the other side of the junction by the electric field (Fig. 5-25). However, carrier generation can take place within the transition region itself. If W is small compared with L_n or L_p, band-to-band generation of EHPs within the transition region is not important compared with generation in the neutral regions. However, the lack of free carriers within the space charge of the transition region can create a current due to the net generation of carriers by *emission from recombination centers*. Of the four generation–recombination processes depicted in Fig. 5-26, the two capture rates R_n and R_p are negligible within W because of the very small carrier concentrations in the reverse-bias space charge region. Therefore, a recombination level E_r near the center of the band gap can provide carriers through the ther-

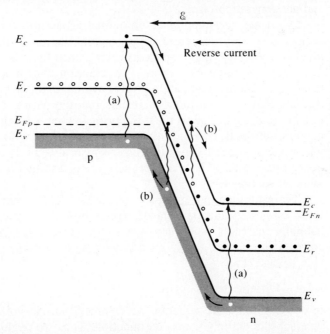

Figure 5-25
Current in a reverse-biased p-n junction due to thermal generation of carriers by (a) band-to-band EHP generation, and (b) generation from a recombination level.

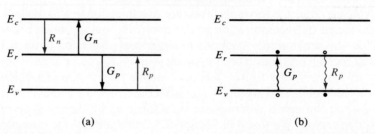

Figure 5-26   Capture and generation of carriers at a recombination center: (a) capture and generation of electrons and holes; (b) hole capture and generation processes redrawn in terms of valence band electron excitation to E_r (hole generation) and electron deexcitation from E_r to E_v (hole capture by E_r).

mal generation rates G_n and G_p. Each recombination center alternately emits an electron and a hole; physically, this means that an electron at E_r is thermally excited to the conduction band (G_n) and a valence band electron is subsequently excited thermally to the empty state on the recombination level, leaving a hole behind in the valence band (G_p). The process can then be repeated over and over, providing electrons for the conduction band and holes for the valence band. Normally, these emission processes are exactly balanced by the corresponding capture processes R_n and R_p. However, in the reverse-bias transition region, generated carriers are swept out before recombination can occur, and net generation results.

Of course, the importance of thermal generation within W depends on the temperature and the nature of the recombination centers. A level near the middle of the band gap is most effective, since for such centers neither G_n nor G_p requires thermal excitation of an electron over more than about half the band gap. If no recombination level is available, this type of generation is negligible. However, in most materials recombination centers exist near the middle of the gap due to trace impurities or lattice defects. Generation from centers within W is most important in materials with large band gaps, for which band-to-band generation in the neutral regions is small. Thus for Si, generation within W is generally more important than for a narrower band gap material such as Ge.

The saturation current due to generation in the neutral regions was found to be essentially independent of reverse bias. However, generation within W naturally increases as W increases with reverse bias. As a result, the reverse current can increase almost linearly with W, or with the square root of reverse-bias voltage (Fig. 5-27).

5.6.3   Ohmic Losses

In deriving the diode equation we assumed that the voltage applied to the device appears entirely across the junction. Thus we neglected any voltage drop

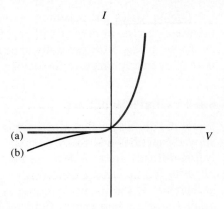

Figure 5-27
Comparison of reverse current mechanisms: (a) carrier generation in the neutral regions; (b) carrier generation within the transition region.

in the neutral regions or at the external contacts. For most devices this is a valid assumption; the doping is usually fairly high, so that the resistivity of each neutral region is low, and the area of a typical diode is large compared with its length. However, some devices do exhibit ohmic effects, which cause significant deviation from the expected I–V characteristic.

We can seldom represent ohmic losses in a diode accurately by including a simple resistance in series with the junction. The effects of voltage drops outside the transition region are complicated by the fact that the voltage drop depends on the current, which in turn is dictated by the voltage across the junction. For example, if we represent the series resistance of the p and n regions by R_p and R_n, respectively, we can write the junction voltage V as

$$V = V_a - I[R_p(I) + R_n(I)] \qquad (5\text{-}73)$$

where V_a is the external voltage applied to the device. As the current increases, there is an increasing voltage drop in R_p and R_n, and the junction voltage V decreases. This reduction in V lowers the level of injection so that the current increases more slowly with increased bias (Fig. 5-28). A further complication in calculating the ohmic loss is that the conductivity of each neutral region increases with increasing carrier injection. Since the effects of Eq. (5-73) are most pronounced at high injection levels, this *conductivity modulation* by the injected excess carriers can reduce R_p and R_n significantly.

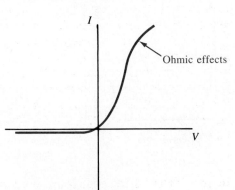

Figure 5-28
Effects of ohmic losses at high injection in a p-n junction diode.

Ohmic losses are purposely avoided in properly designed devices by appropriate choices of doping and geometry. Therefore, deviations of the current as shown in Fig. 5-28 generally appear only for very high currents, outside the normal operating range of the device.

5.6.4   Graded Junctions

While the abrupt junction approximation accurately describes the properties of alloyed junctions and many epitaxial structures, it is often inadequate in analyzing diffused junction devices. For shallow diffusions, in which the diffused impurity profile is very steep (Fig. 5-29a), the abrupt approximation is usually acceptable. If the impurity profile is spread out into the sample, however, a graded junction can result (Fig. 5-29b). Several of the expressions we have derived for the abrupt junction must be modified for this case.

The graded junction problem can be solved analytically if, for example, we make a linear approximation of the net impurity distribution near the junction (Fig. 5-29c). We assume that the graded region can be described approximately by

$$N_d - N_a = Gx \tag{5-74}$$

where G is a grade constant giving the slope of the net impurity distribution.

In Gauss's law [Eq. (5-14)], we can use our linear approximation to write

$$\frac{d\mathscr{E}}{dx} = \frac{q}{\epsilon}(p - n + N_d^+ - N_a^-) \simeq \frac{q}{\epsilon}Gx \tag{5-75}$$

within the transition region. In this approximation we assume complete ionization of the impurities and neglect the carrier concentrations in the transition region, as before. The net space charge varies linearly over W, and the electric field distribution is therefore parabolic. The expressions for contact potential and junction capacitance are different from the abrupt junction case (Fig. 5-30 and Prob. 5.21), since the electric field is no longer linear on each side of the junction.

Figure 5-29
Approximations to diffused junctions: (a) shallow diffusion (abrupt); (b) deep drive-in diffusion with source removed (graded); (c) linear approximation to the graded junction.

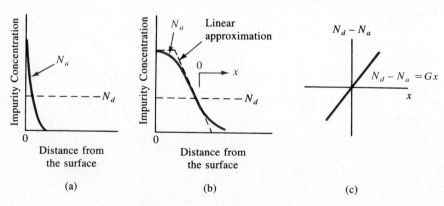

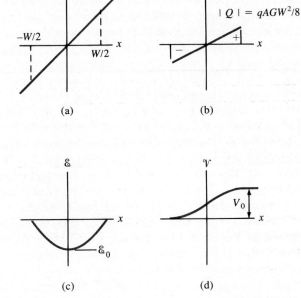

Figure 5-30
Properties of the graded junction transition region: (a) net impurity profile; (b) net charge distribution; (c) electric field; (d) electrostatic potential.

In a graded junction the usual depletion approximation is often inaccurate. If the grade constant G is small, the carrier concentrations $(p - n)$ can be important in Eq. (5-75). Similarly, the usual assumption of negligible space charge outside the transition region is questionable for small G. It would be more accurate to refer to the regions just outside the transition region as quasi-neutral rather than neutral. Thus the edges of the transition region are not sharp as Fig. 5-30 implies but are spread out in x. These effects complicate calculations of junction properties, and a computer must be used in solving the problem accurately.

Most of the conclusions we have made regarding carrier injection, recombination and generation currents, and other properties are qualitatively applicable to graded junctions, with some alterations in the functional form of the resulting equations. Therefore, we can apply most of our basic concepts of junction theory to reasonably graded junctions as long as we remember that certain modifications should be made in accurate computations.

Many of the useful properties of a p-n junction can be achieved by simply forming an appropriate metal–semiconductor contact. This approach is obviously attractive because of its simplicity of fabrication; also, as we shall see in this section, metal–semiconductor junctions are particularly useful when high-speed rectification is required. On the other hand, we also must be able to form nonrectifying (ohmic) contacts to semiconductors. Therefore, this section deals with both rectifying and ohmic contacts.

5.7 METAL-SEMICONDUCTOR JUNCTIONS

5.7.1 Schottky Barriers

In Section 2.2.1 we discussed the work function $q\Phi_m$ of a metal in a vacuum. An energy of $q\Phi_m$ is required to remove an electron at the Fermi level to the vacuum outside the metal. Typical values of Φ_m for very clean surfaces are 4.3V for Al and 4.8V for Au. When negative charges are brought near the metal surface, positive (image) charges are induced in the metal. When this image force is combined with an applied electric field, the effective work function is somewhat reduced. Such barrier lowering is called the *Schottky effect*, and this terminology is carried over to the discussion of potential barriers arising in metal–semiconductor contacts. Although the Schottky effect is only a part of the explanation of metal–semiconductor effects, rectifying contacts are generally referred to as *Schottky barrier diodes.* In this section we shall see how such barriers arise in metal–semiconductor contacts. First we consider barriers in ideal metal–semiconductor junctions, and then in Section 5.7.4 we will include effects which alter the barrier height.

When a metal with work function $q\Phi_m$ is brought in contact with a semiconductor having a work function $q\Phi_s$, charge transfer occurs until the Fermi levels align at equilibrium (Fig. 5-31). For example, when $\Phi_m > \Phi_s$, the semiconductor Fermi level is initially higher than that of the metal before contact is made. To align the two Fermi levels, the electrostatic potential of the semiconductor must be raised (i.e., the electron energies must be lowered) relative to that of the metal. In the n-type semiconductor of Fig. 5-31 a depletion region W is formed near the junction. The positive charge due to uncompensated donor ions within W matches the negative charge on the metal. The electric field and the bending of the bands within W are similar to effects already discussed for p-n junctions. For example, the depletion width W in the semiconductor can be calculated from Eq. (5-21) by using the p^+-n approximation

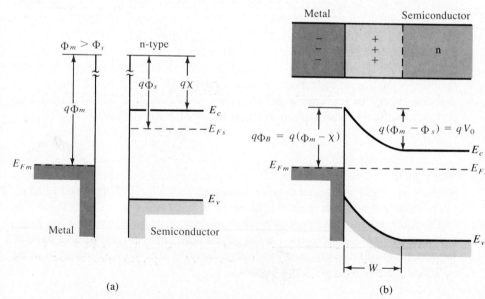

Figure 5-31
A Schottky barrier formed by contacting an n-type semiconductior with a metal having a larger work function: (a) band diagrams for the metal and the semiconductor before joining; (b) equilibrium band diagram for the junction.

(a)   (b)

(i.e., by assuming the negative charge in the dipole is a thin sheet of charge to the left of the junction). Similarly, the junction capacitance is $A\epsilon_s/W$, as in the p^+-n junction.[†]

The equilibrium contact potential V_0, which prevents further net electron diffusion from the semiconductor conduction band into the metal, is the difference in work function potentials $\Phi_m - \Phi_s$. The potential barrier height Φ_B for electron injection from the metal into the semiconductor conduction band is $\Phi_m - \chi$, where $q\chi$ (called the *electron affinity*) is measured from the vacuum level to the semiconductor conduction band edge. The equilibrium potential difference V_0 can be decreased or increased by the application of either forward- or reverse-bias voltage, as in the p-n junction.

Figure 5-32 illustrates a Schottky barrier on a p-type semiconductor, with $\Phi_m < \Phi_s$. In this case aligning the Fermi levels at equilibrium requires a positive charge on the metal side and a negative charge on the semiconductor side of the junction. The negative charge is accommodated by a depletion region W in which ionized acceptors (N_a^-) are left uncompensated by holes. The potential barrier V_0 retarding hole diffusion from the semiconductor to the metal is $\Phi_s - \Phi_m$, and as before this barrier can be raised or lowered by the application of voltage across the junction. In visualizing the barrier for holes, we recall from Fig. 5-7 that the electrostatic potential barrier for positive charge is opposite to the barrier on the electron energy diagram.

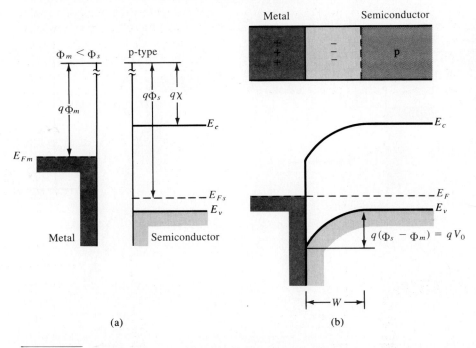

(a)

(b)

Figure 5-32
Schottky barrier between a p-type semiconductor and a metal having a smaller work function: (a) band diagrams before joining; (b) band diagram for the junction at equilibrium.

[†]While the properties of the Schottky barrier depletion region are similar to the p^+-n, it is clear that the analogy does not include forward-bias hole injection, which is dominant for the p^+-n but not for the contact of Fig. 5-31.

The two other cases of ideal metal-semiconductor contacts ($\Phi_m < \Phi_s$ for n-type semiconductors, and $\Phi_m > \Phi_s$ for p-type) result in nonrectifying contacts. We will save treatment of these cases for Section 5.7.3, where ohmic contacts are discussed.

5.7.2   Rectifying Contacts

When a forward-bias voltage V is applied to the Schottky barrier of Fig. 5-31b, the contact potential is reduced from V_0 to $V_0 - V$ (Fig. 5-33a). As a result, electrons in the semiconductor conduction band can diffuse across the depletion region to the metal. This gives rise to a forward current (metal to semiconductor) through the junction. Conversely, a reverse bias increases the barrier to $V_0 + V_r$, and electron flow from semiconductor to metal becomes negligible. In either case flow of electrons from the metal to the semiconductor is retarded by the barrier $\Phi_m - \chi$. The resulting diode equation is similar in form to that of the p-n junction

$$I = I_0(e^{qV/kT} - 1) \tag{5-76}$$

as Fig. 5-33c suggests. In this case the reverse saturation current I_0 is not simply derived as it was for the p-n junction. One important feature we can predict intuitively, however, is that the saturation current should depend upon the size

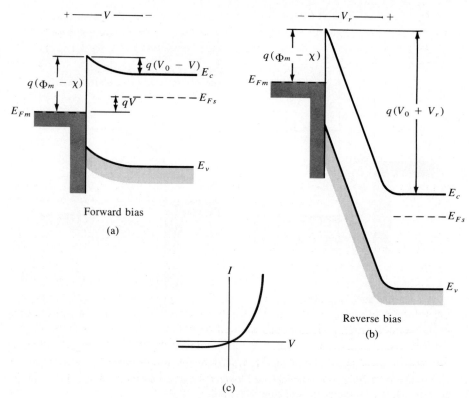

Figure 5-33
Effects of forward and reverse bias on the junction of Fig. 5-31: (a) forward bias; (b) reverse bias; (c) typical current–voltage characteristic.

of the barrier Φ_B for electron injection from the metal into the semiconductor. This barrier (which is $\Phi_m - \chi$ for the ideal case shown in Fig. 5-33) is unaffected by the bias voltage. We expect the probability of an electron in the metal surmounting this barrier to be given by a Boltzmann factor. Thus

$$I_0 \propto e^{-q\Phi_B/kT} \tag{5-77}$$

The diode equation (5-76) applies also to the metal–p-type semiconductor junction of Fig. 5-32. In this case forward voltage is defined with the semiconductor biased positively with respect to the metal. Forward current increases as this voltage lowers the potential barrier to $V_0 - V$ and holes flow from the semiconductor to the metal. Of course, a reverse voltage increases the barrier for hole flow and the current becomes negligible.

In both of these cases the Schottky barrier diode is rectifying, with easy current flow in the forward direction and little current in the reverse direction. We also note that the forward current in each case is due to the injection of *majority* carriers from the semiconductor into the metal. The absence of minority carrier injection and the associated storage delay time is an important feature of Schottky barrier diodes. Although some minority carrier injection occurs at high current levels, these are essentially majority carrier devices. Their high-frequency properties and switching speed are therefore generally better than typical p-n junctions.

In the early days of semiconductor technology, rectifying contacts were made simply by pressing a wire against the surface of the semiconductor. In modern devices, however, the metal–semiconductor contact is made by depositing an appropriate metal film on a clean semiconductor surface and defining the contact pattern photolithographically. Schottky barrier devices are particularly well suited for use in densely packed integrated circuits, because fewer photolithographic masking steps are required compared to p-n junction devices.

5.7.3   Ohmic Contacts

In many cases we wish to have an *ohmic* metal–semiconductor contact, having a linear *I–V* characteristic in both biasing directions. For example, the surface of a typical integrated circuit is a maze of p and n regions, which must be contacted and interconnected. It is important that such contacts be ohmic, with minimal resistance and no tendency to rectify signals.

Ideal metal–semiconductor contacts are ohmic when the charge induced in the semiconductor in aligning the Fermi levels is provided by majority carriers (Fig. 5-34). For example, in the $\Phi_m < \Phi_s$ (n-type) case of Fig. 5-34a, the Fermi levels are aligned at equilibrium by transferring electrons from the metal to the semiconductor. This raises the semiconductor electron energies (lowers the electrostatic potential) relative to the metal at equilibrium (Fig. 5-34b). In this case the barrier to electron flow between the metal and the semiconductor is small and easily overcome by a small voltage. Similarly, the case $\Phi_m > \Phi_s$ (p-type) results in easy hole flow across the junction (Fig. 5-34d). Unlike the

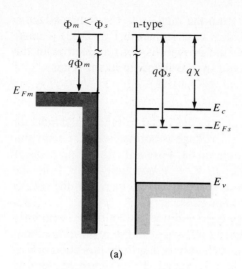

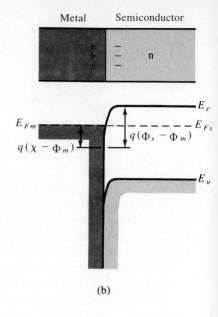

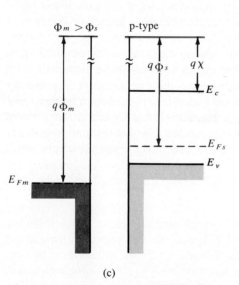

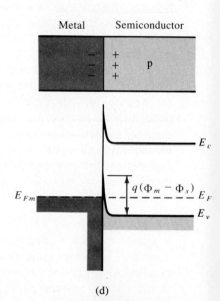

Figure 5-34
Ohmic metal–semiconductor contacts: (a) $\Phi_m < \Phi_s$ for an n-type semiconductor, and (b) the equilibrium band diagram for the junction; (c) $\Phi_m > \Phi_s$ for a p-type semiconductor, and (d) the junction at equilibrium.

(a)

(b)

(c)

(d)

rectifying contacts discussed above, no depletion region occurs in the semiconductor in these cases since the electrostatic potential difference required to align the Fermi levels at equilibrium calls for accumulation of majority carriers in the semiconductor.

A practical method for <u>forming ohmic contacts is by doping the semiconductor heavily in the contact region</u>. Thus if a barrier exists at the interface, the depletion width is small enough to allow carriers to tunnel through the barrier. For example, Au containing a small percentage of Sb can be alloyed to n-type

Si, forming an n$^+$ layer at the semiconductor surface and an excellent ohmic contact. Similarly, p-type material requires a p$^+$ surface layer in contact with the metal. In the case of Al on p-type Si, the metal contact also provides the acceptor dopant. Thus the required p$^+$ surface layer is formed during a brief heat treatment of the contact after the Al is deposited. As we shall see in the discussion of integrated circuits in Chapter 9, Al makes good ohmic contact to p-type Si, but n-type material requires an n$^+$ diffusion or implantation before contacting by Al.

5.7.4   Typical Schottky Barriers

The discussion of ideal metal–semiconductor contacts does not include certain effects of the junction between the two dissimilar materials. Unlike a p-n junction, which occurs within a single crystal, a Schottky barrier junction includes a termination of the semiconductor crystal. The semiconductor surface contains *surface states* due to incomplete covalent bonds and other effects, which can lead to charges at the metal–semiconductor interface. Furthermore, the contact is seldom an atomically sharp discontinuity between the semiconductor crystal and the metal. There is typically a thin interfacial layer, which is neither semiconductor nor metal. For example, silicon crystals are covered by a thin (10–20 Å) oxide layer even after etching or cleaving in atmospheric conditions. Therefore, deposition of a metal on such a Si surface leaves a glassy interfacial layer at the junction. Although electrons can tunnel through this thin layer, it does affect the barrier to current transport through the junction.

Because of surface states, the interfacial layer, microscopic clusters of metal–semiconductor phases, and other effects, it is difficult to fabricate junctions with barriers near the ideal values predicted from the work functions of the two isolated materials. Therefore, measured barrier heights are used in device design. In compound semiconductors the interfacial layer introduces states in the semiconductor band gap that pin the Fermi level at a fixed position, regardless of the metal used (Fig. 5-35). For example, a collection of interface

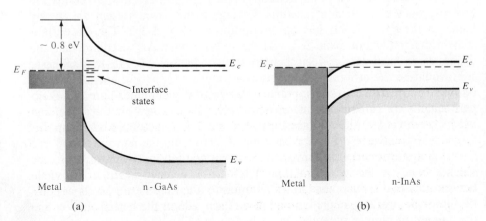

Figure 5-35
Fermi level pinning by interface states in compound semiconductors: (a) E_F is pinned near $E_C - 0.8$ eV in n-type GaAs, regardless of the choice of metal; (b) E_F is pinned above E_C in n-type InAs, providing an excellent ohmic contact.

states located $0.7 \sim 0.9$ eV below the conduction band pins E_F at the surface of n-type GaAs, and the Schottky barrier height is determined from this pinning effect rather than by the work function of the metal. An interesting case is n-type InAs (Fig. 5-35b), in which E_F at the interface is pinned *above* the conduction band edge. As a result, ohmic contact to n-type InAs can be made by depositing virtually any metal on the surface. For Si, good Schottky barriers are formed by various metals, such as Au or Pt. In the case of Pt, heat treatment results in a platinum silicide layer, which provides a reliable Schottky barrier with $\Phi_B \simeq 0.85$ V on n-type Si.

A full treatment of Schottky barrier diodes results in a forward current equation of the form

$$I = ABT^2 e^{-q\Phi_B/kT} e^{qV/nkT} \qquad (5\text{-}78)$$

where B is a constant containing parameters of the junction properties and **n** is a number between 1 and 2, similar to the ideality factor in Eq. (5-71) but arising from different reasons. The mathematics of this derivation is similar to that of *thermionic emission,* and the factor B corresponds to an effective Richardson constant in the thermionic problem.

**5.8
HETEROJUNCTIONS**

Thus far we have discussed p-n junctions formed within a single semiconductor (*homojunctions*) and junctions between a metal and a semiconductor. The third important class of junctions consists of those between two lattice-matched semiconductors with different band gaps (*heterojunctions*). We discussed lattice-matching in Section 1.4.1. The interface between two such semiconductors may be virtually free of defects, and continuous crystals containing single or multiple heterojunctions can be formed. The availability of heterojunctions and multilayer structures in compound semiconductors opens a broad range of possibilities for device development. We will discuss many of these applications in later chapters, including heterojunction bipolar transistors, field-effect transistors, and semiconductor lasers.

When semiconductors of different band gaps and electron affinities are brought together to form a junction, we expect discontinuities in the energy bands as the Fermi levels line up at equilibrium (Fig. 5-36). The discontinuities in the conduction band ΔE_c and the valence band ΔE_v accommodate the difference in band gap between the two semiconductors ΔE_g. In an ideal case, ΔE_c would be the difference in electron affinities $q(\chi_2 - \chi_1)$, and ΔE_v would be found from $\Delta E_g - \Delta E_c$. In practice, the band discontinuities are found experimentally for particular semiconductor pairs. For example, in the commonly used system GaAs–AlGaAs (see Figs. 3-6 and 3-13), the direct band gap difference ΔE_g^Γ between the wider band gap AlGaAs and the narrower band gap GaAs is apportioned approximately $\frac{2}{3}$ in the conduction band and $\frac{1}{3}$ in the valence band for the heterojunction. The built-in contact potential is divided between the two semiconductors as required to align the Fermi levels at equilibrium. The resulting depletion region on each side of the heterojunction and the amount of built-in potential on each side (making up the contact potential V_0) are found by solving Poisson's equation with the boundary condition of

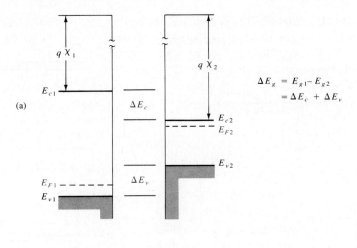

$$\Delta E_g = E_{g1} - E_{g2}$$
$$= \Delta E_c + \Delta E_v$$

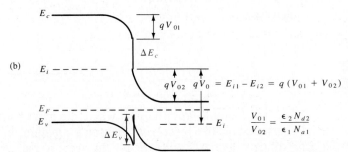

$$qV_0 = E_{i1} - E_{i2} = q(V_{01} + V_{02})$$

$$\frac{V_{01}}{V_{02}} = \frac{\epsilon_2 N_{d2}}{\epsilon_1 N_{a1}}$$

Figure 5-36
An ideal heterojunction between a p-type, wide band gap semiconductor and an n-type narrower band gap semiconductor: (a) band diagrams before joining; (b) band discontinuities and band bending at equilibrium.

continuous electric flux density, $\epsilon_1 \mathscr{E}_1 = \epsilon_2 \mathscr{E}_2$ at the junction. The barrier that electrons must overcome in moving from the n side to the p side may be quite different from the barrier for holes moving from p to n. The depletion region on each side is analogous to that described in Eq. (5-23), except that we must account for the different dielectric constants in the two semiconductors.

To draw the band diagram for a heterojunction accurately, we must not only use the proper values for the band discontinuities but also account for the band bending in the junction. To do this, we must solve Poisson's equation across the heterojunction, taking into account the details of doping and space charge, which generally requires a computer solution. We can, however, sketch an approximate diagram without a detailed calculation. Given the experimental band offsets ΔE_v and ΔE_c, we can proceed as follows:

1. Align the Fermi level with the two semiconductor bands separated. Leave space for the transition region.

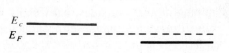

2. The metallurgical junction ($x = 0$) is located near the more heavily doped side. At $x = 0$ put ΔE_v and ΔE_c, separated by the appropriate band gaps.

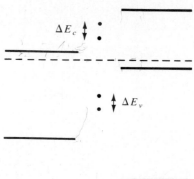

3. Connect the conduction band and valence band regions, keeping the band gap constant in each material.

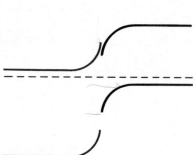

Steps 2 and 3 of this procedure are where the exact band bending is important and must be obtained by solving Poisson's equation. In step 2 we must use the band offset values ΔE_c and ΔE_v for the specific pair of semiconductors in the heterojunction.

EXAMPLE 5-6 For heterojunctions in the GaAs–AlGaAs system, the direct (Γ) band gap difference ΔE_g^Γ is accommodated approximately $\frac{2}{3}$ in the conduction band and $\frac{1}{3}$ in the valence band. For an Al composition of 0.3, the AlGaAs is direct (see Fig. 3-6) with $E_g^\Gamma = 1.85$ eV. Sketch the band diagrams for two heterojunction cases: N^+-Al$_{0.3}$Ga$_{0.7}$As on n-type GaAs, and N^+-Al$_{0.3}$Ga$_{0.7}$As on p^+-GaAs.[†]

SOLUTION Taking $\Delta E_g = 1.85 - 1.43 = 0.42$ eV, the band offsets are $\Delta E_c = 0.28$ eV and $\Delta E_v = 0.14$ eV. In each case we draw the equilibrium Fermi level, add the appropriate bands far from the junction, add the band offsets while estimating the relative amounts of band bending and position of $x = 0$ for the particular doping on the two sides, and finally sketch the band edges so that E_g is maintained in each separate semiconductor right up to the heterojunction at $x = 0$.

[†]In discussing heterojunctions, it is common to use capital N or P to designate the wide band gap material.

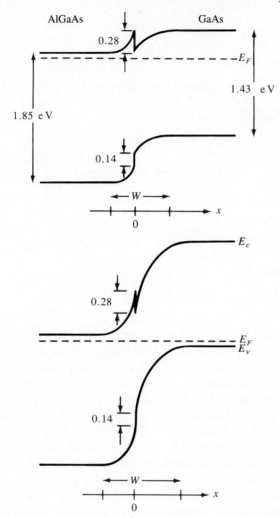

A particularly important example of a heterojunction is shown in Figure 5-37, in which heavily n-type AlGaAs is grown on lightly doped GaAs. In this example the discontinuity in the conduction band allows electrons to spill over from the N^+-AlGaAs into the GaAs, where they become trapped in the potential well. As a result, electrons collect on the GaAs side of the heterojunction and move the Fermi level above the conduction band in the GaAs near the interface. These electrons are confined in a narrow potential well in the GaAs conduction band. If we construct a device in which conduction occurs parallel to the interface, the electrons in such a potential well form a *two-dimensional electron gas* with very interesting device properties. As we shall see in Chapter 8, electron conduction in such a potential well can result in very high mobility electrons. This high mobility is due to the fact that the electrons

Figure 5-37
A heterojunction
between N^+-AlGaAs
and lightly doped
GaAs, illustrating
the potential well
for electrons
formed in the GaAs
conduction band. If
this well is
sufficiently thin,
discrete states
(such as E_1 and E_2)
are formed, as
discussed in
Section 2.4.3.

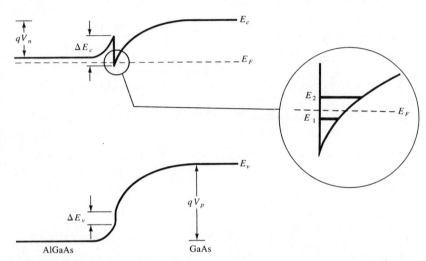

in this well come from the AlGaAs, and not from doping in the GaAs. As a result, there is negligible impurity scattering in the GaAs well, and the mobility is controlled almost entirely by lattice scattering (phonons). At low temperatures, where phonon scattering is low, the mobility in this region can be very high. If the band-bending in the GaAs conduction band is strong enough, the potential well may be extremely narrow, so that discrete states such as E_1 and E_2 in Fig. 5-37 are formed. We will return to this example in Section 8.2.2.

Another obvious feature of Fig. 5-37 is that the concept of a contact potential barrier qV_0 for both electrons and holes in a homojunction is no longer valid for the heterojunction. In Fig. 5-37 the barrier for electrons qV_n is smaller than the barrier for holes qV_p. This property of a heterojunction can be used to alter the relative injection of electrons and holes, as we shall see in Section 7.9.

PROBLEMS

5.1. (a) A Si sample is doped with 10^{21} cm^{-3} phosphorus atoms. Is it possible to form a p-n junction in this sample by alloying or diffusion? If so, what impurity would you recommend for the counterdoping? *Hint:* see Appendix V.

(b) A Ge sample contains 10^{19} cm^{-3} Sb atoms. Is In an appropriate metal for forming an alloy junction in this material? How about In + Ga?

5.2. The mathematics used in deriving Eq. (5-7) can be avoided if we note from Eq. (3-58) that the Fermi level must be invariant across the junction ($E_{Fn} = E_{Fp}$) at equilibrium. Draw a band diagram such as in Fig. 5-8, and show that Eq. (5-7) follows automatically from this condition.

5.3. In a p^+-n junction the hole diffusion current in the neutral n material is given by Eq. (5-32). What are the electron diffusion and electron drift components of current at point x_n in the neutral n region?

5.4. An abrupt Si p-n junction has $N_a = 10^{18}$ cm^{-3} on one side and $N_d = 5 \times 10^{15}$ cm^{-3} on the other.
   (a) Calculate the Fermi level positions at 300 K in the p and n regions.
   (b) Draw an equilibrium band diagram for the junction and determine the contact potential V_0 from the diagram.
   (c) Compare the results of part (b) with V_0 as calculated from Eq. (5-8).

5.5. The junction described in Prob. 5.4 has a circular cross section with diameter of 10 μm. Calculate x_{n0}, x_{p0}, Q_+, and $\mathscr{E}_0$ for this junction at equilibrium (300 K). Sketch $\mathscr{E}(x)$ and charge density to scale, as in Fig. 5-9.

5.6. The hole injection efficiency of a junction is I_p/I at $x_n = 0$.
   (a) Assuming the junction follows the simple diode equation, express I_p/I in terms of the diffusion constants, diffusion lengths, and equilibrium minority carrier concentrations.
   (b) Show that I_p/I can be written as $[1 + L_p^n n_n \mu_n^p / L_n^p p_p \mu_p^n]^{-1}$, where the superscripts refer to the n and p regions. What should be done to increase the hole injection efficiency of a junction?

5.7. A Si p^+-n junction has a donor doping of 5×10^{16} cm^{-3} on the n side and a cross-sectional area of 10^{-3} cm^2. If $\tau_p = 1$ μs and $D_p = 10$ cm^2/s, calculate the current with a forward bias of 0.5 V at 300 K.

5.8. Draw a band diagram for a forward-biased p-n junction, including a sketch of the quasi-Fermi levels $F_n(x)$ and $F_p(x)$ throughout the junction and for several diffusion lengths on either side of the junction. Explain qualitatively the variations in F_n and F_p.

5.9. (a) Explain physically why the charge storage capacitance is unimportant for reverse-biased junctions.
   (b) Assuming that a GaAs junction is doped to equal concentrations on the n and p sides, would you expect electron or hole injection to dominate in forward bias? Explain.

5.10. (a) A Si p^+-n junction 10^{-2} cm^2 in area has $N_d = 10^{15}$ cm^{-3} doping on the n side. Calculate the junction capacitance with a reverse bias of 10 V.
   (b) An abrupt p^+-n junction is formed in Si with a donor doping of $N_d = 10^{15}$ cm^{-3}. What is the depletion region thickness W just prior to avalanche breakdown?

5.11. Use sketches of band diagrams to illustrate that the tunneling distance d in Fig. 5-17b becomes smaller with an increase in reverse bias. Assume the doping is sufficiently heavy that changes in W are negligible.

5.12. Using Eqs. (5-17) and (5-23), show that the peak electric field in the transition region is controlled by the doping on the more lightly doped side of the junction.

5.13. An abrupt Si p-n junction has the following properties at 300 K:

p side	n side	$A = 10^{-4}$ cm^2
$N_a = 10^{17}$ cm^{-3}	$N_d = 10^{15}$	
$\tau_n = 0.1$ μs	$\tau_p = 10$ μs	
$\mu_p = 200$ cm^2/V-s	$\mu_n = 1300$	
$\mu_n = 700$	$\mu_p = 450$	

(a) Draw an equilibrium band diagram for this junction, including numerical values for the Fermi level position relative to the intrinsic level on each side. Find the contact potential from the diagram and check your answer with the analytical expression for V_0.

(b) Calculate the reverse saturation current I_0, and find the current I with a forward bias of 0.5 V.

5.14. A long p$^+$-n diode is forward biased with current I flowing. The current is suddenly tripled at $t = 0$.

(a) What is the slope of the hole distribution at $x_n = 0$ just after the current is tripled?

(b) Assuming the voltage is always $\gg kT/q$, relate the final junction voltage (at $t = \infty$) to the initial voltage (before $t = 0$).

(c) Assume an exponential distribution for the excess holes at each instant during the transient. Find the expression for the instantaneous junction voltage $v(t)$.

5.15. A p$^+$-n diode is switched from zero bias ($I = 0$) to a forward current I at $t = 0$.

(a) Find the expression for the excess hole charge $Q_p(t)$ during the turn-on transient. Sketch $Q_p(t)$ vs. t.

(b) Assuming $\delta p(x_n)$ is always essentially exponential, find $\Delta p_n(t)$ and the voltage $v(t)$.

5.16. Assume that the doping concentration N_a on the p side of an abrupt junction is the same as N_d on the n side. Each side is many diffusion lengths long. Find the expression for the hole current I_p in the p-type material.

5.17. Assume that an abrupt Ge p-n junction with area 10^{-4} cm^2 has $N_a = 5 \times 10^{17}$ cm^{-3} on the p side and $N_d = 5 \times 10^{17}$ cm^{-3} on the n side. The diode has a forward bias of 0.3 V. Using mobility values from Fig. 3-23 and assuming that $\tau_n = \tau_p = 10 \ \mu s$, plot I_p and I_n versus distance on a diagram such as Fig. 5-15, including both sides of the junction. Neglect recombination within W.

5.18. For the diode of Prob. 5-17, plot $\delta n(x_p)$ and $\delta p(x_n)$.

5.19. For the diode of Prob. 5-18, plot F_n and F_p through the junction and for several diffusion lengths on each side.

5.20. In this problem we wish to derive the high-injection diode equation, Eq. (5-70). Assume space charge neutrality in the n and p regions, $\Delta n_p = \Delta p_p$ and $\Delta p_n = \Delta n_n$.
(a) Use Eq. (5-69) to solve for Δp_n and Δn_p in terms of equilibrium quantities and the applied voltage.
(b) Show that Eq. (5-70) follows from

$$I = qA\left(\frac{D_p \, \Delta p_n}{L_p} + \frac{D_n \, \Delta n_p}{L_n}\right)$$

5.21. Assume a linearly graded junction as in Fig. 5-30, with a doping distribution described by Eq. (5-74). The doping is symmetrical, so that $x_{p0} = x_{n0} = W/2$.
(a) Integrate Eq. (5-75) to show that

$$\mathscr{E}(x) = \frac{q}{2\epsilon} G\left[x^2 - \left(\frac{W}{2}\right)^2\right]$$

(b) Show that the width of the depletion region is

$$W = \left[\frac{12\epsilon(V_0 - V)}{qG}\right]^{1/3}$$

(c) Show that the junction capacitance is

$$C_j = A\left[\frac{qG\epsilon^2}{12(V_0 - V)}\right]^{1/3}$$

5.22. When impurities are diffused into a sample from an unlimited source such that the surface concentration N_0 is held constant, the impurity distribution (profile) is given by

$$N(x, t) = N_0 \ \mathrm{erfc}\left(\frac{x}{2\sqrt{Dt}}\right)$$

where D is the diffusion coefficient for the impurity, t is the diffusion time, and erfc is the complementary error function.

If a certain number of impurities are placed in a thin layer on the surface before diffusion, and if no impurities are added and none escape during diffusion, a gaussian distribution is obtained:

$$N(x, t) = \frac{N_s}{\sqrt{\pi Dt}} e^{-(x/2\sqrt{Dt})^2}$$

where N_s is the quantity of impurity placed on the surface (atoms/cm^2) prior to $t = 0$. Notice that this expression differs from Eq. (4-44) by a factor of two. Why?

Figure P5-22 gives curves of the complementary error function and gaussian factors for the variable u, which in our case is $x/2\sqrt{Dt}$. Assume that boron is diffused into n-type Si (uniform $N_d = 5 \times 10^{16}$ cm^{-3}) at 1000°C for 30 min. The diffusion coefficient for B in Si at this temperature is $D = 3 \times 10^{-14}$ cm^2/s.

Figure P5-22

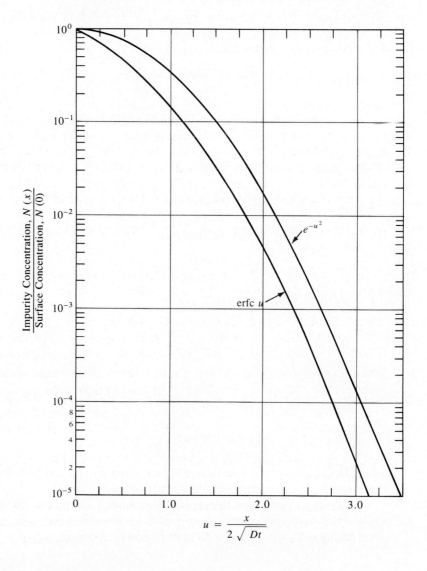

(a) Plot $N_a(x)$ after the diffusion, assuming that the surface concentration is held constant at $N_0 = 5 \times 10^{20}$ cm^{-3}. Locate the position of the junction below the surface.

(b) Plot $N_a(x)$ after diffusion, assuming that B is deposited in a thin layer on the surface prior to diffusion ($N_s = 5 \times 10^{13}$ cm^{-2}), and no additional B atoms are available during the diffusion. Locate the junction for this case.

Hint: Plot the curves on five-cycle semilog paper, with an abscissa varying from zero to $\frac{1}{2} \mu$m. In plotting $N_a(x)$, choose values of x that are simple multiples of $2\sqrt{Dt}$.

5.23. GaAs in implanted with Be ions at 100 keV to a dose of 2.5×10^{13} ions/cm^2. For this implant the range is 0.28 μm and the straggle is 0.1 μm. Plot the Be impurity profile on semilog paper, as in Fig. 5-5(a).

5.24. An ideal metal–semiconductor contact is formed on n-type Si having $N_d = 5 \times 10^{15}$ cm^{-3}. The metal work function is 3.8 eV, and the Si electron affinity is 4 eV. Draw the equilibrium band diagram, with values calculated for appropriate barriers.

5.25. A Schottky barrier is formed between a metal having a work function of 4.3 eV and p-type Si (electron affinity = 4 eV). The acceptor doping in the Si is 10^{17} cm^{-3}.

(a) Draw the equilibrium band diagram, showing a numerical value for qV_0.

(b) Draw the band diagram with 0.3 V forward bias. Repeat for 2 V reverse bias.

5.26. Design an ohmic contact for n-type GaAs using InAs, with an intervening graded InGaAs region (see Fig. 5-35).

5.27. Sketch the band diagrams for Al$_{0.3}$Ga$_{0.7}$As on GaAs for (a) P$^+$-AlGaAs, n$^+$-GaAs, (b) N$^+$-AlGaAs, intrinsic GaAs, (c) N$^+$-AlGaAs, p-GaAs. This composition of AlGaAs is direct, with $E_g = 1.85$ eV. Assume $\Delta E_c = 2/3 \, \Delta E_g$.

*5.28. (a) Using Eq. (5-8), calculate the contact potential V_0 of a Si p-n junction operating at 300 K for $N_a = 10^{14}$ and 10^{19} cm^{-3}, with $N_d = 10^{14}$, 10^{15}, 10^{16}, 10^{17}, 10^{18}, and 10^{19} cm^{-3} in each case and plot vs. N_d.

(b) Plot the maximum electric field $\mathscr{E}_0$ vs. N_d for the junctions described in (a).

*5.29. Calculate and plot the I–V characteristics from 0 to 0.65 V for an abrupt Si diode of area 100 μm^2 with $N_a = 10^{19}$ cm^{-3} on one side and $N_d = 10^{19}$ cm^{-3} on the other side and for $\tau_n = \tau_p = 1 \mu$s, using:

(a) The ideal diode equation,

(b) The high-level injection case [equation (5-70)],

(c) High-level injection plus series resistances $R_p = R_n = 100 \ \Omega$.

*5.30. The ideality factor **n** can be used to describe the relative importance of recombination within the transition region and the neutral region. Calculate and plot the I–V characteristics of the diode in problem 5.29 using Eq. (5-71) for values of the ideality factor of 1.0, 1.2, 1.4, 1.6, 1.8, and 2.0.

READING LIST

Bauer, R. S., and G. Margaritondo. "Probing Semiconductor-Semiconductor Interfaces." *Physics Today* 40, no. 1 (January 1987): 27–34.

Capasso, F., and G. Margaritondo, eds. *Heterojunction Band Discontinuities: Device Physics and Applications*. New York: North Holland, 1987.

Dohler, G. H. "Solid State Superlattices." *Scientific American* 249, no. 5 (November 1983): 144–51.

Green, M., J. E. E. Baglin, G. Y. Chin, H. W. Deckman, W. Mayo, and D. Narasinham, eds. *Semiconductor-Based Heterostructures: Interfacial Structure and Stability*. Warrendale, Pa.: The Metallurgical Society, Inc., 1986.

Grove, A. S. *Physics and Technology of Semiconductor Devices*. New York: John Wiley, 1967.

Muller, R. S., and T. I. Kamins. *Device Electronics for Integrated Circuits*. 2d ed. New York: John Wiley, 1986.

Narayanamurti, V. "Crystalline Semiconductor Heterostructures." *Physics Today* 37, no. 10 (October 1984): 24–32.

Neudeck, G. W. *The Modular Series on Solid State Devices. Vol. II: The p-n Junction Diode,* 2d ed. Reading, Mass.: Addison-Wesley, 1989.

Picraux, S. T., and P. S. Peercy. "Ion Implantation of Surfaces." *Scientific American* 252, no. 3 (March 1985): 102–13.

Ryssel, H., and I. Ruge. *Ion Implantation*. New York: John Wiley, 1986.

Sze, S. M. *Physics of Semiconductor Devices*. New York: John Wiley, 1981.

Weaver, J. H. "Metal-Semiconductor Interfaces." *Physics Today* 39, no. 1 (January 1986): 24–33.

Wilmsen, C. W. *Physics and Chemistry of III–V Compound Semiconductor Interfaces*. New York: Plenum Press, 1985.

Wolfe, C. M., N. Holonyak, and G. E. Stillman. *Physical Properties of Semiconductors*. Englewood Cliffs, NJ: Prentice Hall, 1989.

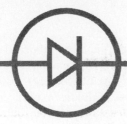

chapter 6

p-n JUNCTION DIODES

In this chapter we shall investigate some of the important electronic devices built with p-n junctions. The devices discussed here are primarily single-junction structures (diodes), with transistors and other multijunction devices left for later chapters. The diodes discussed in this chapter illustrate many of the fundamental applications of junction properties such as rectification, variable capacitance, tunneling, and light emission and detection. Since detailed descriptions of applications can be found in various texts on electronic circuits, we shall limit discussion of circuit applications to examples which illustrate particular device properties.

We have investigated many of the properties of junction diodes in Chapter 5. In this section we discuss the use of these diodes for rectification and other circuit applications. In some cases, diodes are designed to exploit specific junction properties, such as capacitance or charge storage. Therefore, we shall extend the treatment beyond the usual applications of diodes in rectification and switching to include those devices which depend on "secondary" junction properties.

**6.1
THE JUNCTION
DIODE**

6.1.1 Rectifiers

The most obvious property of a p-n junction is its *unilateral* nature; that is, to a good approximation it conducts current in only one direction. We can think of an *ideal diode* as a short circuit when forward biased and as an open circuit when reverse biased (Fig. 6-1a). The p-n junction diode does not quite fit this description, but the $I-V$ characteristics of many junctions can be approximated by the ideal diode in series with other circuit elements to form an equivalent circuit. For example, most forward-biased diodes exhibit an *offset voltage* E_0 (see Fig. 5-23), which can be approximated in a circuit model by a battery in series with the ideal diode (Fig. 6-1b). The series battery in the model keeps the ideal diode turned off for applied voltages less than E_0. From Section 5.6.1 we expect E_0 to be approximately the contact potential of the junction. In some cases the approximation to the actual diode characteristic is improved by adding a series resistor R to the circuit equivalent (Fig. 6-1c). The equivalent circuit approximations illustrated in Fig. 6-1 are called *piecewise-linear equivalents,* since the approximate characteristics are linear over specific ranges of voltage and current.

An ideal diode can be placed in series with an a-c voltage source to provide *rectification* of the signal. Since current can flow only in the forward direction through the diode, only the positive half-cycles of the input sine wave are passed. The output voltage is a *half-rectified sine wave.* Whereas the input sinusoid has no average value, the rectified signal has a positive average value and therefore contains a d-c component. By appropriate filtering, this d-c level can be extracted from the rectified signal.

The unilateral nature of diodes is useful for many other circuit applications that require *waveshaping.* This involves alteration of a-c signals by passing only certain portions of the signal while blocking other portions.

Junction diodes designed for use as rectifiers should have $I-V$ characteristics as close as possible to that of the ideal diode. The reverse current should be negligible, and the forward current should exhibit little voltage dependence (negligible *forward resistance* R). The reverse breakdown voltage should be large, and the offset voltage E_0 in the forward direction should be small. Unfortunately, not all of these requirements can be met by a single device; compromises must be made in the design of the junction to provide the best diode for the intended application.

Figure 6-1
Piecewise-linear approximations of junction diode characteristics: (a) the ideal diode; (b) ideal diode with an offset voltage; (c) ideal diode with an offset voltage and a resistance to account for slope in the forward characteristic.

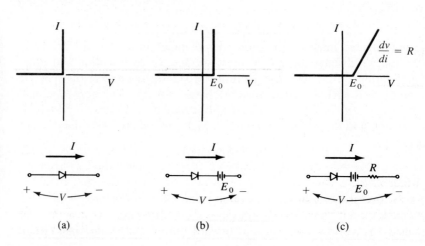

From the theory derived in Chapter 5 we can easily list the various requirements for good rectifier junctions. *Band gap* is obviously an important consideration in choosing a material for rectifier diodes. Since n_i is small for large band gap materials, the reverse saturation current (which depends on thermally generated carriers) decreases with increasing E_g. A rectifier made with a wide band gap material can be operated at higher temperatures, because thermal excitation of EHPs is reduced by the increased band gap. Such temperature effects are critically important in rectifiers, which must carry large currents in the forward direction and are thereby subjected to appreciable heating. On the other hand, the contact potential and offset voltage E_0 generally increase with E_g. This drawback is usually outweighed by the advantages of low n_i; for example, Si is generally preferred over Ge for power rectifiers because of its wider band gap, lower leakage current, and higher breakdown voltage, as well as its more convenient fabrication properties.

The *doping concentration* on each side of the junction influences the avalanche breakdown voltage, the contact potential, and the series resistance of the diode. If the junction has one highly doped side and one lightly doped side (such as a p^+-n junction), the lightly doped region determines many of the properties of the junction. From Fig. 5-19 we see that a high-resistivity region should be used for at least one side of the junction to increase the breakdown voltage V_{br}. However, this approach tends to increase the forward resistance R of Fig. 6-1c, and therefore contributes to the problems of thermal effects due to I^2R heating. To reduce the resistance of the lightly doped region, it is necessary to make its area large and reduce its length. Therefore, the physical *geometry* of the diode is another important design variable. Limitations on the practical area for a diode include problems of obtaining uniform starting material and junction processing over large areas. Localized flaws in junction uniformity can cause premature reverse breakdown in a small region of the device. Similarly, the lightly doped region of the junction cannot be made arbitrarily short. One of the primary problems with a short, lightly doped region is an effect called *punch-through*. Since the transition region width W increases with reverse bias and extends primarily into the lightly doped region, it is possible for W to increase until it fills the entire length of this region (Prob. 6.4). The result of punch-through is a breakdown below the value of V_{br} expected from Fig. 5-19.

In devices designed for use at high reverse bias, care must be taken to avoid premature breakdown across the edge of the sample. This effect can be reduced by *beveling* the edge or by diffusing a *guard ring* to isolate the junction from the edge of the sample (Fig. 6-2). The electric field is lower at the beveled edge of the sample in Fig. 6-2b than it is in the main body of the device. Similarly, the junction at the lightly doped p guard ring of Fig. 6-2c breaks down at higher voltage than the p^+-n junction. Since the depletion region is wider in the p ring than in the p^+ region, the average electric field is smaller at the ring for a given diode reverse voltage.

In fabricating a p^+-n or a p-n^+ junction, it is common to terminate the lightly doped region with a heavily doped layer of the same type (Fig. 6-3a),

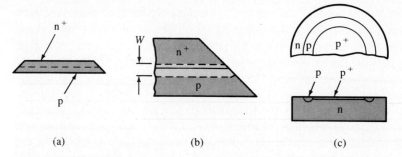

(a)　　　　　　(b)　　　　　　(c)

Figure 6-2　Beveled edge and guard ring to prevent edge breakdown under reverse bias: (a) diode with beveled edge; (b) closeup view of edge, showing reduction of depletion region near the bevel; (c) guard ring.

to ease the problem of making ohmic contact to the device. The result is a p^+-n-n^+ structure with the p^+-n layer serving as the active junction, or a p^+-p-n^+ device with an active p-n^+ junction. The lightly doped center region determines the avalanche breakdown voltage. If this region is short compared with the minority carrier diffusion length, the excess carrier injection for large forward currents can increase the conductivity of the region significantly. This type of *conductivity modulation,* which reduces the forward resistance R, can be very useful for high-current devices. On the other hand, a short, lightly doped center region can also lead to punch-through under reverse bias, as in Fig. 6-3c.

Figure 6-3
A p^+-n-n^+ junction diode: (a) device configuration; (b) zero-bias condition; (c) reverse-biased to punch-through.

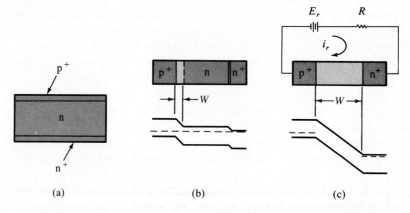

(a)　　　　　　(b)　　　　　　(c)

The mounting of a rectifier junction is critical to its ability to handle power. For diodes used in low-power circuits, glass or plastic encapsulation or a simple header mounting is adequate. However, high-current devices that must dissipate large amounts of heat require special mountings to transfer thermal energy away from the junction. A typical Si power rectifier is mounted on a molybdenum or tungsten disk to match the thermal expansion properties of the Si. This disk is fastened to a large stud of copper or other thermally conductive material that can be bolted to a heat sink with an appropriate cooling facility. The contrast in size and type of mounting between a power rectifier and low-current diodes is shown in Fig. 6-4.

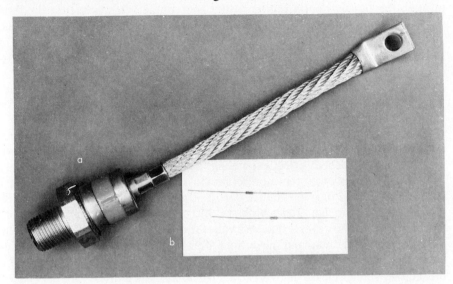

Figure 6-4
Rectifier mountings: (a) power rectifier; (b) low-current diodes. The power rectifier is rated at 250 A, with 2800 V maximum reverse voltage; the small diodes are rated at less than 1 A, with several hundred volts reverse-bias breakdown. (Photograph courtesy of Delco Electronics Corp.)

6.1.2 Switching Diodes

In discussing rectifiers we emphasized the importance of minimizing the reverse-bias current and the power losses under forward bias. In many applications, time response can be important as well. If a junction diode is to be used to switch rapidly from the conducting to the nonconducting state and back again, special consideration must be given to its charge control properties. We have discussed the equations governing the turn-on time and the reverse recovery time of a junction in Section 5.5. From Eqs. (5-47) and (5-54) it is clear that a diode with fast switching properties must either store very little charge in the neutral regions for steady forward currents, or have a very short carrier lifetime, or both.

A common technique for improving the switching speed of a diode is adding efficient recombination centers to the bulk material. For Si diodes, Au doping is useful for this purpose. To a good approximation the carrier lifetime varies with the reciprocal of the recombination center concentration. Thus, for example, a p^+-n Si diode may have $\tau_p = 1$ μs and a reverse recovery time of 0.1 μs before Au doping. If the addition of 10^{14} Au atoms/cm^3 reduces the lifetime to 0.1 μs and t_{sd} to 0.01 μs, 10^{15} cm^{-3} Au atoms could reduce τ_p to 0.01 μs and t_{sd} to 1 ns (10^{-9} s). This process cannot be continued indefinitely, however. The reverse current due to generation of carriers from the Au centers in the depletion region becomes appreciable with large Au concentration (Section 5.6.2). In addition, as the Au concentration approaches the lightest doping of the junction, the equilibrium carrier concentration of that region can be affected.

A second approach to improving the diode switching time is to make the lightly doped neutral region shorter than a minority carrier diffusion length.

This is the *narrow base diode* (Prob. 6.5). In this case the stored charge for forward conduction is very small, since most of the injected carriers diffuse through the lightly doped region to the end contact. When such a diode is switched to reverse conduction, very little time is required to eliminate the stored charge in the narrow neutral region. The mathematics involved in Prob. 6.5 is particularly interesting, because it closely resembles the calculations we shall make in analyzing the bipolar junction transistor in Chapter 7.

6.1.3   The Breakdown Diode

As we discussed in Section 5.4, the reverse-bias breakdown voltage of a junction can be varied by choice of junction doping concentrations. The breakdown mechanism is the Zener effect (tunneling) for abrupt junctions with extremely heavy doping; however, the more common breakdown is avalanche (impact ionization), typical of more lightly doped or graded junctions. By varying the doping we can fabricate diodes with specific breakdown voltages ranging from less than one volt to several hundred volts. If the junction is well designed, the breakdown will be sharp and the current after breakdown will be essentially independent of voltage (Fig. 6-5a). When a diode is designed for a specific breakdown voltage, it is called a *breakdown diode*. Such diodes are also called *Zener diodes*, despite the fact that the actual breakdown mechanism is usually the avalanche effect. This error in terminology is due to an early mistake in identifying the first observations of breakdown in p-n junctions.

Breakdown diodes can be used as *voltage regulators* in circuits with varying inputs. The 15-V breakdown diode of Fig. 6-5 holds the circuit output voltage v_0 constant at 15 V, while the input varies at voltages greater than 15 V. For example, if v_s is a rectified and filtered signal composed of a 17-V d-c component and a 1-V ripple variation above and below 17 V, the output v_0 will remain constant at 15 V. More complicated voltage regulator circuits can be designed using breakdown diodes, depending on the type of signal being regulated and the nature of the output load. In a similar application, such a device

Figure 6-5
A breakdown diode: (a) *I–V* characteristic; (b) application as a voltage regulator.

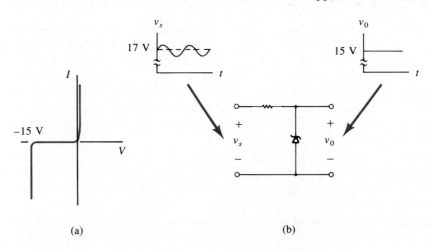

(a)   (b)

can be used as a *reference diode;* since the breakdown voltage of a particular diode is known, the voltage across it during breakdown can be used as a reference in circuits that require a known value of voltage.

6.1.4   The Varactor Diode

The term *varactor* is a shortened form of *variable reactor,* referring to the voltage-variable capacitance of a reverse-biased p-n junction. The equations derived in Section 5.5.3 indicate that junction capacitance depends on the applied voltage and the design of the junction. In some cases a junction with fixed reverse bias may be used as a capacitance of a set value. More commonly the varactor diode is designed to exploit the voltage-variable properties of the junction capacitance. For example, a varactor (or a set of varactors) may be used in the tuning stage of a radio receiver to replace the bulky variable plate capacitor. The size of the resulting circuit can be greatly reduced, and its dependability is often improved. Other applications of varactors include use in harmonic generation, microwave frequency multiplication, and active filters.

If the p-n junction is abrupt, the capacitance varies as the square root of the reverse bias V_r [Eq. (5-61)]. In a graded junction, however, the capacitance can usually be written in the form

$$C_j \propto V_r^{-n} \qquad \text{for } V_r \gg V_0 \qquad (6\text{-}1)$$

For example, in a linearly graded junction the exponent **n** is one-third (Prob. 5-21). Thus the voltage sensitivity of C_j is greater for an abrupt junction than for a linearly graded junction. For this reason, varactor diodes are often made by alloying or epitaxial growth techniques, or by ion implantation. The epitaxial layer and the substrate doping profile can be designed to obtain junctions for which the exponent **n** in Eq. (6-1) is greater than one-half. Such junctions are called *hyperabrupt junctions*.

In the set of doping profiles shown in Fig. 6-6, the junction is assumed p^+-n so that the depletion layer width W extends primarily into the n side. Three

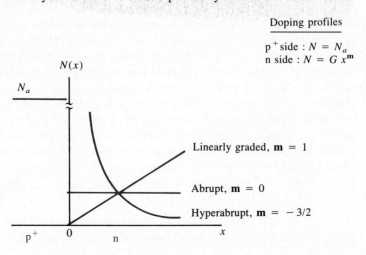

Doping profiles

p^+ side : $N = N_a$
n side : $N = G\,x^{\mathbf{m}}$

Figure 6-6
Graded junction profiles: linearly graded, abrupt, hyperabrupt.

$N(x)$

N_a

Linearly graded, **m** = 1

Abrupt, **m** = 0

Hyperabrupt, **m** = − 3/2

p^+ 0   n x

types of doping profiles on the n side are illustrated, with the donor distribution $N_d(x)$ given by Gx^m, where G is a constant and the exponent **m** is 0, 1, or $-\frac{3}{2}$. We can show (Prob. 6.7) that the exponent **n** in Eq. (6-1) is $1/(m + 2)$ for the p^+-n junction. Thus for the profiles of Fig. 6-6, **n** is $\frac{1}{2}$ for the abrupt junction and $\frac{1}{3}$ for the linearly graded junction. The hyperabrupt junction with $\mathbf{m} = -\frac{3}{2}^\dagger$ is particularly interesting for certain varactor applications, since for this case $\mathbf{n} = 2$ and the capacitance is proportional to V_r^{-2}. When such a capacitor is used with an inductor L in a resonant circuit, the resonant frequency varies linearly with the voltage applied to the varactor

$$\omega_r = \frac{1}{\sqrt{LC}} \propto \frac{1}{\sqrt{V_r^{-n}}} \propto V_r \qquad \text{for } \mathbf{n} = 2 \qquad (6\text{-}2)$$

Because of the wide variety of C_j vs. V_r dependencies available by choosing doping profiles, varactor diodes can be designed for specific applications. For some high-frequency applications, varactors can be designed to exploit the forward-bias charge storage capacitance.

6.2 TUNNEL DIODES

The tunnel diode is a p-n junction device that operates in certain regions of its I–V characteristic by the quantum mechanical tunneling of electrons through the potential barrier of the junction (see Sections 2.4.4 and 5.4.1). The tunneling process for reverse current is essentially the Zener effect, although negligible reverse bias is needed to initiate the process in tunnel diodes. This device can be used in many applications, including high-speed switching and logic circuits. As we shall see in this section, the tunnel diode (often called the *Esaki diode* after L. Esaki, who in 1973 received the Nobel prize for his work on the effect) exhibits the important feature of *negative resistance* over a portion of its I–V characteristic.

6.2.1 Degenerate Semiconductors

Thus far we have discussed the properties of relatively pure semiconductors; any impurity doping represented a small fraction of the total atomic density of the material. Since the few impurity atoms were so widely spaced throughout the sample, we could be confident that no charge transport could take place within the donor or acceptor levels themselves. What happens, however, if we continue to dope a semiconductor with impurities of either type? As might be expected, a point is reached at which the impurities become so closely packed within the lattice that interactions between them cannot be ignored. For example, donors present in high concentrations (e.g., 10^{20} donors/cm^3) are so close together that we can no longer consider the donor level as being composed of discrete, noninteracting energy states. Instead, the donor states form a

---

†It is clear that $N_d(x)$ cannot become arbitrarily large at $x = 0$. However, the $\mathbf{m} = -\frac{3}{2}$ profile can be approximated a short distance away from the junction.

band, which may overlap the bottom of the conduction band. If the conduction band electron concentration n exceeds the effective density of states N_c, the Fermi level is no longer within the band gap but lies within the conduction band.) When this occurs, the material is called *degenerate* n-type. The analogous case of degenerate p-type material occurs when the acceptor concentration is very high and the Fermi level lies in the valence band. We recall that the energy states below E_F are mostly filled and states above E_F are empty, except for a small distribution dictated by the Fermi statistics. Thus in a degenerate n-type sample the region between E_c and E_F is for the most part filled with electrons, and in degenerate p-type the region between E_v and E_F is almost completely filled with holes.

6.2.2 Tunnel Diode Operation

A p-n junction between two degenerate semiconductors is illustrated in terms of energy bands in Fig. 6-7a. This is the equilibrium condition, for which the Fermi level is constant throughout the junction. We notice that E_{Fp} lies below the valence band edge on the p side and E_{Fn} is above the conduction band edge on the n side. Thus the bands must overlap on the energy scale in order for E_F to be constant. This overlapping of bands is very important; it means that with a small forward or reverse bias, filled states and empty states appear opposite each other, separated by essentially the width of the depletion region. If the metallurgical junction is sharp, as in an alloyed junction, the depletion region will be very narrow for such high doping concentrations, and the electric field at the junction will be quite large. Thus the conditions for electron tunneling are met — filled and empty states separated by a narrow potential barrier of finite height.

As mentioned above, the filled and empty states are distributed about E_F according to the Fermi distribution function; thus there are some filled states above E_{Fp} and some empty states below E_{Fn}. In Fig. 6-7 the bands are shown filled to the Fermi level for convenience of illustration, with the understanding that a distribution is implied.

Since the bands overlap under equilibrium conditions, a small reverse bias (Fig. 6-7b) allows electron tunneling from the filled valence band states below E_{Fp} to the empty conduction band states above E_{Fn}. This condition is similar to the Zener effect except that no bias is required to create the condition of overlapping bands. As the reverse bias is increased, E_{Fn} continues to move down the energy scale with respect to E_{Fp}, placing more filled states on the p side opposite empty states on the n side. Thus the tunneling of electrons from p to n increases with increasing reverse bias. The resulting conventional current is opposite to the electron flow, that is, from n to p. At equilibrium (Fig. 6-7a) there is equal tunneling from n to p and from p to n, given a zero net current.

When a small forward bias is applied (Fig. 6-7c), E_{Fn} moves up in energy with respect to E_{Fp} by the amount qV. Thus electrons below E_{Fn} on the n side are placed opposite empty states above E_{Fp} on the p side. Electron tunneling

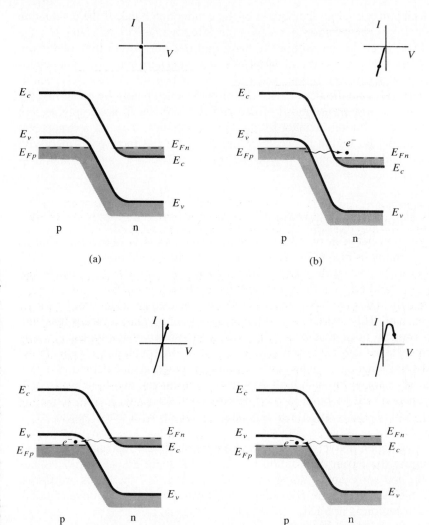

Figure 6-7
Tunnel diode band diagrams and I–V characteristics for various biasing conditions: (a) equilibrium (zero bias) condition, no net tunneling; (b) small reverse bias, electron tunneling from p to n; (c) small forward bias, electron tunneling from n to p; (d) increased forward bias, electron tunneling from n to p decreases as bands pass by each other.

occurs from n to p as shown, with the resulting conventional current from p to n. This forward tunneling current continues to increase with increased bias as more filled states are placed opposite empty states. However, as E_{Fn} continues to move up with respect to E_{Fp}, a point is reached at which the bands begin to pass by each other. When this occurs, the number of filled states opposite empty states decreases. The resulting decrease in tunneling current is illustrated in Fig. 6-7d. This region of the I–V characteristic is important in that the *decrease* of tunneling current with *increased* bias produces a region of negative slope; that is, the *dynamic resistance* dV/dI is negative. This negative resistance region is useful in a number of applications.

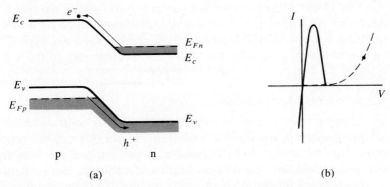

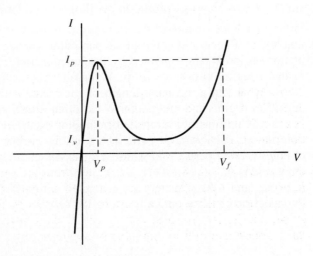

Band diagram (a) and *I–V* characteristic (b) for the tunnel diode beyond the tunnel current region. In (b) the tunneling component of current is shown by the solid curve and the diffusion current component is dashed.

Figure 6-8

If the forward bias is increased beyond the negative resistance region, the current begins to increase again (Fig. 6-8). Once the bands have passed each other, the characteristic resembles that of a conventional diode. The forward current is now dominated by the diffusion current — electrons surmounting the potential barrier from n to p and holes surmounting their potential barrier from p to n. Of course, the diffusion current is present in the forward tunneling region, but it is negligible compared to the tunneling current.

The total tunnel diode characteristic (Fig. 6-9) has the general shape of an *N* (if a little imagination is applied); therefore, it is common to refer to this characteristic as exhibiting a *type N negative resistance*. It is also called a *voltage-controlled negative resistance,* meaning that the current decreases rapidly at some critical voltage (in this case the *peak voltage* V_p, taken at the point of maximum forward tunneling).

Figure 6-9
Total tunnel diode characteristic.

The values of *peak tunneling current* I_p and *valley current* I_v (Fig. 6-9) determine the magnitude of the negative resistance slope for a diode of given material. For this reason, their ratio I_p/I_v is often used as a figure of merit for the tunnel diode. Similarly, the ratio V_p/V_f is a measure of the voltage spread between the two positive resistance regions.

6.2.3   Circuit Applications

The negative resistance of the tunnel diode can be used in a number of ways to achieve switching, oscillation, amplification, and other circuit functions. This wide range of applications, coupled with the fact that the tunneling process does not present the time delays of drift and diffusion, makes the tunnel diode a natural choice for certain high-speed circuits. However, the tunnel diode has not achieved widespread application, because of its relatively low current operation and competition from other devices.

**6.3
PHOTODIODES**

In Section 4.3.4 we saw that bulk semiconductor samples can be used as photoconductors by providing a change in conductivity proportional to an optical generation rate. Often, junction devices can be used to improve the speed of response and sensitivity of detectors of optical or high-energy radiation. Two-terminal devices designed to respond to photon absorption are called *photodiodes*. Some photodiodes have extremely high sensitivity and response speed. Since modern electronics often involves optical as well as electrical signals, photodiodes serve important functions as electronic devices. In this section, we shall investigate the response of p-n junctions to optical generation of EHPs and discuss a few typical *photodiode detector* structures. We shall also consider the very important use of junctions as *solar cells,* which convert absorbed optical energy into useful electrical power.

6.3.1   Current and Voltage in an Illuminated Junction

In Chapter 5 we identified the current due to drift of minority carriers across a junction as a generation current. In particular, carriers generated within the depletion region W are separated by the junction field, electrons being collected in the n region and holes in the p region. Also, minority carriers generated thermally within a diffusion length of each side of the junction diffuse to the depletion region and are swept to the other side by the electric field. If the junction is uniformly illuminated by photons with $hv > E_g$, an added generation rate g_{op} (EHP/cm^3-s) participates in this current (Fig. 6-10). The number of holes created per second within a diffusion length of the transition region on the n side is $AL_p g_{op}$. Similarly $AL_n g_{op}$ electrons are generated per second within L_n of x_{p0}, and $AW g_{op}$ carriers are generated within W. The resulting current due to collection of these optically generated carriers by the junction is

$$I_{op} = qAg_{op}(L_p + L_n + W) \qquad (6\text{-}3)$$

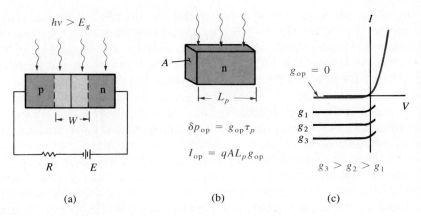

$hv > E_g$

$g_{op} = 0$

g_1

g_2

g_3

$g_3 > g_2 > g_1$

$\delta p_{op} = g_{op}\tau_p$

$I_{op} = qAL_p g_{op}$

(a)   (b)   (c)

Figure 6-10
Optical generation of carriers in a p-n junction: (a) absorption of light by the device; (b) current I_{op} resulting from EHP generation within a diffusion length of the junction on the n side; (c) I–V characteristics of an illuminated junction.

Although Eq. (6-3) illustrates the concept of photocurrent, it is uncommon to have a uniform optical generation rate g_{op} throughout the device. More commonly, the optical excitation enters from one side of the junction, and the absorption is described by Eq. (4-2).

Since the current of Eq. (6-3) is directed from n to p, it subtracts from the total current from p to n, and the diode equation [Eq. (5-36)] must be modified accordingly. Neglecting recombination and generation within W to be consistent with Eq. (5-36), the I–V relationship of an illuminated diode is

$$I = qA\left(\frac{L_p}{\tau_p}p_n + \frac{L_n}{\tau_n}n_p\right)(e^{qV/kT} - 1) - qAg_{op}(L_p + L_n) \qquad (6-4)$$

Thus the I–V curve is lowered by an amount proportional to the generation rate (Fig. 6-10c). This equation can be considered in two parts — the current described by the usual diode equation, and the current due to optical generation.

When the device is short circuited ($V = 0$), the terms from the diode equation cancel in Eq. (6-4), as expected. However, there is a short-circuit current from n to p equal to I_{op}. Thus the I–V characteristics of Fig. 6-10c cross the I-axis at negative values proportional to g_{op}. When there is an open circuit across the device, $I = 0$ and the voltage $V = V_{oc}$ is

$$V_{oc} = \frac{kT}{q}\ln\left[\frac{L_p + L_n}{(L_p/\tau_p)p_n + (L_n/\tau_n)n_p}\cdot g_{op} + 1\right] \qquad (6-5)$$

For the special case of a symmetrical junction, $p_n = n_p$ and $\tau_p = \tau_n$, we can rewrite Eq. (6-5) in terms of the thermal generation rate $p_n/\tau_n = g_{th}$ and the optical generation rate g_{op}:

$$V_{oc} \simeq \frac{kT}{q}\ln\frac{g_{op}}{g_{th}} \qquad \text{for } g_{op} \gg g_{th} \qquad (6-6)$$

Actually, the term $g_{th} = p_n/\tau_n$ represents the *equilibrium* thermal generation–recombination rate. As the minority carrier concentration is increased by optical generation of EHPs, the lifetime τ_n becomes shorter, and p_n/τ_n becomes larger (p_n is fixed, for a given N_d and T). Therefore, V_{oc} cannot increase indefinitely with increased generation rate; in fact, the limit on V_{oc} is the equilibrium contact potential V_0 (Fig. 6-11). This result is to be expected, since the contact potential is the maximum forward bias that can appear across a junction. The appearance of a forward voltage across an illuminated junction is known as the *photovoltaic effect*.

Depending on the intended application, the photodiode of Fig. 6-10 can be operated in either the third or fourth quarters of its I–V characteristic. As Fig. 6-12 illustrates, power is delivered to the device from the external circuit when the current and junction voltage are both positive or both negative (first or third quadrants). In the fourth quadrant, however, the junction voltage is positive and the current is negative. In this case power is delivered from the junction to the external circuit (notice that in the fourth quadrant the current flows from the negative side of V to the positive side, as in a battery).

If power is to be extracted from the device, the fourth quadrant is used; on the other hand, in applications as a photodetector we usually reverse bias the

Figure 6-11
Effects of illumination on the open circuit voltage of a junction: (a) junction at equilibrium; (b) appearance of a voltage V_{oc} with illumination.

Figure 6-12
Operation of an illuminated junction in the various quadrants of its I–V characteristic; in (a) and (b) power is delivered to the device by the external circuit; in (c) the device delivers power to the load.

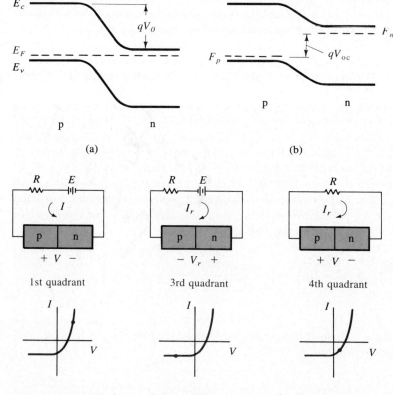

junction and operate it in the third quadrant. We shall investigate these applications more closely in the discussion to follow.

6.3.2 Solar Cells

Since power can be delivered to an external circuit by an illuminated junction, it is possible to convert solar energy into electrical energy. If we consider the fourth quadrant of Fig. 6-12c, it appears doubtful that much power can be delivered by an individual device. The voltage is restricted to values less than the contact potential, which in turn is generally less than the band gap voltage E_g/q. For Si the voltage V_{oc} is less than about 1 V. The current generated depends on the illuminated area, but typically I_{op} is in the 10-100 mA range for a junction with an area of about 1 cm^2. However, if many such devices are used, the resulting power can be significant. In fact, arrays of p-n junction solar cells are currently used to supply electrical power for many space satellites. Solar cells can supply power for the electronic equipment aboard a satellite over a long period of time, which is a distinct advantage over batteries. The array of junctions can be distributed over the surface of the satellite or can be contained in solar cell "paddles" attached to the main body of the satellite (Fig. 6-13).

To utilize a maximum amount of available optical energy, it is necessary to design a solar cell with a large area junction located near the surface of the device (Fig. 6-14). The planar junction is formed by diffusion or ion implantation, and the surface is coated with appropriate materials to reduce reflection and to decrease surface recombination. Many compromises must be made in solar cell design. In the device shown in Fig. 6-14, for example, the junction depth d must be less than L_p in the n material to allow holes generated near the

Figure 6-13
Solar cell arrays attached to the Mars fly-by satellite *Mariner*. (Provided through the courtesy of the National Aeronautics and Space Administration, California Institute of Technology, Jet Propulsion Laboratory.)

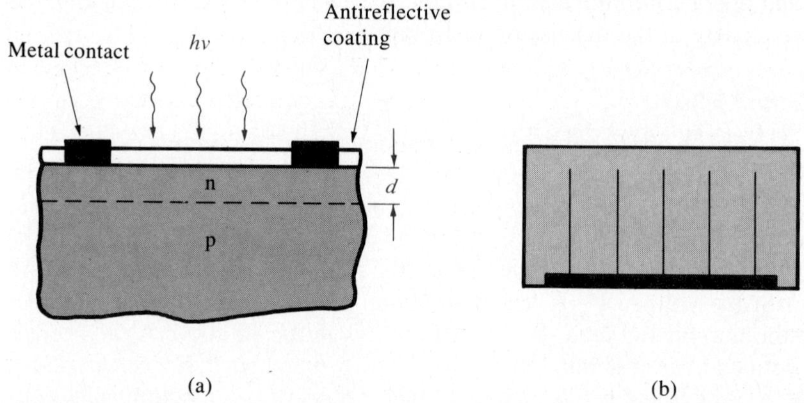

Metal contact hv Antireflective coating

n

d

p

(a)

(b)

Figure 6-14
Configuration of a solar cell: (a) enlarged view of the planar junction; (b) top view, showing metal contact "fingers."

surface to diffuse to the junction before they recombine; similarly, the thickness of the p region must be such that electrons generated in this region can diffuse to the junction before recombination takes place. This requirement implies a proper match between the electron diffusion length L_n, the thickness of the p region, and the mean optical penetration depth $1/\alpha$ [see Eq. (4-2)]. It is desirable to have a large contact potential V_0 to obtain a large photovoltage, and therefore heavy doping is indicated; on the other hand, long lifetimes are desirable and these are reduced by doping too heavily. It is important that the series resistance of the device be very small so that power is not lost to heat due to ohmic losses in the device itself. A series resistance of only a few ohms can seriously reduce the output power of a solar cell (Prob. 6.14). Since the area is large, the resistance of the p-type body of the device can be made small. However, contacts to the thin n region require special design. If this region is contacted at the edge, current must flow along the thin n region to the contact, resulting in a large series resistance. To prevent this effect, the contact can be distributed over the n surface by providing small contact "fingers" as in Fig. 6-14b. These narrow contacts serve to reduce the series resistance without interfering appreciably with the incoming light.

Figure 6-15 shows the fourth-quadrant portion of a solar cell characteristic, with I_r plotted upward for convenience of illustration. The open-circuit voltage

Figure 6-15
I–V characteristic of an illuminated solar cell. The maximum power rectangle is shaded.

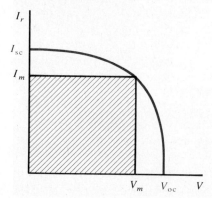

V_{oc} and the short-circuit current I_{sc} are determined for a given light level by the cell properties. The maximum power delivered to a load by this solar cell occurs when the product VI_r is a maximum. Calling these values of voltage and current V_m and I_m, we can see that the maximum delivered power illustrated by the shaded rectangle in Fig. 6-15 is less than the $I_{sc}V_{oc}$ product. The ratio $I_m V_m / I_{sc} V_{oc}$ is called the *fill factor,* and is a figure of merit for solar cell design.

Applications of solar cells are not restricted to outer space. It is possible to obtain useful power from the sun in terrestrial applications using solar cells, even though the solar intensity is reduced by the atmosphere. About 1 kW/m² is available in a particularly sunny location, but not all of this solar power can be converted to electricity. Much of the photon flux is at energies less than the cell band gap, and is not absorbed. High-energy photons are strongly absorbed, and the resulting EHPs may recombine at the surface. A well-made Si cell can have about 10 percent efficiency for solar energy conversion, providing approximately 100 W/m² of electrical power under full illumination. This is a modest amount of power per unit solar cell area, considering the effort involved in fabricating a large area of Si cells. One approach to obtaining more power per cell is to focus considerable light onto the cell using mirrors. Although Si cells lose efficiency at the resulting high temperatures, GaAs and related compounds can be used at 100°C or higher. In such solar concentrator systems more effort and expense can be put into the solar cell fabrication, since fewer cells are required. For example, a GaAs–AlGaAs heterojunction cell provides good conversion efficiency and operates at the elevated temperatures common in solar concentrator systems.

6.3.3   Photodetectors

When the photodiode is operated in the third quadrant of its *I–V* characteristic (Fig. 6-12b), the current is essentially independent of voltage but is proportional to the optical generation rate. Such a device provides a useful means of measuring illumination levels or of converting time-varying optical signals into electrical signals.

In most optical detection applications the detector's speed of response is critical. For example, if the photodiode is to respond to a series of light pulses 1 ns apart, the photogenerated minority carriers must diffuse to the junction and be swept across to the other side in a time much less than 1 ns. The carrier diffusion step in this process is time consuming and should be eliminated if possible. Therefore, it is desirable that the width of the depletion region W be large enough so that most of the photons are absorbed within W rather than in the neutral p and n regions. When an EHP is created in the depletion region, the electric field sweeps the electron to the n side and the hole to the p side. Since this carrier drift occurs in a very short time, the response of the photodiode can be quite fast. When the carriers are generated primarily within the depletion layer W, the detector is called a *depletion layer photodiode*. Obviously, it is desirable to dope at least one side of the junction lightly so that W can be made large. The appropriate width for W is chosen as a compromise between sensitivity and speed of response. If W is wide, most of the incident photons will be absorbed in the depletion region. Also, a wide W results in a small junction capacitance [see Eq. (5-62)], thereby reducing the RC time constant of the detector circuit. On the other hand, W must not be so wide that the time required for drift of photogenerated carriers out of the depletion region is excessive.

One convenient method of controlling the width of the depletion region is to build a *p-i-n photodetector* (Fig. 6-16). The "i" region need not be truly intrinsic, as long as the resistivity is high. It can be grown epitaxially on the n-type substrate, and the p region can be obtained by diffusion. When this device is reverse biased, the applied voltage appears almost entirely across the i region. If the carrier lifetime within the i region is long compared with the drift time, most of the photogenerated carriers will be collected by the n and p regions.

If low-level optical signals are to be detected, it is often desirable to operate the photodiode in the avalanche region of its characteristic. In this mode each photogenerated carrier results in a significant change in the current because of avalanche multiplication. In the *avalanche photodiode* the junction must be uniform, and a guard ring (Fig. 6-2) is generally used to ensure against edge breakdown. With proper design a Si avalanche photodiode can have high sensitivity to low-level optical signals, and the response time is in the neighborhood of 1 ns. These devices are particularly useful in fiber optic communication systems (Section 6.4.2).

The type of photodiode described here is sensitive to photons with energies near the band gap energy (*intrinsic* detectors). If $h\nu$ is less than E_g, the photons will not be absorbed; on the other hand, if the photons are much more energetic than E_g, they will be absorbed very near the surface, where the recombination rate is high. Therefore, it is necessary to choose a photodiode material with a band gap corresponding to a particular region of the spectrum. Detectors sensitive to longer wavelengths can be designed such that photons can excite electrons into or out of impurity levels (*extrinsic* detectors). However, the sensitivity of such extrinsic detectors is much less than intrinsic detectors, where electron–hole pairs are generated by excitation across the band gap.

By using lattice-matched multilayers of compound semiconductors, the band gap of the absorbing region can be tailored to match the wavelength of

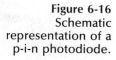

Figure 6-16
Schematic representation of a p-i-n photodiode.

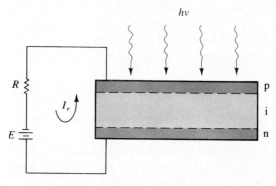

light being detected. Wider band gap material can then be used as a window through which the light is transmitted to the absorbing region (Fig. 6-17). For example, we saw in Fig. 1-15 that InGaAs with an In mole fraction of 53 percent can be grown epitaxially on InP with excellent lattice matching. This composition of InGaAs has a band gap of about 0.75 eV, which is sensitive to a useful wavelength for fiber optic systems. (1.55 μm), as we shall see in Section 6.4.2. In making a photodiode using InGaAs as the active material, it is possible to bring the light through the wider band gap InP (1.35 eV), thus greatly reducing surface recombination effects. In the case of avalanche photodiodes requiring narrow band gap material, it is often advantageous to absorb the light in the narrow-gap semiconductor (e.g., InGaAs) and transport the resulting carriers to a junction made in wider band gap material (e.g., InP), where the avalanche multiplication takes place (Fig. 6-17b). Such a separation of the absorption and multiplication regions avoids the excessive leakage currents typical of reverse-biased junctions in narrow-gap materials.

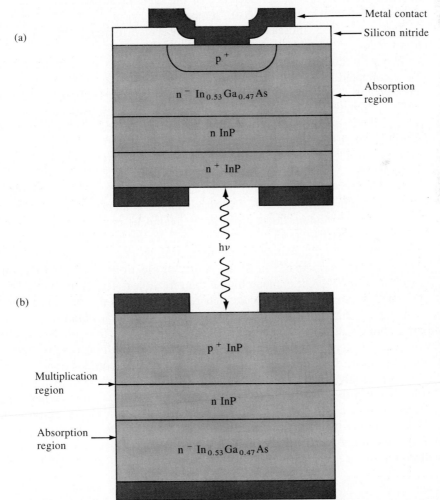

(a)

Metal contact
Silicon nitride

p^+

n^- In$_{0.53}$Ga$_{0.47}$As

Absorption region

n InP

n^+ InP

$h\nu$

(b)

Multiplication region

Absorption region

p^+ InP

n InP

n^- In$_{0.53}$Ga$_{0.47}$As

Figure 6-17
Use of multilayer heterojunctions to enhance photodiode operation: (a) a p-i-n photodiode in which light near 1.55 μm is absorbed in a narrow band gap material (InGaAs, E_g = 0.75 eV) after passing through a wider-gap material (InP, E_g = 1.35 eV); (b) an avalanche photodiode in which light is absorbed in the InGaAs and holes are swept to an InP junction, where the avalanche multiplication takes place. This separation of the absorption and multiplication regions reduces the junction leakage current. In this figure, n$^-$ refers to lightly doped n-type material.

6.3.4 Noise and Bandwidth of Photodetectors

In optical communication systems the sensitivity of the photodetector and its response time are of critical importance. Unfortunately, these two properties are generally difficult to optimize without making compromises between them. For example, in a photoconductor the gain depends on the ratio of carrier lifetime to transit time (see Prob. 4-11). On the other hand, the frequency response (and therefore the bandwidth) varies inversely with carrier lifetime. As a result, trade-offs must be made between these two desirable characteristics. It is common to express the *gain-bandwidth product* as a figure of merit for detectors. Designs which increase gain tend to decrease bandwidth and vice versa. Another important property of detectors is the *signal-to-noise ratio,* which is the amount of usable information compared with the background noise in the detector.

In the case of photoconductors the major source of noise is random fluctuations in the dark current (called *Johnson noise*). The noise current increases with temperature and the conductance of the material in the dark. Therefore, the photoconductor noise at a given temperature can be reduced by increasing the dark resistance. Increased dark resistance also increases the gain of the photoconductor, thereby decreasing the bandwidth.

In a p-i-n diode there is no gain mechanism, since at most one electron-hole pair is collected by the junction for each photon absorbed. Thus the gain is essentially unity, and the gain-bandwidth product is determined by the bandwidth, or frequency response. In a p-i-n the response time is dependent on the width of the depletion region, and the main source of noise is random thermal generation of EHPs within this region (called *shot noise*). The noise in a p-i-n device is considerably lower than that in a photoconductor, which compensates for the lack of gain in the p-i-n.

Avalanche photodiodes have the advantage of providing gain through the avalanche multiplication effect. The disadvantage is increased noise relative to the p-i-n, due to random fluctuations in the avalanche process. This noise is reduced if the impact ionization in the high field region is due to only one type of carrier, since more fluctuations in the ionization process occur when both electrons and holes participate. In Si the ability of electrons to create EHPs in an impact ionization event is much higher than for holes. Therefore, Si avalanche photodiodes can be operated with high gain and relatively low noise. On the other hand, ionization rates for the two carriers are fairly equal in most compound semiconductors. One creative way of overcoming this problem is shown in Fig. 6-18. By using a multilayer heterostructure with successive layers of AlGaAs graded from GaAs to wider band gap AlGaAs compositions, it is possible to take advantage of the wider step in the conduction band than in the valence band in the GaAs-AlGaAs heterojunction (see Fig. 3-13). When a large reverse bias is applied to this diode, electrons enter each GaAs region high in the conduction band with enough energy to cause impact ionization. In contrast, holes enter the GaAs regions with less energy because the valence band discontinuity is less. As a result, the avalanche multiplication in this diode is

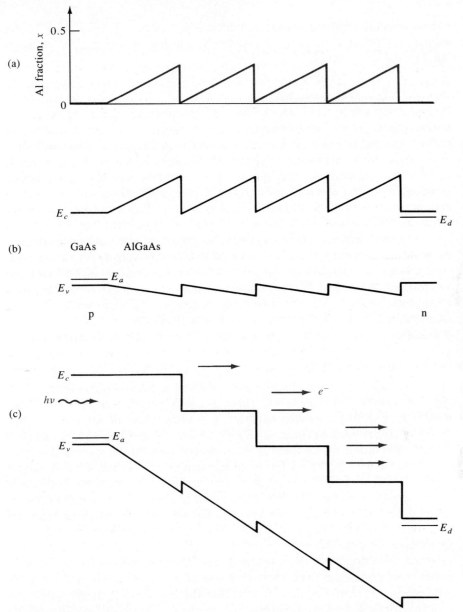

Figure 6-18
A multilayer structure for an avalanche photodiode, which is designed to selectively enhance impact ionization by electrons: (a) variation of the composition in the $Al_xGa_{1-x}As$ regions; (b) the resulting band diagram shows ramps in the AlGaAs conduction bands, with much smaller variations in the valence bands; (c) the conduction bands under strong reverse bias resemble successive steps over which electrons are swept at high energy, resulting in successive regions of avalanche multiplication. The multiplication by holes (swept in the opposite direction) is much less in this example.

due primarily to electrons. In Fig. 6-18, an electron entering the first GaAs layer causes multiplication, and the process continues through the structure. This device is similar to the vacuum-tube *photomultiplier,* in which incoming photons eject electrons from a metal electrode, and these electrons are accelerated to a second electrode where they eject more electrons, and so forth. Obviously, the gain in the device of Fig. 6-18 can be high, and the noise is reduced compared with ordinary avalanche photodiodes; however, the frequency response (and thus the bandwidth) is reduced, as expected. It takes time for the avalanche process to build up and to decay, and higher gain results in longer buildup and decay times.

6.4
LIGHT-EMITTING
DIODES AND
LASERS

When carriers are injected across a forward-biased junction, the current is usually accounted for by recombination in the transition region and in the neutral regions near the junction. In a semiconductor with an indirect band gap, such as Si or Ge, the recombination releases heat to the lattice. On the other hand, in a material charaterized by direct recombination, considerable light may be given off from the junction under forward bias. This effect, called *injection electroluminescence* (Section 4.2.3), provides an important application of diodes as generators of light. The use of light-emitting diodes (LEDs) in digital displays is well known. There are also other important applications in communications and other areas. Another important device making use of radiative recombination in a forward-biased p-n junction is the *semiconductor laser*. As we shall see in Chapter 10, lasers emit coherent light in much narrower wavelength bands than LEDs, and with more collimation (directionality). We will leave the details of laser operation for Chapter 10, but at this point we can discuss both LEDs and lasers as applications of injection electroluminescence.

6.4.1 Light-Emitting Materials

The band gaps of various binary compound semiconductors are illustrated in Fig. 4-4 relative to the spectrum. There is a wide variation in band gaps and, therefore, in available photon energies, extending from the ultraviolet (ZnS, 3.6 eV) into the infrared (InSb, 0.18 eV). In fact, by utilizing ternary and quaternary compounds the number of available energies can be increased significantly (see Figs 1-15 and 3-6). A good example of the variation in photon energy obtainable from the compound semiconductors is the ternary alloy gallium arsenide–phosphide, which is illustrated in Fig. 6-19. When the percentage of As is reduced and P is increased in this material, the resulting band gap varies from the direct 1.43-eV gap of GaAs (infrared) to the indirect 2.26-eV gap of GaP (green). The band gap of $GaAs_{1-x}P_x$ varies almostly linearly with x until the 0.45 composition is reached, and electron–hole recombination is direct over this range. The most common alloy composition used in LED displays is $x \simeq 0.4$. For this composition the band gap is direct, since the Γ minimum (at $\mathbf{k} = 0$) is the lowest part of the conduction band. This results in efficient radiative recombination, and the emitted photons (~ 1.9 eV) are in the red portion of the spectrum. This composition is used for red LEDs common in calculators and other displays.

For $GaAs_{1-x}P_x$ with P concentrations above 45 percent, the band gap is due to the indirect X minimum. Radiative recombination in such indirect materials is generally unlikely, because electrons in the conduction band have different momentum from holes in the valence band (see Fig. 3-5). Interestingly, however, indirect $GaAs_{1-x}P_x$ (including GaP, $x = 1$) doped with nitrogen can be used in LEDs with light output in the yellow to green portions of the spectrum. This is possible because the nitrogen impurity binds an electron very tightly.

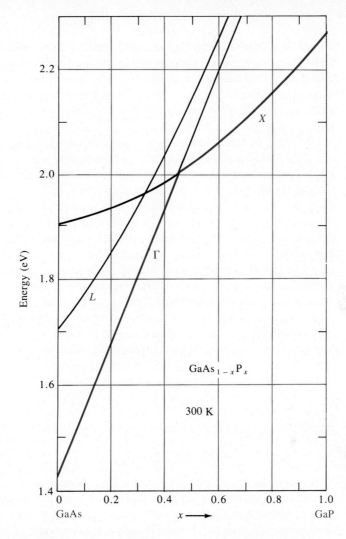

Figure 6-19
Conduction band energies as a function of alloy composition for $GaAs_{1-x}P_x$.

This confinement in real space (Δx) means that the electron momentum is spread out in momentum space Δp by the Heisenberg uncertainty principle (see Eq. 2-18). As a result, the momentum conservation rules, which generally prevent radiative recombination in indirect materials, are circumvented. Thus nitrogen doping of $GaAs_{1-x}P_x$ is not only useful technologically, but also provides an interesting and practical illustration of the uncertainty principle.

In many applications light from a laser or an LED need not be visible to the eye. Infrared emitters such as GaAs, InP, and mixed alloys of these compounds are particularly well suited to optical communication systems. For example, a laser or light-emitting diode can be used in conjunction with a photodiode or other photosensitive device to transmit information optically between locations. By varying the current through the diode, the light output can be modulated such that analog or digital information appears in the optical signal directed at the detector. Alternatively, the information may be introduced between the source and detector. For example, a semiconductor laser-photodetector arrangement can be used in a compact disc system for reading digital information from the spinning disc. A light emitter and a photodiode form an *optoelectronic pair,* which provides complete electrical isolation between input and output, since the only link between the two devices is optical. In an *optoelectronic isolator,* both devices may be mounted on a ceramic substrate and packaged together to form a unit that passes information while maintaining isolation.

In view of the broad range of applications requiring semiconductor lasers and LEDs with visible and infrared wavelengths, the wide variety of available III–V materials is extremely useful. Even more wavelengths would be accessible if II–VI materials could be used in LEDs and lasers.

Unfortunately, the II-VI compound semiconductors cannot generally be doped p-type or n-type at will (see Appendix III). In these materials, crystal defects are formed in the doping process, such that self-compensation occurs. For example, CdS and CdSe can be made n-type but cannot be converted to p-type by typical counterdoping with acceptor atoms. In the process of counter-doping, a lattice defect with donor properties is created for each acceptor atom introduced. Thus each acceptor state is automatically compensated by a donor state, and the Fermi level remains essentially unchanged by the doping process. This difficulty makes it impossible to form the usual p-n junction structure necessary for a luminescent diode. In some materials which do not allow p-n junction formation, a junction can be formed between the sample and a metal or another semiconductor such that avalanche breakdown or tunneling of carriers through a thin barrier can occur. With such methods enough carriers can be injected into the semiconductor to cause luminescence, although the efficiency of the resulting device is generally poor compared with a p-n junction.

6.4.2 Fiber Optic Communications

The transmission of optical signals from source to detector can be greatly enhanced if an *optical fiber* is placed between the light source and the detector. An optical fiber is essentially a "light pipe" or waveguide for optical frequencies. The fiber is typically drawn from a boule of glass to a diameter of ~ 25 μm. The fine glass fiber is relatively flexible and can be used to guide optical signals over distances of kilometers without the necessity of perfect alignment between source and detector. This significantly increases the applications of optical communication in areas such as telephone and data transmission.

One type of optical fiber has an outer layer of very pure fused silica (SiO$_2$), with a core of germanium doped glass having a higher index of refraction (Fig. 6-20a).[†] Such a *step-index* fiber maintains the light beam primarily in the central core with little loss at the surface. The light is transmitted along the length of the fiber by internal reflection at the step in the refractive index.

[†]The *index of refraction* (or *refractive index*) **n** compares the velocity of light **v** in the material to its velocity c in a vacuum, $\mathbf{n} = c/\mathbf{v}$. Thus if $\mathbf{n}_1 > \mathbf{n}_2$ in Fig. 6-20a, the light velocity is greater in material 2 than in 1. The value of **n** varies somewhat with the wavelength of light.

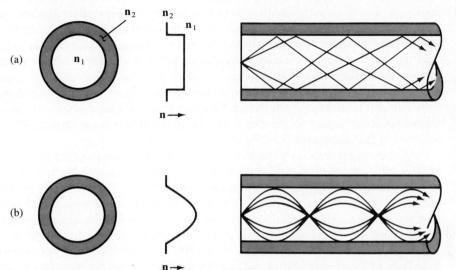

Figure 6-20
Two examples of multimode fibers: (a) *step-index*, having a core with slightly larger refractive index **n**; (b) *graded-index*, having in this case a parabolic grading of **n** in the core. The figure illustrates the cross section (left) of the fiber, its index of refraction profile (center), and typical mode patterns (right).

Losses in the fiber at a given wavelength can be described by an attenuation coefficient α [similar to the absorption coefficient of Eq. (4-3)]. The intensity of the signal at a distance x along the fiber is then related to the starting intensity by the usual expression,

$$\mathbf{I}(x) = \mathbf{I}_0 e^{-\alpha x} \qquad (6\text{-}7)$$

The attenuation is not the same for all wavelengths, however, and it is therefore important to choose a signal wavelength carefully. A plot of α vs. λ for a typical silica glass fiber is shown in Fig. 6-21. It is clear that dips in α near 1.3 and 1.55 μm provide "windows" in the attenuation, which can be exploited to reduce the degradation of signals. The overall decrease in absorption with increasing wavelength is due to the reduced scattering from small random inhomogenieties which result in fluctuations of the refractive index on a scale

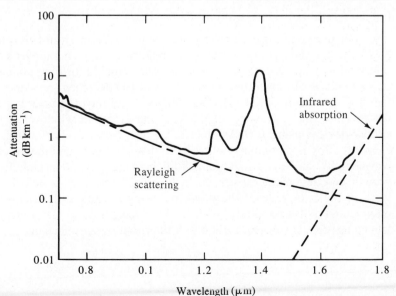

Figure 6-21
Typical plot of attenuation coefficient α vs. wavelength λ for a fused silica optical fiber. Peaks are due primarily to OH$^-$ impurities.

comparable to the wavelength. This type of attenuation, called *Rayleigh scattering,* decreases with the fourth power of wavelength. This effect is observed at sunrise and sunset, when attenuation of short wavelength blue and green light results in red and orange sunlight. Obviously, Rayleigh scattering encourages operation at long wavelengths in fiber optic systems. However, a competing process of infrared absorption dominates for wavelengths longer than about 1.7 μm, due to vibrational excitation of the atoms making up the glass. Therefore, a useful minimum in absorption for silica fibers occurs at about 1.55 μm, where epitaxial layers in the (In, Ga) (As, P) system can be grown lattice-matched to InP substrates (see Fig. 1-15).

Another consideration in choice of operating wavelength is the *pulse dispersion,* or spreading of data pulses as they propagate down the fiber. This effect can be caused by the frequency dependence of the refractive index, causing different optical frequencies to travel down the fiber with slightly different velocities. This effect, called *chromatic dispersion,* is much less pronounced at the 1.3 μm window in Fig. 6-21. Another cause of dispersion is the fact that different modes propagate with different path lengths (Fig. 6-20a). This type of dispersion can be reduced by grading the refractive index of the core (Fig. 6-20b) such that various modes are continually refocused, reducing the differences in path lengths.

In early optoelectronic systems for fiber optics, it was most convenient to use the well-established GaAs–AlGaAs system for making lasers and LEDs. These light sources are very efficient, and good detectors can be made using Si p-i-n or avalanche photodiodes. However, these sources operate in the wavelength range near 0.9 μm, where the attenuation is greater than for longer wavelengths. Modern systems, therefore, operate near the 1.3- and 1.55-μm minima in Fig. 6-21. At these wavelengths, sources can be made using InGaAs or InGaAsP grown on InP, and detectors can be made of the same materials or using Ge.

6.4.3 Multilayer Heterojunctions for LEDs

The light source in a fiber optic system may be a laser or an LED. In the case of a laser, the light is of essentially a single frequency and allows a very large information bandwidth. Semiconductor lasers suitable for fiber optic communications will be discussed in Chapter 10. An LED designed for a fiber optic system is illustrated in Fig. 6-22. The LED is a multilayer structure of GaAs and AlGaAs. To take advantage of the 1.3- and 1.55-μm windows in Fig. 6-21, similar devices using InGaAs or InGaAsP can be used. The quaternary (four-element) alloy is particularly suitable, in that band gap (and therefore emission wavelength) can be adjusted along with choosing lattice constants for epitaxial growth on convenient substrates. In Fig. 6-22 the fiber is held in an etched well on the back side of the diode by an epoxy resin. This configuration, often called a "Burrus diode" after its developer, is particularly convenient for launching signals from an LED into a fiber, with good mechanical stability.

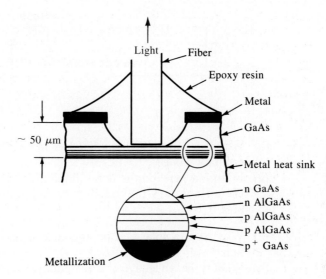

Figure 6-22
Cross section of a
GaAs–AlGaAs LED
for fiber–optic
applications. [After
C. A. Burrus and
B. I. Miller, *Optics
Communications,*
vol. 4, p. 307
(1971).]

Although LEDs are less suited to transmission of digital information than are lasers, they are easily modulated by analog signals. The optical power emitted by a properly constructed LED varies linearly with the input current over a wide range. An LED is an *incoherent* light source, in that photons are emitted randomly from the junction in all directions and not in phase with each other. Therefore, transmission of LED-generated signals inherently involves many modes, as in Fig. 6-20. *Multimode* fibers are larger ($\sim$25 μm in diameter) than are *single-mode* fibers ($\sim$5 μm), which transmit a coherent laser beam.

By forming numerous optical fibers into a bundle, with an appropriate jacket for mechanical strength, an enormous amount of information can be transmitted over long distances.[†] Depending upon the losses in the fibers, repeater stations may be required periodically along the path. Thus many photodetectors and LED or laser sources are required in a fiber optic system. Semiconductor device development, including appropriate binary, ternary, and quaternary compounds for both emitters and detectors, is therefore crucial to the successful implementation of such optical communication systems.

PROBLEMS

6.1. Assume $v_s = 3 \sin \omega t$ in the circuits shown in Fig. P6-1. The diodes are ideal, with *I–V* characteristics given by Fig. 6-1a. Sketch $v_0(t)$ over a full cycle for each circuit.

[†]Transmission rates of many G-bit/s have been achieved. As a convenient calibration of this rate, it is worth noting that the human eye is able to transmit about one G-bit/s to the brain.

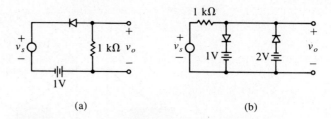

(a)   (b)

Figure P6-1

6.2. (a) Why is a narrow base diode faster in switching operations than an ordinary diode?

(b) In a p^+-n junction the donor doping on the n side varies as $N_d = Gx^2$. How does the junction capacitance C_j vary with reverse bias V_r?

6.3. (a) Why must a solar cell be operated in the fourth quadrant of the junction I–V characteristic?

(b) What is the advantage of a quaternary alloy in fabricating LEDs for fiber optics?

(c) Why is a reverse-biased GaAs p-n junction not a good photodetector for light of $\lambda = 1$ μm?

6.4. It is found experimentally that the avalanche breakdown voltage V_{br} is 60 V for an abrupt Si p^+-n junction with $N_d = 10^{16}$ cm^{-3}. What is the minimum thickness of the n region (between the metallurgical junction and the ohmic contact) required to ensure avalanche breakdown rather than punch-through?

6.5. Assume that a p^+-n diode is built with an n region width l smaller than a hole diffusion length ($l < L_p$). This is the so-called *narrow base diode*. Since for this case holes are injected into a short n region under forward bias, we cannot use the assumption $\delta p(x_n = \infty) = 0$ in Eq. (4-35). Instead, we must use as a boundary condition the fact that $\delta p = 0$ at $x_n = l$.

(a) Solve the diffusion equation to obtain

$$\delta p(x_n) = \frac{\Delta p_n [e^{(l-x_n)/L_p} - e^{(x_n - l)/L_p}]}{e^{l/L_p} - e^{-l/L_p}}$$

(b) Show that the current in the diode is

$$I = \left(\frac{qAD_p p_n}{L_p} \text{ ctnh } \frac{l}{L_p}\right)(e^{qV/kT} - 1)$$

6.6. Given the narrow base diode result (Prob. 6.5), (a) calculate the current due to recombination in the n region, and (b) show that the current due to recombination at the ohmic contact is

$$I(\text{ohmic contact}) = \left(\frac{qAD_p p_n}{L_p} \text{ csch } \frac{l}{L_p}\right)(e^{qV/kT} - 1)$$

6.7.   Assume that a p^{+}-n junction is built with a graded n region in which the doping is described by $N_d(x) = Gx^m$. The depletion region ($W \cong x_{n0}$) extends from essentially the junction at $x = 0$ to a point W within the n region. The singularity at $x = 0$ for negative **m** can be neglected.

   (a) Integrate Gauss's law across the depletion region to obtain the maximum value of the electric field $\mathcal{E}_0 = -qGW^{(m+1)}/\epsilon(m+1)$.

   (b) Find the expression for $\mathcal{E}(x)$ and use the result to obtain $V_0 - V = qGW^{(m+2)}/\epsilon(m+2)$.

   (c) Find the charge Q due to ionized donors in the depletion region; write Q explicitly in terms of $(V_0 - V)$.

   (d) Using the results of (c), take the derivative $dQ/d(V_0 - V)$ to show that the capacitance is

$$C_j = A \left[\frac{qG\epsilon^{(m+1)}}{(m + 2)(V_0 - V)} \right]^{1/(m+2)}$$

6.8.   What determines the peak tunneling voltage V_p of a tunnel diode? Explain.

6.9.   If a large density of trapping centers is present in a tunnel diode (Fig. P6-9), tunneling can occur from the n-side conduction band to the trapping level (A–B). Then the electrons may drop to the valence band on the p side (B–C), thereby completing a two-step process of charge transport across the junction. In fact, if the density of trapping centers is large, it is possible to observe an increase in current as the states below E_{Fn} pass by the trapping level with increased bias. In Fig. P6-9, the trapping level E_t is located 0.3 eV above the valence band. Assume $E_g = 1$ eV, and $E_{Fn} - E_c$ on the n side equals $E_v - E_{Fp}$ on the p side, equals 0.1 eV.

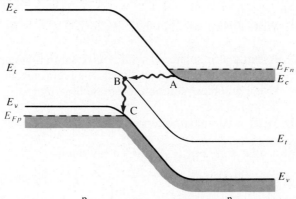

Figure P6-9

   (a) Calculate the minimum forward bias at which tunneling through E_t occurs.

   (b) Calculate the maximum forward bias for tunneling via E_t.

   (c) Sketch the I–V curve for this tunnel diode. Assume the maximum tunneling current via E_t is about one-third of the peak band-to-band tunneling current.

6.10. For steady state optical excitation, we can write the hole diffusion equation as

$$D_p \frac{d^2 \, \delta p}{dx^2} = \frac{\delta p}{\tau_p} - g_{op}$$

Assume that a long p^+-n diode is uniformly illuminated by an optical signal, resulting in g_{op} EHP/cm^3-s.

(a) Show that the excess hole distribution in the n region is

$$\delta p(x_n) = \left[p_n(e^{qV/kT} - 1) - g_{op} \frac{L_p^2}{D_p} \right] e^{-x_n/L_p} + \frac{g_{op} L_p^2}{D_p}$$

(b) Calculate the hole diffusion current $I_p(x_n)$ and evaluate it at $x_n = 0$. Compare the result with Eq. (6-4) evaluated for a p^+-n junction.

6.11. A Si solar cell has a short-circuit current of 100 mA and an open-circuit voltage of 0.8 V under full solar illumination. The fill factor is 0.7. What is the maximum power delivered to a load by this cell?

6.12. The maximum power delivered by a solar cell can be found by maximizing the I–V product.

(a) Show that maximizing the power leads to the expression

$$\left(1 + \frac{q}{kT} V_{mp} \right) e^{qV_{mp}/kT} = 1 + \frac{I_{sc}}{I_{th}}$$

where V_{mp} is the voltage for maximum power, I_{sc} is the magnitude of the short-circuit current, and I_{th} is the thermally induced reverse saturation current.

(b) Write this equation in the form $\ln x = C - x$ for the case $I_{sc} \gg I_{th}$, and $V_{mp} \gg kT/q$.

(c) Assume a Si solar cell with a dark saturation current I_{th} of 1.5 nA is illuminated such that the short-circuit current is $I_{sc} = 100$ mA. Use a graphical solution to obtain the voltage V_{mp} at maximum delivered power.

(d) What is the maximum power output of the cell at this illumination?

6.13. For a solar cell, Eq. (6-4) can be rewritten

$$V = \frac{kT}{q} \ln \left(1 + \frac{I_{sc} + I}{I_{th}} \right)$$

Given the cell parameters of Prob. 6.12, plot the I–V curve as in Fig. 6-15 and draw the maximum power rectangle. Remember that I is a negative number but is plotted positive as I_r in the figure. I_{th} and I_{sc} are positive magnitudes in the equation.

6.14. A major problem with solar cells is internal resistance, generally in the thin region at the surface which must be only partially contacted, as in Fig. 6-14. Assume that the cell of Prob. 6.12 has a series resistance of 1 Ω, so that the cell voltage is reduced by the IR drop. Replot the I–V curve for this case and compare with the cell of Prob. 6.13.

6.15. Based upon Fig. 1-15, what ternary alloy, composition, and binary substrate can be used for an LED at the 1.55-μm optical fiber window? What type of epitaxial layer/substrate combination would you use for an LED with emission at 1.3 μm?

*6.16. Solar cells are severely degraded by unwanted series resistance. For the cell described in Prob. 6-14, include a series resistance R, which reduces the cell voltage by the amount IR. Calculate and plot the fill factor for a series resistance R from 0 to 5 Ω, and comment on the effect of R on cell efficiency.

READING LIST

Capasso, F. "Band-Gap Engineering: From Physics and Materials to New Semiconductor Devices." *Science* 235 (January 9, 1987): 172–6.

Coutts, T. J., and J. D. Meakin. *Current Topics in Photovoltaics.* New York: Academic, 1985.

Drummond, T. J., P. L. Gourley, and T. E. Zipperian. "Quantum-Tailored Solid-State Devices." *IEEE Spectrum* (June 1988): 33–37.

Esaki, L. "Discovery of the Tunnel Diode." *IEEE Transactions on Electron Devices.* ED-23, no. 7 (July 1976): 644–7.

Fischetti, M. A. "Photovoltaic-cell Technologies Joust for Position." *IEEE Spectrum* 21, no. 3 (March 1984): 40–47.

Forrest, S. R. "Optical Detectors: Three Contenders." *IEEE Spectrum* 23, no. 5 (May 1986): 76–84.

Miller, S. E., and I. P. Kaminow, eds. *Optical Fiber Telecommunications II.* San Diego: Academic Press, 1988.

Special Issue on Optoelectronics, *Physics Today,* 38, no. 5 (May 1985): 23–64.

Senior, J. *Optical Fiber Communications.* Englewood Cliffs, N.J.: Prentice Hall International, 1985.

Sze, S. M. *Physics of Semiconductor Devices.* New York: John Wiley, 1981.

Sze, S. M. *Semiconductor Devices.* New York: John Wiley, 1985.

Williams, E. W. *Luminescence and the Light-Emitting Diode.* Oxford: Pergamon Press, 1978.

Wilson, J. and J. F. B. Hawkes. *Optoelectronics.* 2nd ed. Cambridge: Prentice Hall International, 1989.

Yang, E. S. *Microelectronic Devices.* New York: McGraw-Hill, 1988.

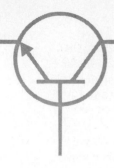

chapter 7

BIPOLAR JUNCTION TRANSISTORS

The modern era of semiconductor electronics was ushered in by the invention of the bipolar transistor in 1948 by Bardeen, Brattain, and Shockley at the Bell Telephone Laboratories. This device, along with its field-effect counterpart, has had an enormous impact on virtually every area of modern life. In this chapter we will learn about the operation, applications, and fabrication of this basic electronic device.

In Chapter 5 we found that two dominant features of p-n junctions are the injection of minority carriers with forward bias and the variation of the depletion width W with reverse bias. These two p-n junction properties are used in two important types of transistors. The *bipolar junction transistor (BJT)* discussed in this chapter uses the injection of minority carriers across a forward-biased junction, and the *junction field effect transistor (JFET)* discussed in Chapter 8 depends on control of a junction depletion width under reverse bias. The FET is a majority carrier device, and is therefore often called a *unipolar* transistor. The BJT, on the other hand, operates by the injection and collection of *minority* carriers. Since the action of both electrons and holes is important in this device, it is called a *bipolar* transistor.

We begin this chapter with a general discussion of amplification and switching, the basic circuit functions performed by transistors. This is followed by a qualitative discussion of charge transport in a BJT, to establish a sound physical understanding of its operation. Then we shall investigate carefully the charge distributions in the transistor and relate the three terminal currents to the physical characteristics of the device. Our aim is to gain a solid understanding of the current flow and control of the transistor and to discover the most important secondary effects that influence its operation. We shall discuss the properties of the transistor with proper biasing for amplification and then consider the effects of more general biasing, as encountered in switching circuits.

In this chapter we shall use the p-n-p transistor for most illustrations. The main advantage of the p-n-p for discussing transistor action is that hole flow and current are in the same direction. This makes the various mechanisms of charge transport somewhat easier to visualize in a preliminary explanation. Once these basic ideas are established for the p-n-p device, it is simple to relate them to the more widely used transistor, the n-p-n.

The transistor is a three-terminal device with the important feature that the current through two terminals can be controlled by small changes we make in the current or voltage at the third terminal. This control feature allows us to amplify small a-c signals or to switch the device from an *on* state to an *off* state and back. These two operations, *amplification* and *switching,* are the basis of a host of electronic functions. This section provides a brief introduction to these operations, as a foundation for understanding both bipolar and field-effect transistors.

7.1
**AMPLIFICATION
AND SWITCHING**

7.1.1 The Load Line

Consider a two-terminal device that has a nonlinear *I–V* characteristic, as in Fig. 7-1. We might determine this curve experimentally by measuring the current for various applied voltages, or by using an oscilloscope called a *curve tracer,* which varies *I* and *V* repetitively and displays the resulting curve. When such a device is biased with the simple battery–resistor combination shown in the figure, steady state values of I_D and V_D are attained. To find these values we begin by writing a loop equation around the circuit:[†]

$$E = i_D R + v_D \tag{7-1}$$

This gives us one equation describing the circuit, but it contains two unknowns (i_D and v_D). Fortunately, we have another equation of the form $i_D = f(v_D)$ in the

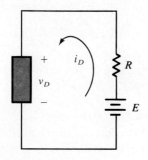

(a)

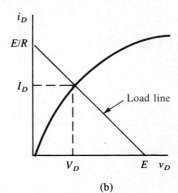

(b)

Figure 7-1
A two-terminal nonlinear device: (a) biasing circuit; (b) *I–V* characteristic and load line.

[†]We use i_D to symbolize the total current, I_D for the d-c value, and i_d for the a-c component. A similar scheme is used for other currents and voltages.

curve of Fig. 7-1b, giving us two equations with two unknowns. The steady state current and voltage are found by a simultaneous solution of these two equations. However, since one equation is analytical and the other is graphical, we must first put them into the same form. It is easy to make the linear equation (7-1) graphical, so we plot it on Fig. 7-1b to find the simultaneous solution. The end points of the line described by Eq. (7-1) are at E when $i_D = 0$ and at E/R when $v_D = 0$. The two graphs cross at $v_D = V_D$ and $i_D = I_D$, the steady state values of current and voltage for the device with this biasing circuit.

Now let's add a third terminal which somehow controls the I–V characteristic of the device. For example, assume that the device current–voltage curve can be moved up the current axis by increasing the control current i_T as in Fig. 7-2b. This results in a family of i_D–v_D curves, depending upon the choice of i_T. We can still write the loop equation (7-1) and draw it on the set of curves, but now the simultaneous solution depends on the value of i_T. In the example of Fig. 7-2, I_T is 0.1 mA and the d-c values of I_D and V_D are found at the intersection to be 10 mA and 5 V, respectively. Whatever the value of the control current i_T at the third terminal, values of I_D and V_D are obtained from points along the line representing Eq. (7-1). This is called the *load line*.

7.1.2 Amplification

If an a-c source is added to the control current, we can achieve large variations in i_D by making small changes in i_T. For example, as i_T varies about its d-c value by 0.05 mA in Fig. 7-2, i_d varies about its d-c value I_D by 2 mA. Thus the amplification of the a-c signal is $2/0.05 = 40$. If the curves for equal changes in i_T are equally spaced on the i_D axis, a faithful amplified version of the small control signal can be obtained. This type of current-controlled amplification is typical of bipolar transistors, and is widely used in transistor circuits.

Figure 7-2
A three-terminal nonlinear device that can be controlled by the current at the third terminal i_T: (a) biasing circuit; (b) *I–V* characteristic and load line. If $I_T = 0.1$ mA, the d-c values of I_D and V_D are as shown by the dashed lines.

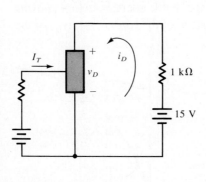

(a)

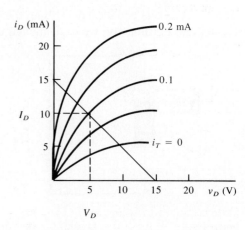

(b)

7.1.3   Switching

Another important circuit function of transistors is the controlled switching of the device off and on. In the example of Fig. 7-2, we can switch from the bottom of the load line ($i_D = 0$) to almost the top ($i_D \simeq E/R$) by appropriate changes in i_T. This type of switching with control at a third terminal is particularly useful in digital circuits. Later in this chapter we will deal with the details of such switching with bipolar transistors.

The bipolar transistor is basically a simple device, and this section is devoted to a simple and largely qualitative view of BJT operation. We will deal with the details of these transistors in following sections, but first we must define some terms and gain physical understanding of how carriers are transported through the device. Then we can discuss how the current through two terminals can be controlled by small changes in the current at a third terminal.

7.2

FUNDAMENTALS OF BJT OPERATION

7.2.1   Charge Transport in a BJT

Let us begin the discussion of bipolar transistors by considering the reverse-biased p-n junction diode of Fig. 7-3. According to the theory of Chapter 5, the reverse saturation current through this diode depends on the rate at which minority carriers are generated in the neighborhood of the junction. We found, for example, that the reverse current due to holes being swept from n to p is essentially independent of the size of the junction $\mathscr{E}$ field and hence is independent of the reverse bias. The reason given was that the hole current depends on how often minority holes are generated by EHP creation within a diffusion length of the junction—not upon how fast a particular hole is swept across the depletion layer by the field. As a result, it is possible to increase the reverse current through the diode by increasing the rate of EHP generation (Fig. 7-3b).

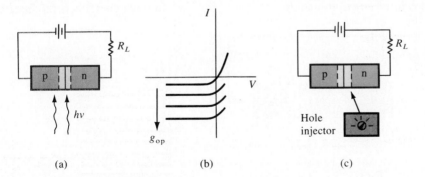

(a)   (b)   (c)

External control of the current in a reverse-biased p-n junction: (a) optical generation; (b) junction I–V characteristics as a function of EHP generation; (c) minority carrier injection by a hypothetical device.

Figure 7-3

One convenient method for accomplishing this is optical excitation of EHPs with light ($hv > E_g$), as in the photodetector of Section 6.3. With steady photoexcitation the reverse current will still be essentially independent of bias voltage, and if the dark saturation current is negligible, the reverse current is directly proportional to the optical generation rate g_{op} [see Eq. (6-3)].

The example of external control of current through a junction by optical generation raises an interesting question: Is it possible to inject minority carriers into the neighborhood of the junction *electrically* instead of optically? If so, we could control the junction reverse current simply by varying the rate of minority carrier injection. For example, let us consider a hypothetical *hole injection device* as in Fig. 7-3c. If we can inject holes at a predetermined rate into the n side of the junction, the effect on the junction current will resemble the effects of optical generation. The current from n to p will depend on the hole injection rate and will be essentially independent of the bias voltage. There are several obvious advantages to such external control of a current; for example, the current through the reverse-biased junction would vary very little if the load resistor R_L were changed, since the magnitude of the junction voltage is relatively unimportant. Therefore, such an arrangement should be a good approximation to a controllable constant current source.

A convenient hole injection device is a forward-biased p^+-n junction. According to Section 5.3.2, the current in such a junction is due primarily to holes injected from the p^+ region into the n material. If we make the n side of the forward-biased junction the same as the n side of the reverse-biased junction, the p^+-n-p structure of Fig. 7-4 results. With this configuration, injection of holes from the p^+-n junction into the center n region supplies the minority carrier holes to participate in the reverse current through the n-p junction. Of course, it is important that the injected holes do not recombine in the n region before they can diffuse to the depletion layer of the reverse-biased junction. Thus we must make the n region narrow compared with a hole diffusion length.

The structure we have described is a p-n-p bipolar junction transistor. The forward-biased junction which injects holes into the center n region is called the *emitter junction,* and the reverse-biased junction which collects the injected

Figure 7-4
A p-n-p transistor:
(a) schematic
representation of a
p-n-p device with a
forward-biased
emitter junction
and a
reverse-biased
collector junction;
(b) *I–V*
characteristics of
the reverse-biased
n-p junction as a
function of emitter
current.

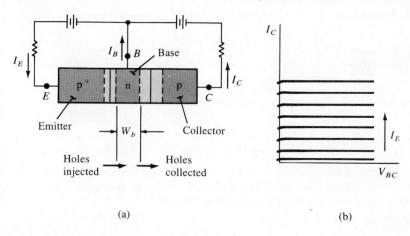

(a)   (b)

holes is called the *collector junction*. The p^+ region, which serves as the source of injected holes, is called the *emitter*, and the p region into which the holes are swept by the reverse-biased junction is called the *collector*. The center n region is called the *base*, for reasons which will become clear in Section 7.3, when we discuss the historical development of transistor fabrication. The biasing arrangement in Fig. 7-4 is called the *common base* configuration, since the base electrode B is common to the emitter and collector circuits.

To have a good p-n-p transistor, we would prefer that almost all the holes injected by the emitter into the base be collected. Thus the n-type base region should be narrow, and the hole lifetime τ_p should be long. This requirement is summed up by specifying $W_b \ll L_p$, where W_b is the length of the *neutral* n material of the base (measured between the depletion regions of the emitter and collector junctions), and L_p is the diffusion length for holes in the base $(D_p\tau_p)^{1/2}$. With this requirement satisfied, an average hole injected at the emitter junction will diffuse to the depletion region of the collector junction without recombination in the base. A second requirement is that the current I_E crossing the emitter junction should be composed almost entirely of holes injected into the base, rather than electrons crossing from base to emitter. This requirement is satisfied by doping the base region lightly compared with the emitter, so that the p^+-n emitter junction of Fig. 7-4 results.

It is clear that current I_E flows into the emitter of a properly biased p-n-p transistor and that I_C flows out at the collector, since the direction of hole flow is from emitter to collector. However, the base current I_B requires a bit more thought. In a good transistor the base current will be very small since I_E is essentially hole current, and the collected hole current I_C is almost equal to I_E. There must be some base current, however, due to requirements of electron flow into the n-type base region (Fig. 7-5). We can account for I_B physically by three dominant mechanisms:

(a) There must be some recombination of injected holes with electrons in the base, even with $W_b \ll L_p$. The electrons lost to recombination must be resupplied through the base contact.

(b) Some electrons will be injected from n to p in the forward biased emitter junction, even if the emitter is heavily doped compared to the base. These electrons must also be supplied by I_B.

(c) Some electrons are swept into the base at the reverse-biased collector junction due to thermal generation in the collector. This small current reduces I_B by supplying electrons to the base.

The dominant sources of base current are (a) recombination in the base and (b) injection into the emitter region. Both of these effects can be greatly reduced by device design, as we shall see. In a well-designed transistor, I_B will be a very small fraction (perhaps one-hundredth) of I_E.

In an n-p-n transistor the three current directions are reversed, since electrons flow from emitter to collector and holes must be supplied to the base.

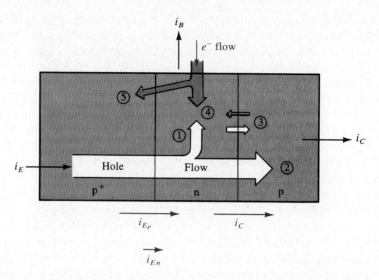

Figure 7-5 Summary of hole and electron flow in a p-n-p transistor with proper biasing: (1) injected holes lost to recombination in the base; (2) holes reaching the reverse-biased collector junction; (3) thermally generated electrons and holes making up the reverse saturation current of the collector junction; (4) electrons supplied by the base contact for recombination with holes; (5) electrons injected across the forward-biased emitter junction.

The physical mechanisms for operation of the n-p-n can be understood simply by reversing the roles of electrons and holes in the p-n-p discussion.

7.2.2 Amplification with BJTs

In this section we shall discuss rather simply the various factors involved in transistor amplification. Basically, the transistor is useful in amplifiers because the currents at the emitter and collector are controllable by the relatively small base current. The essential mechanisms are easy to understand if various secondary effects are neglected. We shall use total current (d-c plus a-c) in this discussion, with the understanding that the simple analysis applies only to d-c and to small-signal a-c at low frequencies. We can relate the terminal currents of the transistor i_E, i_B, and i_C by several important factors. In this introduction we shall neglect the saturation current at the collector (Fig. 7-5, component 3) and such effects as recombination in the transition regions. Under these assumptions, the collector current is made up entirely of those holes injected at the emitter which are not lost to recombination in the base. Thus i_C is proportional to the hole component of the emitter current i_{Ep}:

$$i_C = Bi_{Ep} \tag{7-2}$$

The proportionality factor B is simply the fraction of injected holes which make it across the base to the collector; B is called the *base transport factor*. The total emitter current i_E is made up of the hole component i_{Ep} and the electron component i_{En}, due to electrons injected from base to emitter (component 5 in Fig. 7-5). The *emitter injection efficiency* γ is

$$\gamma = \frac{i_{Ep}}{i_{En} + i_{Ep}} \qquad\qquad (7\text{-}3)$$

For an efficient transistor we would like B and γ to be very near unity; that is, the emitter current should be due mostly to holes ($\gamma \simeq 1$), and most of the injected holes should eventually participate in the collector current ($B \simeq 1$). The relation between the collector and emitter currents is

$$\frac{i_C}{i_E} = \frac{Bi_{Ep}}{i_{En} + i_{Ep}} = B\gamma \equiv \alpha \qquad\qquad (7\text{-}4)$$

The product $B\gamma$ is defined as the factor α, called the *current transfer ratio*, which represents the emitter-to-collector current amplification. There is no real amplification between these currents, since α is smaller than unity. On the other hand, the relation between i_C and i_B is more promising for amplification.

In accounting for the base current, we must include the rates at which electrons are lost from the base by injection across the emitter junction (i_{En}) and the rate of electron recombination with holes in the base. In each case, the lost electrons must be resupplied through the base current i_B. If the fraction of injected holes making it across the base *without* recombination is B, then it follows that $(1 - B)$ is the fraction *recombining* in the base. Thus the base current is

$$i_B = i_{En} + (1 - B)i_{Ep} \qquad\qquad (7\text{-}5)$$

neglecting the collector saturation current. The relation between the collector and base currents is found from Eqs. (7-2) and (7-5):

$$\frac{i_C}{i_B} = \frac{Bi_{Ep}}{i_{En} + (1 - B)i_{Ep}} = \frac{B[i_{Ep}/(i_{En} + i_{Ep})]}{1 - B[i_{Ep}/(i_{En} + i_{Ep})]}$$

$$\frac{i_C}{i_B} = \frac{B\gamma}{1 - B\gamma} = \frac{\alpha}{1 - \alpha} \equiv \beta \qquad\qquad (7\text{-}6)$$

The factor β relating the collector current to the base current is the *base-to-collector current amplification factor.*[†] Since α is near unity, it is clear that β can be large for a good transistor, and the collector current is large compared with the base current.

It remains to be shown that the collector current i_C can be controlled by variations in the small current i_B. In the discussion to this point, we have indicated the control of i_C by the emitter current i_E, with the base current characterized as a small side effect. In fact, we can show from space charge neutrality

---

[†]α is also called the *common-base current gain*. β is also called the *common-emitter current gain*.

arguments that i_B can indeed be used to determine the magnitude of i_C. Let us consider the transistor of Fig. 7-6, in which i_B is determined by a biasing circuit. For simplicity, we shall assume unity emitter injection efficiency and negligible collector saturation current. Since the n-type base region is electrostatically neutral between the two transition regions, the presence of excess holes in transit from emitter to collector calls for compensating excess electrons from the base contact. However, there is an important difference in the times which electrons and holes spend in the base. The average excess hole spends a time τ_t, defined as the *transit time* from emitter to collector. Since the base width W_b is made small compared with L_p, this transit time is much less than the average hole lifetime τ_p in the base.[†] On the other hand, an average excess electron supplied from the base contact spends τ_p seconds in the base supplying space charge neutrality during the lifetime of an average excess

Figure 7-6
Example of amplification in a common-emitter transistor circuit: (a) biasing circuit; (b) addition of an a-c variation of base current i_b to the d-c value of I_B, resulting in an a-c component i_c.

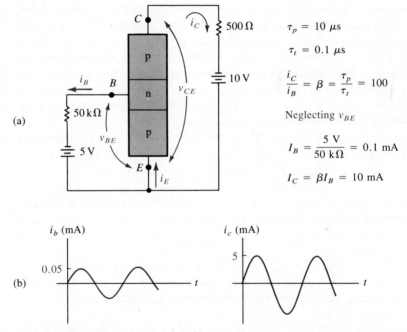

$$\tau_p = 10 \ \mu s$$

$$\tau_t = 0.1 \ \mu s$$

$$\frac{i_C}{i_B} = \beta = \frac{\tau_p}{\tau_t} = 100$$

Neglecting v_{BE}

$$I_B = \frac{5 \text{ V}}{50 \text{ k}\Omega} = 0.1 \text{ mA}$$

$$I_C = \beta I_B = 10 \text{ mA}$$

[†]This difference between average hole lifetime before recombination (τ_p) and the average time a hole spends in transit across the base (τ_t) may be confusing at first. How can the lifetime be longer than the time a hole actually spends in transit? The answer depends on the fact that holes are indistinguishable in the recombination kinetics. Think of an analogy with a shooting gallery, in which a good marksman fires slowly at a line of quickly moving ducks. Although many individual ducks make it across the firing line without being hit, the lifetime of an *average* duck within the firing line is determined by the time between shots. We can speak of the lifetime of an average duck because they are essentially indistinguishable. Similarly, the rate of recombination in the base (and therefore i_B) depends on the average lifetime τ_p and the distribution of the indistinguishable holes in the base region.

hole. While the average electron waits τ_p seconds for recombination, many individual holes can enter and leave the base region, each with an average transit time τ_t. In particular, for each electron entering from the base contact, τ_p/τ_t holes can pass from emitter to collector while maintaining space charge neutrality. Thus the ratio of collector current to base current is simply

$$\frac{i_C}{i_B} = \beta = \frac{\tau_p}{\tau_t} \tag{7-7}$$

for $\gamma = 1$ and negligible collector saturation current.

If the electron supply to the base (i_B) is restricted, the traffic of holes from emitter to base is correspondingly reduced. This can be argued simply by supposing that the hole injection does continue despite the restriction on electrons from the base contact. The result would be a net buildup of positive charge in the base and a loss of forward bias (and therefore a loss of hole injection) at the emitter junction. Clearly, the supply of electrons through i_B can be used to raise or lower the hole flow from emitter to collector.

The base current is controlled independently in Fig. 7-6. This is called a *common-emitter* circuit, since the emitter electrode is common to the base and collector circuits. The emitter junction is clearly forward biased by the battery in the base circuit. The voltage drop in the forward-biased emitter junction is small, however, so that almost all of the voltage from collector to emitter appears across the reverse-biased collector junction. Since v_{BE} is small for the forward-biased junction, we can neglect it and approximate the base current as 5 V/50 kΩ = 0.1 mA. If $\tau_p = 10$ μs and $\tau_t = 0.1$ μs, β for the transistor is 100 and the collector current I_C is 10 mA. It is important to note that i_C is determined by β and the base current, rather than by the battery and resistor in the collector circuit (as long as these are of reasonable values to maintain a reverse-biased collector junction). In this example 5 V of the collector circuit battery voltage appears across the 500 Ω resistor, and 5 V serves to reverse bias the collector junction.

If a small a-c current i_b is superimposed on the steady state base current of Fig. 7-6a, a corresponding a-c current i_C appears in the collector circuit. The time-varying portion of the collector current will be i_b multiplied by the factor β, and current gain results.

We have neglected a number of important properties of the transistor in this introductory discussion, and many of these properties will be treated in detail below. We have established, however, the fundamental basis of operation for the bipolar transistor and have indicated in a simplified way how it can be used to produce current gain in an electronic circuit.

(a) Show that Eq. (7-7) is valid from arguments of the steady state replacement of stored charge. Assume that $\tau_n = \tau_p$.

(b) What is the steady state charge $Q_n = Q_p$ due to excess electrons and holes in the neutral base region for the transistor of Fig. 7-6?

EXAMPLE 7-1

SOLUTION

(a) In steady state there are excess electrons and holes in the base. The charge in the electron distribution Q_n is replaced every τ_p seconds. Thus $i_B = Q_n/\tau_p$. The charge in the hole distribution Q_p is collected every τ_t seconds, and $i_C = Q_p/\tau_t$. For space charge neutrality, $Q_n = Q_p$, and

$$\frac{i_C}{i_B} = \frac{Q_n/\tau_t}{Q_n/\tau_p} = \frac{\tau_p}{\tau_t}$$

(b) $Q_n = Q_p = i_C\tau_t = i_B\tau_p = 10^{-9}$ C.

■

7.3
BJT FABRICATION

The first commercial bipolar junction transistors were fabricated by alloying techniques. For example, alloyed p-type regions on opposite sides of an n-type Ge sample result in a p^+-n-p^+ structure. In the early versions of alloyed transistors, it was common practice to mount the semiconductor sample on a header and connect the emitter and collector alloyed regions by wires to the appropriate header posts. In this configuration the material common to the base region of the transistor provides mechanical support for the structure. In the point contact transistor, which preceded alloyed structures, the emitter and collector were sharp wires pressed upon the surface of a semiconductor which served as the base of the device. This is the origin of the term "base" as applied to the central region of a transistor. Although the base region is not used to provide mechanical support in modern transistors, the terminology is retained.

Although alloyed junction transistors served a useful purpose for many years, they have now been replaced with transistors fabricated by the more precise and convenient processes of diffusion and ion implantation. Most transistors are now made in Si, utilizing photolithography and oxide masking. The most common type of BJT is the n-p-n, which takes advantage of the higher mobility for electrons compared with holes. A simple example of an n-p-n diffused transistor is shown in Fig. 7-7. Beginning with an n-type Si substrate, an oxide layer is grown on the surface and a window is opened by the photolithographic techniques described in Section 5.1.3. The sample is placed in a furnace for a boron diffusion, forming the p-type base region. After reoxidation, a new window is opened in the oxide for a phosphorus diffusion, forming the n^+ emitter region. The time and temperature of each diffusion can be controlled quite accurately to ensure reproducible junction depths and base region thickness. Alternatively, ion implantation can be used to form either or both of these regions. After opening windows to the top surface of the p and n^+ regions, aluminum is evaporated onto the wafer. The final metallization pattern is defined by photolithography, and the unwanted Al is etched away. Several thousand such transistors can be made on a single wafer of Si and then separated by scribing and breaking into individual devices. In this geometry the n-type substrate is alloyed to a header to provide contact to the collector region, and Au or Al wires are bonded to the metallized regions to provide leads

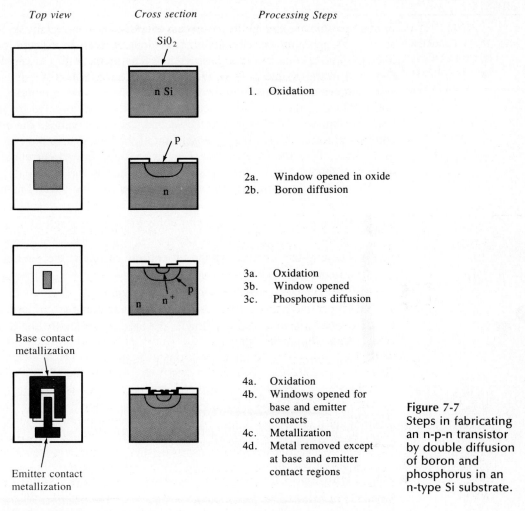

Top view	Cross section	Processing Steps
	SiO$_2$ n Si	1. Oxidation
	p n	2a. Window opened in oxide 2b. Boron diffusion
	n n$^+$ p	3a. Oxidation 3b. Window opened 3c. Phosphorus diffusion
Base contact metallization Emitter contact metallization		4a. Oxidation 4b. Windows opened for base and emitter contacts 4c. Metallization 4d. Metal removed except at base and emitter contact regions

Figure 7-7
Steps in fabricating an n-p-n transistor by double diffusion of boron and phosphorus in an n-type Si substrate.

to the emitter and base.[†] The final encapsulation can be a hermetically sealed header or a molded plastic case, depending on the power rating and environmental specifications.

Many alterations can be made in this basic process to improve the transistor characteristics. For example, the n-type collector region is often an epitaxial layer grown on an n$^+$ substrate. This reduces the series resistance of the collector. Special requirements exist for high-power and high-frequency devices (Sections 7.7 and 7.8). Further changes, such as ion implantation, are required in fabricating transistors in integrated circuits. In fact, many discrete (single-unit) transistors are made using the techniques of IC technology because of the requirements of mass production. We shall discuss these processes in Chapter 9, where IC fabrication will be treated in detail.

[†]Look ahead to Fig. 7-26(a).

In this section we examine the operation of a BJT in more detail. We begin our analysis by applying the techniques of previous chapters to the problem of hole injection into a narrow n-type base region. The mathematics is very similar to that used in the problem of the narrow base diode (Prob. 6.5). Basically, we assume holes are injected into the base at the forward-biased emitter, and these holes diffuse to the collector junction. The first step is to solve for the excess hole distribution in the base, and the second step is to evaluate the emitter and collector currents (I_E, I_C) from the gradient of the hole distribution on each side of the base. Then the base current (I_B) can be found from a current summation or from a charge control analysis of recombination in the base.

We shall at first simplify the calculations by making several assumptions:

1. Holes diffuse from emitter to collector; drift is negligible in the base region.
2. The emitter current is made up entirely of holes; the emitter injection efficiency is $\gamma = 1$.
3. The collector saturation current is negligible.
4. The active part of the base and the two junctions are of uniform cross-sectional area A; current flow in the base is essentially one-dimensional from emitter to collector.
5. All currents and voltages are steady state.

In later sections we shall consider the implications of imperfect injection efficiency, drift due to nonuniform doping in the base, structural effects such as different areas for the emitter and collector junctions, and capacitance and transit time effects in a-c operation.

7.4.1 Solution of the Diffusion Equation in the Base Region

Since the injected holes are assumed to flow from emitter to collector by diffusion, we can evaluate the currents crossing the two junctions by techniques used in Chapter 5. Neglecting recombination in the two depletion regions, the hole current entering the base at the emitter junction is the current I_E, and the hole current leaving the base at the collector is I_C. If we can solve for the distribution of excess holes in the base region, it is simple to evaluate the gradient of the distribution at the two ends of the base to find the currents. We shall consider the simplified geometry of Fig. 7-8, in which the base width is W_b between the two depletion regions, and the uniform cross-sectional area is A. The excess hole concentration at the edge of the emitter depletion region Δp_E and the corresponding concentration on the collector side of the base Δp_C are found from Eq. (5-29):

$$\Delta p_E = p_n(e^{qV_{EB}/kT} - 1) \tag{7-8a}$$

$$\Delta p_C = p_n(e^{qV_{CB}/kT} - 1) \tag{7-8b}$$

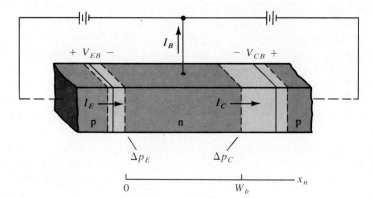

Figure 7-8
Simplified p-n-p transistor geometry used in the calculations.

If the emitter junction is strongly forward biased ($V_{EB} \gg kT/q$) and the collector junction is strongly reverse biased ($V_{CB} \ll 0$), these excess concentrations simplify to

$$\Delta p_E \simeq p_n e^{qV_{EB}/kT} \tag{7-9a}$$

$$\Delta p_C \simeq -p_n \tag{7-9b}$$

We can solve for the excess hole concentration as a function of distance in the base $\delta p(x_n)$ by using the proper boundary conditions in the diffusion equation, Eq. (4-34b):

$$\frac{d^2 \, \delta p(x_n)}{dx_n^2} = \frac{\delta p(x_n)}{L_p^2} \tag{7-10}$$

The solution of this equation is

$$\delta p(x_n) = C_1 e^{x_n/L_p} + C_2 e^{-x_n/L_p} \tag{7-11}$$

where L_p is the diffusion length of holes in the base region. Unlike the simple problem of injection into a long n region, we cannot eliminate one of the constants by assuming the excess holes disappear for large x_n. In fact, since $W_b \ll L_p$ in a properly designed transistor, most of the injected holes reach the collector at W_b. The solution is very similar to that of the narrow base diode problem. In this case the appropriate boundary conditions are

$$\delta p(x_n = 0) = C_1 + C_2 = \Delta p_E \tag{7-12a}$$

$$\delta p(x_n = W_b) = C_1 e^{W_b/L_p} + C_2 e^{-W_b/L_p} = \Delta p_C \tag{7-12b}$$

Solving for the parameters C_1 and C_2 we obtain

$$C_1 = \frac{\Delta p_C - \Delta p_E e^{-W_b/L_p}}{e^{W_b/L_p} - e^{-W_b/L_p}} \tag{7-13a}$$

$$C_2 = \frac{\Delta p_E e^{W_b/L_p} - \Delta p_C}{e^{W_b/L_p} - e^{-W_b/L_p}} \tag{7-13b}$$

These parameters applied to Eq. (7-11) give the full expression for the excess hole distribution in the base region. For example, if we assume that the collector junction is strongly reverse biased [Eq. (7-9b)] and the equilibrium hole concentration p_n is negligible compared with the injected concentration Δp_E, the excess hole distribution simplifies to

$$\delta p(x_n) = \Delta p_E \frac{e^{W_b/L_p} e^{-x_n/L_p} - e^{-W_b/L_p} e^{x_n/L_p}}{e^{W_b/L_p} - e^{-W_b/L_p}} \qquad \text{(for } \Delta p_C \simeq 0\text{)} \qquad (7\text{-}14)$$

The various terms in Eq. (7-14) are sketched in Fig. 7-9, and the corresponding excess hole distribution in the base region is demonstrated for a moderate value of W_b/L_p. Note that $\delta p(x_n)$ varies almost linearly between the emitter and collector junction depletion regions (Prob. 7.5). As we shall see below, the slight deviation from linearity of the distribution indicates the small value of I_B caused by recombination in the base region.

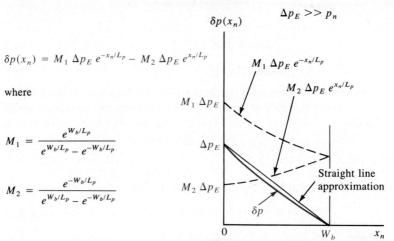

7.4.2 Evaluation of the Terminal Currents

Having solved for the excess hole distribution in the base region, we can evaluate the emitter and collector currents from the gradient of the hole concentration at each depletion region edge. From Eq. (4-22b) we have

$$I_p(x_n) = -qAD_p \frac{d\delta p(x_n)}{dx_n} \qquad (7\text{-}15)$$

This expression evaluated at $x_n = 0$ gives the hole component of the emitter current,

$$I_{Ep} = I_p(x_n = 0) = qA \frac{D_p}{L_p}(C_2 - C_1) \qquad (7\text{-}16)$$

Similarly, if we neglect the electrons crossing from collector to base in the collector reverse saturation current, I_C is made up of entirely of holes entering the collector depletion region from the base. Evaluating Eq. (7-15) at $x_n = W_b$, we have the collector current

$$I_C = I_p(x_n = W_b) = qA\frac{D_p}{L_p}(C_2 e^{-W_b/L_p} - C_1 e^{W_b/L_p}) \qquad (7\text{-}17)$$

When the parameters C_1 and C_2 are substituted from Eqs. (7-13), the emitter and collector currents take a form that is most easily written in terms of hyperbolic functions:

$$I_{Ep} = qA\frac{D_p}{L_p}\left[\frac{\Delta p_E(e^{W_b/L_p} + e^{-W_b/L_p}) - 2\Delta p_C}{e^{W_b/L_p} - e^{-W_b/L_p}}\right]$$

$$I_{Ep} = qA\frac{D_p}{L_p}\left(\Delta p_E\ \mathrm{ctnh}\ \frac{W_b}{L_p} - \Delta p_C\ \mathrm{csch}\ \frac{W_b}{L_p}\right) \qquad (7\text{-}18a)$$

$$I_C = qA\frac{D_p}{L_p}\left(\Delta p_E\ \mathrm{csch}\ \frac{W_b}{L_p} - \Delta p_C\ \mathrm{ctnh}\ \frac{W_b}{L_p}\right) \qquad (7\text{-}18b)$$

Now we can obtain the value of I_B by a current summation, noting that the sum of the base and collector currents leaving the device must equal the emitter current entering. If $I_E \simeq I_{Ep}$ for $\gamma \simeq 1$,

$$I_B = I_E - I_C = qA\frac{D_p}{L_p}\left[(\Delta p_E + \Delta p_C)\left(\mathrm{ctnh}\ \frac{W_b}{L_p} - \mathrm{csch}\ \frac{W_b}{L_p}\right)\right]$$

$$I_B = qA\frac{D_p}{L_p}\left[(\Delta p_E + \Delta p_C)\ \tanh\frac{W_b}{2L_p}\right] \qquad (7\text{-}19)$$

By using the techniques of Chapter 5 we have evaluated the three terminal currents of the transistor in terms of the material parameters, the base width, and the excess concentrations Δp_E and Δp_C. Furthermore, since these excess concentrations are related in a straightforward way to the emitter and collector junction bias voltages by Eq. (7-8), it should be simple to evaluate the transistor performance under various biasing conditions. It is important to note here that Eqs. (7-18) and (7-19) are not restricted to the case of the usual transistor biasing. For example, Δp_C may be $-p_n$ for a strongly reverse-biased collector, or it may be a significant positive number if the collector is positively biased. The generality of these equations will be used in Section 7.5 in considering the application of transistors to switching circuits.

EXAMPLE 7-2

(a) Find the expression for the current I for the transistor connection shown if $\gamma = 1$.

(b) How does the current I divide between the base lead and the collector lead?

SOLUTION

(a) Since $V_{CB} = 0$, Eq. (7-8b) gives $\Delta p_C = 0$.
Thus from Eq. (7-18a),

$$I_E = I = \frac{qAD_p}{L_p} \Delta p_E \text{ ctnh} \frac{W_b}{L_p}$$

similarly,

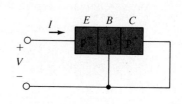

(b)
$$I_C = \frac{qAD_p}{L_p} \Delta p_E \text{ csch} \frac{W_b}{L_p}$$

$$I_B = \frac{qAD_p}{L_p} \Delta p_E \text{ tanh} \frac{W_b}{2L_p}$$

where I_C and I_B are the components in the collector lead and base lead, respectively. Note that these results are analogous to those of Probs. 6.5 and 6.6 for the narrow base diode.

7.4.3 Approximations of the Terminal Currents

The general equations of the previous section can be simplified for the case of normal transistor biasing, and such simplification allows us to gain insight into the current flow. For example, if the collector is reverse biased, $\Delta p_C = -p_n$ from Eq. (7-9b). Furthermore, if the equilibrium hole concentration p_n is small (Fig. 7-10a), we can neglect the terms involving Δp_C. For $\gamma \simeq 1$, the terminal currents reduce to those of Example 7-2:

$$I_E \simeq qA\frac{D_p}{L_p} \Delta p_E \text{ ctnh} \frac{W_b}{L_p} \tag{7-20a}$$

Figure 7-10
Approximate excess hole distributions in the base:
(a) forward-biased emitter, reverse-biased collector;
(b) triangular distribution for $V_{CB} = 0$ or for negligible p_n.

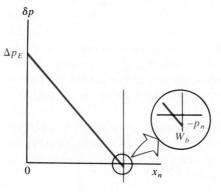

(a)

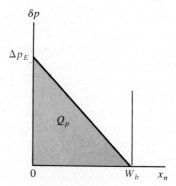

(b)

$$I_C \simeq qA \frac{D_p}{L_p} \Delta p_E \; \text{csch} \; \frac{W_b}{L_p} \tag{7-20b}$$

$$I_B \simeq qA \frac{D_p}{L_p} \Delta p_E \; \text{tanh} \; \frac{W_b}{2L_p} \tag{7-20c}$$

Series expansions of the hyperbolic functions are given in Table 7-1. For small values of W_b/L_p, we can neglect terms above the first order of the argument. It is clear from this table and Eq. (7-20) that I_C is only slightly smaller than I_E, as expected. The first-order approximation of tanh y is simply y, so that the base current is

$$I_B \simeq qA \frac{D_p}{L_p} \Delta p_E \frac{W_b}{2L_p} = \frac{qA W_b \, \Delta p_E}{2\tau_p} \tag{7-21}$$

Table 7-1. Expansions of several pertinent hyperbolic functions.

$$\text{sech} \; y = 1 - \frac{y^2}{2} + \frac{5y^4}{24} - \cdots$$

$$\text{ctnh} \; y = \frac{1}{y} + \frac{y}{3} - \frac{y^3}{45} + \cdots$$

$$\text{csch} \; y = \frac{1}{y} - \frac{y}{6} + \frac{7y^3}{360} - \cdots$$

$$\text{tanh} \; y = y - \frac{y^3}{3} + \cdots$$

The same approximate expression for the base current is found from the difference in the first-order approximations to I_E and I_C:

$$
\begin{aligned}
I_B &= I_E - I_C \\
&\simeq qA \frac{D_p}{L_p} \Delta p_E \left[\left(\frac{1}{W_b/L_p} + \frac{W_b/L_p}{3} \right) - \left(\frac{1}{W_b/L_p} - \frac{W_b/L_p}{6} \right) \right] \\
&\simeq \frac{qA D_p W_b \, \Delta p_E}{2L_p^2} = \frac{qA W_b \, \Delta p_E}{2\tau_p}
\end{aligned}
\tag{7-22}
$$

This expression for I_B accounts for recombination in the base region. We must include injection into the emitter in many BJT devices, as discussed in Section 7.4.4.

If recombination in the base dominates the base current, I_B can be obtained from the charge control model, assuming an essentially straight-line hole distribution in the base (Fig. 7-10b). Since the hole distribution diagram appears as a triangle in this approximation, we have

$$Q_p \simeq \tfrac{1}{2} qA \, \Delta p_E W_b \tag{7-23}$$

If we consider that this stored charge must be replaced every τ_p seconds and relate the recombination rate to the rate at which electrons are supplied by the base current, I_B becomes

$$I_B \simeq \frac{Q_p}{\tau_p} = \frac{qAW_b\,\Delta p_E}{2\tau_p} \tag{7-24}$$

which is the same as that found in Eqs. (7-21) and (7-22).

Since we have neglected the collector saturation current and have assumed $\gamma = 1$ in these approximations, the difference between I_E and I_C is accounted for by the requirements of recombination in the base. In Eq. (7-24) we have a clear demonstration that the base current is reduced for small W_b and large τ_p. We can increase τ_p by using light doping in the base region, which of course also improves the emitter injection efficiency.

The straight-line approximation of the excess hole distribution (Fig. 7-10) is fairly accurate in calculating the base current. On the other hand, it does not give a valid picture of I_E and I_C. If the distribution were perfectly straight, the slope would be the same at each end of the base region. This would imply zero base current, which is not the case. There must be some "droop" to the distribution, as in the more accurate curve of Fig. 7-9. This slight deviation from linearity gives a steeper slope at $x_n = 0$ than at $x_n = W_b$, and the value of I_E is larger than I_C by the amount I_B. The reason we can use the straight-line approximation in the charge control calculation of base current is that the area under the hole distribution curve is essentially the same in the two cases.

7.4.4 Current Transfer Ratio

The value of I_E calculated thus far in this section is more properly designated I_{Ep}, since we have assumed that $\gamma = 1$ (the emitter current due entirely to hole injection). Actually, there is always some electron injection across the forward-biased emitter junction in a real transistor, and this effect is important in calculating the current transfer ratio α. It is easy to show (Prob. 7.7) that the emitter injection efficiency of a p-n-p transistor can be written in terms of the emitter and base properties:

$$\gamma = \left[1 + \frac{L_p^n n_n \mu_n^p}{L_n^p p_p \mu_p^n}\tanh\frac{W_b}{L_p^n}\right]^{-1} \simeq \left[1 + \frac{W_b n_n \mu_n^p}{L_n^p p_p \mu_p^n}\right]^{-1} \tag{7-25}$$

In this equation we use superscripts to indicate which side of the emitter–base junction is referred to. For example, L_p^n is the hole diffusion length in the n-type base region and μ_n^p is the electron mobility in the p-type emitter region. In an n-p-n the superscripts and subscripts would be changed along with the majority carrier symbols. Using Eq. (7-20a) for I_{Ep}, and Eq. (7-20b) for I_C, the base transport factor B is

$$B = \frac{I_C}{I_{Ep}} = \frac{\text{csch } W_b/L_p}{\text{ctnh } W_b/L_p} = \text{sech}\frac{W_b}{L_p} \tag{7-26}$$

and the current transfer ratio α is the product of B and γ as in Eq. (7-4).

Assume that a p-n-p transistor is doped such that the emitter doping is ten times that in the base, the minority carrier mobility in the emitter is one-half that in the base, and the base width is one-tenth the minority carrier diffusion length. The carrier lifetimes are equal. Calculate α and β for this transistor.

EXAMPLE 7-3

From Eqs. (7-25) and (7-26), we have

SOLUTION

$$\alpha = B\gamma = \left[\cosh \frac{W_b}{L_p^n} + \frac{L_p^n n_n \mu_n^p}{L_n^p p_p \mu_p^n} \sinh \frac{W_b}{L_p^n}\right]^{-1}$$

Using the values given, and taking $L_p^n/L_n^p = \sqrt{\mu_p^n/\mu_n^p}$ for equal lifetimes,

$$\alpha = [\cosh 0.1 + \sqrt{2}\,(0.1)\,(0.5)\,\sinh 0.1]^{-1}$$

$$= [1.005 + 0.0707\,(0.1)]^{-1} = 0.988$$

We can find β from Eq. (7-6):

$$\beta = \frac{\alpha}{1 - \alpha} = 82$$

Thus an incremental change in I_B causes a significant change in I_C.

The expressions derived in Section 7.4 describe the terminal currents of the transistor, if the device geometry and other factors are consistent with the assumptions. Real transistors may deviate from these approximations, as we shall see in Section 7.7. The collector and emitter junctions may differ in area, saturation current, and other parameters, so that the proper description of the terminal currents may be more complicated than Eqs. (7-18) and (7-19) suggest. For example, if the roles of emitter and collector are reversed, these equations predict that the behavior of the transistor is symmetrical. Real transistors, on the other hand, are generally not symmetrical between emitter and collector. This is a particularly important consideration when the transistor is not biased in the usual way. We have discussed normal biasing (sometimes called the *normal active* mode), in which the emitter junction is forward biased and the collector is reverse biased. In some applications, particularly in switching, this normal biasing rule is violated. In these cases it is important to account for the differences in injection and collection properties of the two junctions. In this section we shall develop a generalized approach which accounts for transistor operation in terms of a coupled-diode model, valid for all combinations of emitter and collector bias. This model involves four measurable parameters that can be related to the geometry and material properties of the device. Using this model in conjunction with the charge control approach, we can describe the physical operation of a transistor in switching circuits and in other applications.

7.5 GENERALIZED BIASING

7.5.1 The Coupled-Diode Model

If the collector junction of a transistor is forward biased, we cannot neglect Δp_C; instead, we must use a more general hole distribution in the base region. Figure 7-11a illustrates a situation in which the emitter and collector junctions are both forward biased, so that Δp_E and Δp_C are positive numbers. We can handle this situation with Eqs. (7-18) and (7-19) for the symmetrical transistor. It is interesting to note that these equations can be considered as linear superpositions of the effects of injection by each junction. For example, the straight-line hole distribution of Fig. 7-11a can be broken into the two components of Figs. 7-11b and 7-11c. One component (Fig. 7-11b) accounts for the holes injected by the emitter and collected by the collector. We can call the resulting currents (I_{EN} and I_{CN}) the *normal mode* components, since they are due to injection from emitter to collector. The component of the hole distribution illustrated by Fig. 7-11c results in currents I_{EI} and I_{CI}, which describe injection in the *inverted mode* of injection from collector to emitter.[†] Of course, these inverted components will be negative, since they account for hole flow opposite to our original definitions of I_E and I_C.

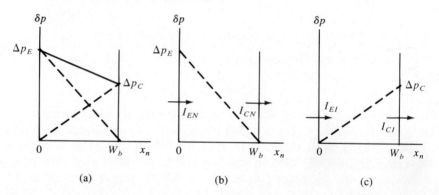

(a)   (b)   (c)

Figure 7-11 Evaluation of a hole distribution in terms of components due to normal and inverted modes: (a) approximate hole distribution in the base with emitter and collector junctions forward biased; (b) component due to injection and collection in the normal mode; (c) component due to the inverted mode.

For the symmetrical transistor, these various components are described by Eqs. (7-18). Defining $a \equiv (qAD_p/L_p)\ \text{ctnh}\ (W_b/L_p)$ and $b \equiv (qAD_p/L_p)\ \text{csch}\ (W_b/L_p)$, we have

$$I_{EN} = a\Delta p_E \quad \text{and} \quad I_{CN} = b\Delta p_E \qquad \text{with } \Delta p_C = 0 \qquad (7\text{-}27a)$$

$$I_{EI} = -b\Delta p_C \quad \text{and} \quad I_{CI} = -a\Delta p_C \qquad \text{with } \Delta p_E = 0 \qquad (7\text{-}27b)$$

The four components are combined by linear superposition in Eq. (7-18):

$$I_E = I_{EN} + I_{EI} = a\Delta p_E - b\Delta p_C$$
$$= \mathsf{A}(e^{qV_{EB}/kT} - 1) - \mathsf{B}(e^{qV_{CB}/kT} - 1) \qquad (7\text{-}28a)$$

[†]Here the words *emitter* and *collector* refer to physical regions of the device rather than to the functions of injection and collection of holes.

$$I_C = I_{CN} + I_{CI} = b\Delta p_E - a\Delta p_C$$
$$= \mathsf{B}(e^{qV_{EB}/kT} - 1) - \mathsf{A}(e^{qV_{CB}/kT} - 1) \qquad (7\text{-}28b)$$

where $\mathsf{A} \equiv ap_n$ and $\mathsf{B} \equiv bp_n$.

We can see from these equations that a linear superposition of the normal and inverted components does give the result we derived previously for the symmetrical transistor. To be more general, however, we must relate the four components of current by factors which allow for asymmetry in the two junctions. For example, the emitter current in the normal mode can be written

$$I_{EN} = I_{ES}(e^{qV_{EB}/kT} - 1), \qquad \Delta p_C = 0 \qquad (7\text{-}29)$$

where I_{ES} is the magnitude of the emitter saturation current in the normal mode. Since we specify $\Delta p_C = 0$ in this mode, we imply that $V_{CB} = 0$ in Eq. (7-8b). Thus we shall consider I_{ES} to be the magnitude of the emitter saturation current with the collector junction short circuited. Similarly, the collector current in the inverted mode is

$$I_{CI} = -I_{CS}(e^{qV_{CB}/kT} - 1), \qquad \Delta p_E = 0 \qquad (7\text{-}30)$$

where I_{CS} is the magnitude of the collector saturation current with $V_{EB} = 0$. As before, the minus sign associated with I_{CI} simply means that in the inverted mode holes are injected opposite to the defined direction of I_C.

The corresponding collected currents for each mode of operation can be written by defining a new α for each case:

$$I_{CN} = \alpha_N I_{EN} = \alpha_N I_{ES}(e^{qV_{EB}/kT} - 1) \qquad (7\text{-}31a)$$

$$I_{EI} = \alpha_I I_{CI} = -\alpha_I I_{CS}(e^{qV_{CB}/kT} - 1) \qquad (7\text{-}31b)$$

where α_N and α_I are the ratios of collected current to injected current in each mode. We notice that in the inverted mode the injected current is I_{CI} and the collected current is I_{EI}.

The total currents can again be obtained by superposition of the components:

$$I_E = I_{EN} + I_{EI} = I_{ES}(e^{qV_{EB}/kT} - 1) - \alpha_I I_{CS}(e^{qV_{CB}/kT} - 1) \qquad (7\text{-}32a)$$

$$I_C = I_{CN} + I_{CI} = \alpha_N I_{ES}(e^{qV_{EB}/kT} - 1) - I_{CS}(e^{qV_{CB}/kT} - 1) \qquad (7\text{-}32b)$$

These relations were derived by J. J. Ebers and J. L. Moll and are referred to as the *Ebers–Moll equations.*[†] While the general form is the same as Eqs. (7-28) for the symmetrical transistor, these equations allow for variations

[†]J. J. Ebers and J. L. Moll, "Large-Signal Behavior of Junction Transistors," *Proceedings of the IRE* 42, pp. 1761–72 (December 1954). In the original paper and in many texts, the terminal currents are all defined as flowing *into* the transistor. This introduces minus signs into the expressions for I_C and I_B as we have developed them here.

in I_{ES}, I_{CS}, α_I, and α_N due to asymmetry between the junctions. Although we shall not prove it here, it is possible to show by reciprocity arguments that

$$\alpha_N I_{ES} = \alpha_I I_{CS} \qquad (7\text{-}33)$$

even for nonsymmetrical transistors.

An interesting feature of the Ebers–Moll equations is that I_E and I_C are described by terms resembling diode relations (I_{EN} and I_{CI}), plus terms which provide coupling between the properties of the emitter and collector (I_{EI} and I_{CN}). This *coupled-diode* property is illustrated by the equivalent circuit of Fig. 7-12. In this figure we take advantage of Eq. (7-8) to write the Ebers–Moll equations in the form

$$I_E = I_{ES}\frac{\Delta p_E}{p_n} - \alpha_I I_{CS}\frac{\Delta p_C}{p_n} = \frac{I_{ES}}{p_n}(\Delta p_E - \alpha_N \Delta p_C) \qquad (7\text{-}34a)$$

$$I_C = \alpha_N I_{ES}\frac{\Delta p_E}{p_n} - I_{CS}\frac{\Delta p_C}{p_n} = \frac{I_{CS}}{p_n}(\alpha_I \Delta p_E - \Delta p_C) \qquad (7\text{-}34b)$$

It is often useful to relate the terminal currents to each other as well as to the saturation currents. We can eliminate the saturation current from the coupling term in each part of Eq. (7-32). For example, by multiplying Eq. (7-32a) by α_N and subtracting the resulting expression from Eq. (7-32b), we have

$$I_C = \alpha_N I_E - (1 - \alpha_N \alpha_I)I_{CS}(e^{qV_{CB}/kT} - 1) \qquad (7\text{-}35)$$

Similarly, the emitter current can be written in terms of the collector current:

$$I_E = \alpha_I I_C + (1 - \alpha_N \alpha_I)I_{ES}(e^{qV_{EB}/kT} - 1) \qquad (7\text{-}36)$$

The terms $(1 - \alpha_N \alpha_I)I_{CS}$ and $(1 - \alpha_N \alpha_I)I_{ES}$ can be abbreviated as I_{CO} and I_{EO}, respectively, where I_{CO} is the magnitude of the collector saturation current with the emitter junction *open* ($I_E = 0$), and I_{EO} is the magnitude of the emitter satu-

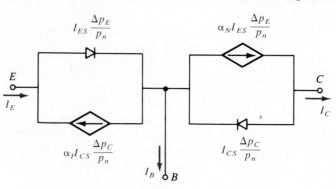

$$I_B = (1 - \alpha_N)\, I_{ES}\frac{\Delta p_E}{p_n} + (1 - \alpha_I)\, I_{CS}\frac{\Delta p_C}{p_n}$$

ration current with the collector open (Prob. 7.19). The Ebers–Moll equations then become

$$I_E = \alpha_I I_C + I_{EO}(e^{qV_{EB}/kT} - 1) \qquad (7\text{-}37\text{a})$$

$$I_C = \alpha_N I_E - I_{CO}(e^{qV_{CB}/kT} - 1) \qquad (7\text{-}37\text{b})$$

and the equivalent circuit is shown in Fig. 7-13a. In this form the equations describe both the emitter and collector currents in terms of a simple diode characteristic plus a current generator proportional to the other current. For example, under normal biasing the equivalent circuit reduces to the form shown in Fig. 7-13b. The collector current is α_N times the emitter current plus the collector saturation current, as expected. The resulting collector characteristics of the transistor appear as a series of reverse diode curves, displaced by increments proportional to the emitter current (Fig. 7-13c).

7.5.2 Charge Control Analysis

The charge control approach is useful in analyzing the transistor terminal currents, particularly in a-c applications. Considerations of transit time effects and charge storage are revealed easily by this method. Following the techniques of the previous section, we can separate an arbitrary excess hole distribution in the base into the normal and inverted distributions of Fig. 7-11. The charge stored in the normal distribution will be called Q_N and the charge under the in-

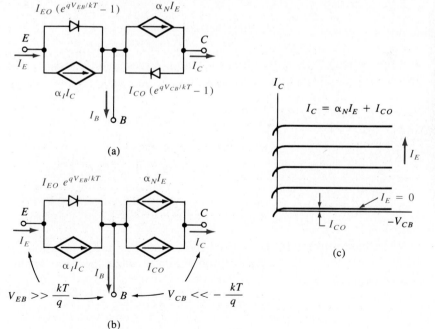

Figure 7-13
Equivalent circuits of the transistor in terms of the terminal currents and the open-circuit saturation currents: (a) synthesis of Eqs. (7-37); (b) equivalent circuit with normal biasing; (c) collector characteristics with normal biasing.

verted distribution will be called Q_I. Then we can evaluate the currents for the normal and inverted modes in terms of these stored charges. For example, the collected current in the normal mode I_{CN} is simply the charge Q_N divided by the mean time required for this charge to be collected. This time is the transit time for the normal mode τ_{tN}. On the other hand, the emitter current must support not only the rate of charge collection by the collector but also the recombination rate in the base Q_N/τ_{pN}. Here we use a subscript N with the transit time and recombination lifetime in the normal mode in contrast to the inverted mode, to allow for possible asymmetries due to imbalance in the transistor structure. With these definitions, the normal components of current become

$$I_{CN} = \frac{Q_N}{\tau_{tN}}, \qquad I_{EN} = \frac{Q_N}{\tau_{tN}} + \frac{Q_N}{\tau_{pN}} \tag{7-38a}$$

Similarly, the inverted components are

$$I_{EI} = -\frac{Q_I}{\tau_{tI}}, \qquad I_{CI} = -\frac{Q_I}{\tau_{tI}} - \frac{Q_I}{\tau_{pI}} \tag{7-38b}$$

where the I subscripts on the stored charge and on the transit and recombination times designate the inverted mode. Combining these equations as in Eq. (7-32) we have the terminal currents for general biasing:

$$I_E = Q_N\left(\frac{1}{\tau_{tN}} + \frac{1}{\tau_{pN}}\right) - \frac{Q_I}{\tau_{tI}} \tag{7-39a}$$

$$I_C = \frac{Q_N}{\tau_{tN}} - Q_I\left(\frac{1}{\tau_{tI}} + \frac{1}{\tau_{pI}}\right) \tag{7-39b}$$

It is not difficult to show that these equations correspond to the Ebers–Moll relations [Eq. (7-34) and Prob. 7.20], where

$$\alpha_N = \frac{\tau_{pN}}{\tau_{tN} + \tau_{pN}}, \qquad \alpha_I = \frac{\tau_{pI}}{\tau_{tI} + \tau_{pI}} \tag{7-40}$$

$$I_{ES} = q_N\left(\frac{1}{\tau_{tN}} + \frac{1}{\tau_{pN}}\right), \qquad I_{CS} = q_I\left(\frac{1}{\tau_{tI}} + \frac{1}{\tau_{pI}}\right)$$

$$Q_N = q_N\frac{\Delta p_E}{p_n}, \qquad Q_I = q_I\frac{\Delta p_C}{p_n}$$

The base current in the normal mode supports recombination, and the base-to-collector current amplification factor β_N takes the form predicted by Eq. (7-7):

$$I_{BN} = \frac{Q_N}{\tau_{pN}}, \qquad \beta_N = \frac{I_{CN}}{I_{BN}} = \frac{\tau_{pN}}{\tau_{tN}} \tag{7-41}$$

This expression for β_N is also obtained from $\alpha_N/(1 - \alpha_N)$. Similarly, I_{BI} is Q_I/τ_{pI}, and the total base current is

$$I_B = I_{BN} + I_{BI} = \frac{Q_N}{\tau_{pN}} + \frac{Q_I}{\tau_{pI}} \qquad (7\text{-}42)$$

This expression for the base current is substantiated by $I_E - I_C$ from Eq. (7-39).

The effects of time dependence of stored charge can be included in these equations by the methods introduced in Section 5.5.1. We can include the proper dependencies by adding a rate of change of stored charge to each of the injection currents I_{EN} and I_{CI}:

$$i_E = Q_N\left(\frac{1}{\tau_{tN}} + \frac{1}{\tau_{pN}}\right) - \frac{Q_I}{\tau_{tI}} + \frac{dQ_N}{dt} \qquad (7\text{-}43\text{a})$$

$$i_C = \frac{Q_N}{\tau_{tN}} - Q_I\left(\frac{1}{\tau_{tI}} + \frac{1}{\tau_{pI}}\right) - \frac{dQ_I}{dt} \qquad (7\text{-}43\text{b})$$

$$i_B = \frac{Q_N}{\tau_{pN}} + \frac{Q_I}{\tau_{pI}} + \frac{dQ_N}{dt} + \frac{dQ_I}{dt} \qquad (7\text{-}43\text{c})$$

We shall return to these equations in Section 7.8, when we discuss the use of transistors at high frequencies.

7.6 SWITCHING

In a switching operation a transistor is usually controlled in two conduction states, which can be referred to loosely as the "on" state and the "off" state. Ideally, a switch should appear as a short circuit when turned on and an open circuit when turned off. Furthermore, it is desirable to switch the device from one state to the other with no lost time in between. Transistors do not fit this ideal description of a switch, but they can serve as a useful approximation in practical electronic circuits. The two states of a transistor in switching can be seen in the simple common-emitter example of Fig. 7-14. In this figure the collector current i_C is controlled by the base current i_B over most of the family of characteristic curves. The load line specifies the locus of allowable $(i_C, -v_{CE})$ points for the circuit, in analogy with Fig. 7-2. If i_B is such that the operating point lies somewhere between the two end points of the load line (Fig. 7-14b), the transistor operates in the normal active mode. That is, the emitter junction is forward biased and the collector is reverse biased, with a reasonable value of i_B flowing out of the base. On the other hand, if the base current is zero or negative, the point C is reached at the bottom end of the load line, and the collector current is negligible. This is the "off" state of the transistor, and the device is said to be operating in the *cutoff* regime. If the base current is positive and sufficiently large, the device is driven to the *saturation* regime, marked S. This is the "on" state of the transistor, in which a large value of i_C flows with only a very small voltage drop v_{CE}. As we shall see below, the beginning of the

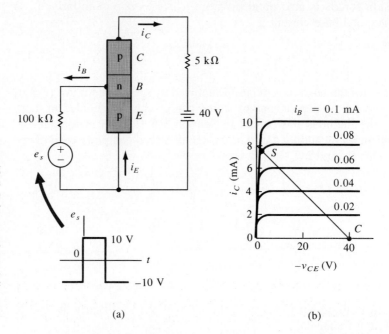

Figure 7-14
Simple switching
circuit for a
transistor in the
common-emitter
configuration:
(a) biasing circuit;
(b) collector
characteristics and
load line for the
circuit, with cutoff
and saturation
indicated.

saturation regime corresponds to the loss of reverse bias across the collector junction. In a typical switching operation the base current swings from positive to negative, thereby driving the device from saturation to cutoff, and vice versa. In this section we shall explore the nature of conduction in the cutoff and saturation regimes; also we shall investigate the factors affecting the speed with which the transistor can be switched between the two states.

7.6.1 Cutoff

If the emitter junction is reverse biased in the cutoff regime (negative i_B), we can approximate the excess hole concentrations at the edges of the reverse-biased emitter and collector junctions as

$$\frac{\Delta p_E}{p_n} \simeq \frac{\Delta p_C}{p_n} \simeq -1 \tag{7-44}$$

which implies $p(x_n) = 0$. With a straight-line approximation, the excess hole distribution in the base appears constant at $-p_n$, as shown in Fig. 7-15a. Actually, there will be some slope to the distribution at each edge to account for the reverse saturation current in the junctions, but Fig. 7-15a is approximately correct. The base current i_B can be approximated for a symmetrical transistor on a charge storage basis as $-qAp_nW_b/\tau_p$. In this calculation a negative excess hole concentration corresponds to *generation* in the same way that a positive distribution indicates recombination. This expression is also obtained by applying Eq. (7-44) to Eq. (7-19) with an approximation from Table 7-1. Physically, a small saturation current flows from n to p in each reverse-biased junction, and

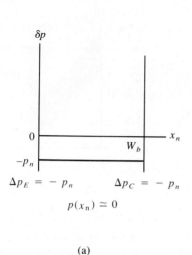

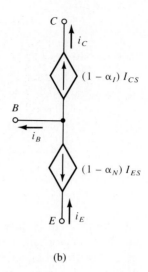

(a)   (b)

Figure 7-15
The cutoff regime of a p-n-p transistor: (a) excess hole distribution in the base region with emitter and collector junctions reverse biased; (b) equivalent circuit corresponding to Eqs. (7-45).

this current is supplied by the base current i_B (which is negative when flowing into the base of a p-n-p device according to our definitions). A more general evaluation of the currents can be obtained from the Ebers–Moll equations by applying Eq. (7-44) to Eq. (7-34):

$$i_E = -I_{ES} + \alpha_I I_{CS} = -(1 - \alpha_N)I_{ES} \qquad (7\text{-}45a)$$

$$i_C = -\alpha_N I_{ES} + I_{CS} = (1 - \alpha_I)I_{CS} \qquad (7\text{-}45b)$$

$$i_B = i_E - i_C = -(1 - \alpha_N)I_{ES} - (1 - \alpha_I)I_{CS} \qquad (7\text{-}45c)$$

If the short-circuit saturation currents I_{ES} and I_{CS} are small and α_N and α_I are both near unity, these currents will be negligible and the cutoff regime will closely approximate the "off" condition of an ideal switch. The equivalent circuit corresponding to Eq. (7-45) is illustrated in Fig. 7-15b.

7.6.2   Saturation

The saturation regime begins when the reverse bias across the collector junction is reduced to zero, and it continues as the collector becomes forward biased. The excess hole distribution in this case is illustrated in Fig. 7-16. The device is saturated when $\Delta p_C = 0$, and forward bias of the collector junction (Fig. 7-16b) leads to a positive Δp_C, driving the device further into saturation. With the load line fixed by the battery and the 5-kΩ resistor in Fig. 7-14, saturation is reached by increasing the base current i_B. We can see how a large value of i_B leads to saturation by applying the reasoning of charge control to Fig. 7-16. Since a certain amount of stored charge is required to accommodate a given i_B (and vice versa), an increase in i_B calls for an increase in the area under the $\delta p(x_n)$ distribution.

In Fig. 7-16a the device has just reached saturation, and the collector junction is no longer reverse biased. The implication of this condition for the cir-

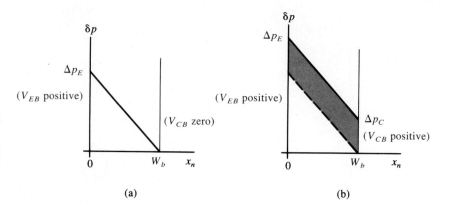

Figure 7-16
Excess hole
distribution in the
base of a saturated
transistor: (a) the
beginning of
saturation;
(b) oversaturation.

cuit of Fig. 7-14 is easy to state. Since the emitter junction is forward biased
and the collector junction has zero bias, very little voltage drop appears across
the device from collector to emitter. The magnitude of $-v_{CE}$ is only a fraction
of a volt. Therefore, almost all of the battery voltage appears across the resistor, and the collector current is approximately 40 V/5 kΩ = 8 mA. As the
device is driven deeper into saturation (Fig. 7-16b), the collector current stays
essentially constant while the base current increases. In this saturation condition the transistor approximates the "on" state of an ideal switch.

Whereas the degree of "oversaturation" (indicated by the shaded area in
Fig. 7-16b) does not affect the value of i_C significantly, it is important in determining the time required to switch the device from one state to the other. For
example, from previous experience we expect the turn-off time (from saturation to cutoff) to be longer for larger values of stored charge in the base. We
can calculate the various charging and delay times from Eq. (7-43). Detailed
calculations are somewhat involved, but we can simplify the problem greatly
with approximations of the type used in Chapter 5 for transient effects in p-n
junctions.

7.6.3 The Switching Cycle

The various mechanisms of a switching cycle are illustrated in Fig. 7-17. If the
device is originally in the cutoff condition, a step increase of base current to I_B
causes the hole distribution to increase approximately as illustrated in Fig. 7-17b.
As in the transient analysis of Chapter 5, we assume for simplicity of calculation that the distribution maintains a simple form in each time interval of the
transient. At time t_s the device enters saturation, and the hole distribution
reaches its final state at t_2. As the stored charge in the base Q_b increases, there
is an increase in the collector current i_C. The collector current does not increase
beyond its value at the beginning of saturation t_s, however. We can approximate this saturated collector current as $I_C \simeq E_{CC}/R_L$, where E_{CC} is the value of
the collector circuit battery and R_L is the load resistor ($I_C \simeq 8$ mA for the example of Fig. 7-14). There is an essentially exponential increase in the collector
current while Q_b rises to its value Q_s at t_s; this rise time serves as one of the

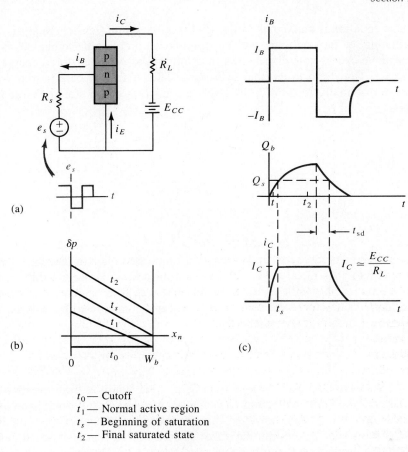

(a)

(b)

t_0 — Cutoff
t_1 — Normal active region
t_s — Beginning of saturation
t_2 — Final saturated state

(c)

Figure 7-17
Switching effects in a common-emitter transistor circuit: (a) circuit diagram; (b) approximate hole distributions in the base during switching from cutoff to saturation; (c) base current, stored charge, and collector current during a turn-on and a turn-off transient.

limitations of the transistor in a switching application. Similarly, when the base current is switched negative (e.g., to the value $-I_B$), the stored charge must be withdrawn from the base before cutoff is reached. While Q_b is larger than Q_s, the collector current remains at the value I_C, fixed by the battery and resistor. Thus there is a storage delay time t_{sd} after the base current is switched and before i_C begins to fall toward zero. After the stored charge is reduced below Q_s, i_c drops exponentially with a characteristic fall time. Once the stored charge is withdrawn, the base current cannot be maintained any longer at its large negative value and must decay to the small cutoff value described by Eq. (7-45c).

7.6.4   Turn-On Transient

We can calculate the various times involved in the switching transient of a symmetrical device by relating the base current to the stored charge and the rate of change of stored charge. For $\gamma = 1$,

$$i_B(t) = \frac{Q_b(t)}{\tau_p} + \frac{dQ_b(t)}{dt} \tag{7-46}$$

This equation is analogous to Eq. (5-47) in the diode transient analysis. In calculating the turn-on time, we can simplify the problem if we neglect the small negative excess hole distribution in cutoff and assume the stored charge Q_b increases from zero to its final value $I_B\tau_p$. Thus when the base current switches from essentially zero to I_B, the Laplace transform of the stored charge relation [Eq. (7-46)] is

$$\frac{I_B}{s} = \frac{Q_b(s)}{\tau_p} + sQ_b(s) \tag{7-47}$$

$$Q_b(s) = \frac{I_B}{s(s + 1/\tau_p)}$$

with the solution

$$Q_b(t) = I_B\tau_p(1 - e^{-t/\tau_p}) \tag{7-48}$$

While the stored charge increases with time, the collector current follows according to $i_C = Q_b(t)/\tau_t$, where τ_t is the transit time. However, this increase in i_C continues only until saturation is reached at t_s, when i_C reaches its maximum value I_C. Therefore, the turn-on time t_s can be calculated by solving for the time at which $Q_b(t)/\tau_t$ reaches I_C:

$$\frac{I_B\tau_p}{\tau_t}(1 - e^{-t_s/\tau_p}) = I_C \qquad \text{when } t_s = \tau_p \ln\frac{1}{1 - I_C/\beta I_B} \tag{7-49}$$

We have neglected an important delay time resulting from the necessity of charging the emitter junction capacitance, which we shall discuss below. It is clear from Eq. (7-49), however, that the time t_s will be short if (1) the lifetime τ_p is short and (2) the limiting value of collector current I_C is small compared with the product of β and the base current drive I_B. The latter condition is met by driving the device into oversaturation.

7.6.5 Turn-Off Transient

If the turn-off transient involves simply the reduction of the base current from its "on" value I_B to zero, the decays of the stored charge and collector current are particularly easy to calculate (Fig. 7-18a). The stored charge at the beginning of the turn-off transient is simply $Q_b = I_B\tau_p$. The stored charge falls exponentially from this value to zero with the time constant τ_p. The collector current remains at its saturated value I_C until Q_b falls to the value for minimum saturation $Q_s = I_C\tau_t$. Therefore, the storage delay time t_{sd} over which the collector current remains constant after the base current has been switched is given by

$$t_{sd} = \tau_p \ln\frac{I_B\tau_p}{I_C\tau_t} = \tau_p \ln\frac{\beta I_B}{I_C} \tag{7-50}$$

After the storage delay time, the collector current falls to zero with the time constant τ_p. We notice that in this case the delay time is increased by the condition of oversaturation ($\beta I_B > I_C$).

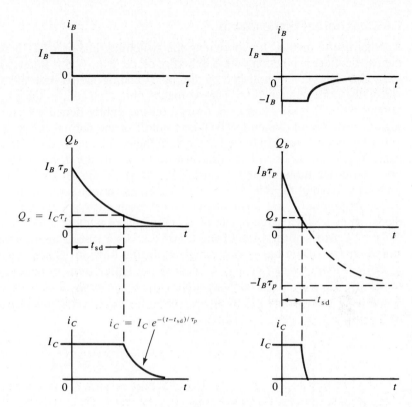

Figure 7-18
Turn-off transients:
(a) base current
switched from a
large positive value
to zero; (b) base
current switched
from positive to
negative.

The decay time can be shortened by driving the device to cutoff, that is, by switching the base current to a negative value instead of simply reducing it to zero. For example, Fig. 7-18b illustrates a case in which the base current is switched from I_B to an equal negative value $-I_B$. In this case the stored charge decays according to (Prob. 7.21)

$$Q_b(t) = I_B \tau_p (2e^{-t/\tau_p} - 1) \qquad (7\text{-}51)$$

Thus Q_b falls from its oversaturation value of $I_B \tau_p$ toward a negative value of equal magnitude. Of course, the stored charge does not reach the final negative value but stops at the negligibly small negative stored charge corresponding to cutoff. The effect, however, is to drive the stored charge (and therefore the collector current) to the cutoff value in a shorter time.

In fabricating transistors for switching applications it is desirable to make the lifetime τ_p in the base region as small as possible, as suggested by Eqs. (7-49), (7-50), and (7-51). One common technique is to increase the density of recombination centers in the base — for example, by doping a Si device with Au. This reduction of τ_p can be tolerated while maintaining the β of the device at an acceptable value by making the base region very narrow.

7.6.6 Schottky Diode Clamp

Another useful method for shortening the switching time of a BJT is to bypass the base-collector junction with a Schottky diode (Fig. 7-19). In an n-p-n transistor this can be accomplished by simply extending the Al base metallization across the collector junction. The Al makes ohmic contact to the p-type base region, but a Schottky barrier is formed on the lightly doped n-type collector region (Fig. 7-19b). When the BJT is in cutoff or the normal active mode, the Schottky diode is reverse biased along with the collector junction. On the other hand, when this junction becomes forward biased as the device enters saturation, the diode turns on and "clamps" V_{BC} to the diode forward voltage. The forward voltage of the Schottky diode for a given current is less than that of the collector junction, and most of the excess base current passes through the diode. Thus a rather large I_B can be accommodated without a large buildup of excess electrons in the base of the transistor. As a result, there is very little stored charge in the base to be extracted during turn-off. Since the Schottky diode is a majority carrier device, it does not suffer from storage delay time problems. Therefore, this configuration, called a *Schottky diode clamped transistor,* can be switched off in about one-tenth the time required for an ordinary BJT.

Figure 7-19
A Schottky diode clamped n-p-n transistor: (a) circuit equivalent, showing a Schottky diode between base and collector; (b) device cross section.

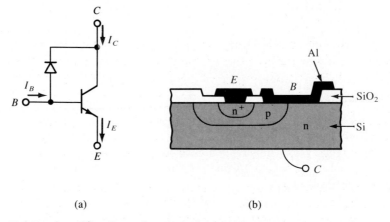

(a)　　　　　　(b)

7.6.7 Specifications for Switching Transistors

As mentioned before, we have simplified this discussion greatly so the basic principles of switching times could be understood apart from secondary effects. Thus we have neglected the effects of asymmetry in the transistor, typified by differing values of τ_{tN}, τ_{tI}, and the other parameters defined in the Ebers–Moll model. The main discrepancy of this analysis with the switching of real devices is the neglect of the charging time of the emitter junction capacitance in going from cutoff to saturation. Since the emitter junction is reverse biased in cutoff, it is necessary for the emitter space charge layer to be charged to the forward bias condition before collector current can flow. Therefore, we should include a *delay time t_d* as in Fig. 7-20 to account for this effect. Typical

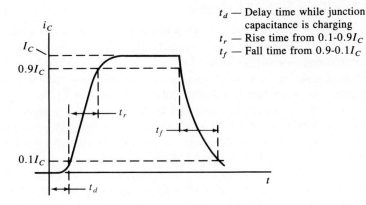

t_d — Delay time while junction
capacitance is charging
t_r — Rise time from 0.1-$0.9I_C$
t_f — Fall time from 0.9-$0.1I_C$

Figure 7-20
Collector current
during switching
transients,
including the delay
time required for
charging the
junction
capacitance;
definitions of the
rise time and fall
time.

values of t_d are given in the specification information of most switching transistors, along with a *rise time* t_r defined as the time required for the collector current to rise from 10 to 90 per cent of its final value. A third specification is the *fall time* t_f required for i_C to fall through a similar fraction of its turn-off excursion.

The approach we have taken in analyzing the properties of transistors has involved a number of simplifying assumptions. Some of the assumptions must be modified in dealing with practical devices. In this section we investigate some common deviations from the basic theory and indicate situations in which each effect is important. Since the various effects discussed here involve modifications of the more straightforward theory, they are labeled "secondary effects." This does not imply that they are unimportant; in fact, the effects described in this section can dominate the conduction in transistors under certain conditions of device geometry and circuit application.

   In this section we shall consider the effects of nonuniform doping in the base region of the transistor. In particular, we shall find that graded doping can lead to a drift component of charge transport across the base, adding to the diffusion of carriers from emitter to collector. We shall discuss the effects of large reverse bias on the collector junction, in terms of widening the space charge region about the junction and avalanche multiplication. We shall see that transistor parameters are affected at high current levels by the degree of injection and by heating effects. We shall consider several structural effects that are important in practical devices, such as asymmetry in the areas of the emitter and collector junctions, series resistance between the base contact and the active part of the base region, and nonuniformity of injection at the emitter junction. All these effects are important in understanding the operation of transistors, and proper consideration of their interactions can contribute greatly to the usefulness of practical transistor circuits.

**7.7
SECONDARY
EFFECTS**

7.7.1 Drift in the Base Region

The assumption of uniform doping in the base is generally valid for an alloyed transistor, in which the base region is made up of a thin layer of the starting material sandwiched between two abrupt alloyed junctions. On the other hand, diffused junction transistors usually involve an appreciable amount of impurity grading; for example, the double-diffused transistor of Fig. 7-7 has a doping profile similar to that sketched in Fig. 7-21. In this example there is a fairly sharp discontinuity in the doping profile, when the donor concentration in the base region becomes smaller than the constant p-type background doping in the collector. Similarly, the emitter is assumed to be a heavily doped (p^+) shallow region, providing a second rather sharp boundary for the base. Within the base region itself, however, the net doping concentration ($N_d - N_a \equiv N$) varies along a profile which decreases from the emitter edge to the collector edge. The most likely doping distribution in the base is a portion of a gaussian (see Section 5.1.3); however, it is often a good approximation to assume that $N(x_n)$ varies exponentially within the base region (Fig. 7-21b).

One important result of a graded base region is that a built-in electric field exists from emitter to collector (for a p-n-p), thereby adding a drift component to the transport of holes across the base. We can demonstrate this effect very simply by considering the required balance of drift and diffusion in the base at equilibrium. If the net donor doping of the base is large enough to allow the usual approximation $n(x_n) \simeq N(x_n)$, the balance of electron drift and diffusion currents at equilibrium requires

$$I_n(x_n) = qA\mu_n N(x_n)\mathscr{E}(x_n) + qAD_n\frac{dN(x_n)}{dx_n} = 0 \qquad (7\text{-}52)$$

Therefore, the built-in electric field is

$$\mathscr{E}(x_n) = -\frac{D_n}{\mu_n}\frac{1}{N(x_n)}\frac{dN(x_n)}{dx_n} = -\frac{kT}{q}\frac{1}{N(x_n)}\frac{dN(x_n)}{dx_n} \qquad (7\text{-}53)$$

Figure 7-21
Graded doping in the base region of a p-n-p transistor: (a) typical doping profile on a semilog plot; (b) approximate exponential distribution of the net donor concentration in the base region on a linear plot.

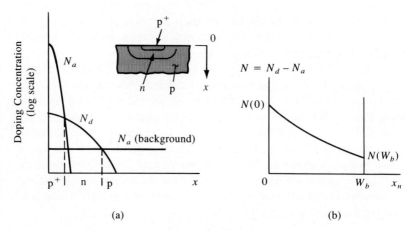

(a)   (b)

For a doping profile $N(x_n)$ that decreases in the positive x_n-direction, this field is positive, directed from emitter to collector.

For the example of an exponential doping profile, the electric field $\mathscr{E}(x_n)$ turns out to be constant with position in the base. We can represent an exponential distribution as

$$N(x_n) = N(0)e^{-ax_n/W_b} \qquad \text{where } a \equiv \ln \frac{N(0)}{N(W_b)} \qquad (7\text{-}54)$$

Taking the derivative of this distribution and substituting in Eq. (7-53), we obtain the constant field

$$\mathscr{E}(x_n) = \frac{kT}{q} \frac{a}{W_b} \qquad (7\text{-}55)$$

Since this field aids the transport of holes across the base region from emitter to collector, the transit time τ_t is reduced below that of a comparable uniform base transistor. Similarly, electron transport in an n-p-n is aided by the built-in field in the base. This shortening of the transit time can be very important in high-frequency devices (Section 7.8.2). Since the base transport factor B for a graded junction is even closer to unity than we indicated for the uniform device, the current transfer ratio α is often determined almost entirely by the emitter injection efficiency.

(a) Find the hole distribution in the base for the transistor of Fig. 7-21b, assuming a hole current I_p throughout the base. **EXAMPLE 7-4**

(b) Sketch the hole distribution for $a = 5$. Where is drift most important in the base?

(a) The combined drift and diffusion in the base, neglecting recombination is **SOLUTION**

$$I_p = qA\mu_p \mathscr{E}p(x_n) - qAD_p \frac{dp(x_n)}{dx_n}$$

Using Eq. (7-55),

$$I_p = qAD_p \left[\frac{a}{W_b}p(x_n) - \frac{dp(x_n)}{dx_n} \right] \qquad \text{since } D_p = \mu_p \frac{kT}{q}$$

Thus we have

$$\frac{dp(x_n)}{dx_n} - \frac{a}{W_b}p(x_n) + \frac{I_p}{qAD_p} = 0$$

The solution is

$$p(x_n) = \left(\frac{W_b}{a} \right) \frac{I_p}{qAD_p} + Ce^{ax_n/W_b}$$

If the collector junction is reverse biased, $p(W_b) = 0$; thus

$$C = -\frac{W_b I_p}{aqAD_p}e^{-a} \quad \text{and} \quad p(x_n) = \frac{I_p W_b}{qAD_p a}[1 - e^{a(x_n/W_b - 1)}]$$

(b) For $a = 5$, $p(x_n) = \text{(const.)} \frac{1}{5}[1 - e^{5(x_n/W_b - 1)}]$. As the figure shows, $p(x_n)$ is fairly constant in the first half of the base region, then drops rapidly in the second half. Thus the current is due mostly to drift for the first part of the transit.

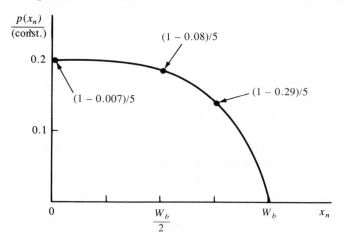

7.7.2 Base Narrowing

In the discussion of transistors thus far, we have assumed that the effective base width W_b is essentially independent of the bias voltages applied to the collector and emitter junctions. This assumption is not always valid; for example, the p$^+$-n-p$^+$ transistor of Fig. 7-22 is affected by the reverse bias applied to the collector. If the base region is lightly doped, the depletion region at the reverse-biased collector junction can extend significantly into the n-type base region. As the collector voltage is increased, the space charge layer takes up more of the metallurgical width of the base L_b, and as a result, the effective base width W_b is decreased. This effect is variously called *base narrowing, base-width modulation,* and the *Early effect* after J. M. Early, who first interpreted it. The effects of base narrowing are apparent in the collector characteristics for the common-emitter configuration (Fig. 7-22b). The decrease in W_b causes β to increase. As a result, the collector current I_C increases with collector voltage rather than staying constant as predicted from the simple treatment. The slope introduced by the Early effect is almost linear with I_C, and the common-emitter characteristics extrapolate to an intersection with the voltage axis at V_A, called the Early voltage.

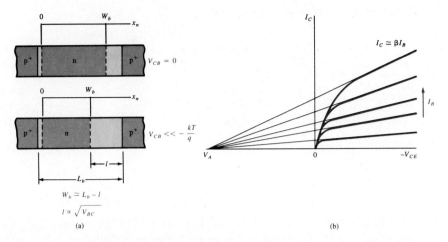

The effects of base narrowing on the characteristics of a p^+-n-p^+ transistor: (a) decrease in the effective base width as the reverse bias on the collector junction is increased; (b) common-emitter characteristics showing the increase in I_C with increased collector voltage. The black lines in (b) indicate the extrapolation of the curves to the Early voltage V_A.

Figure 7-22

For the p^+-n-p^+ device of Fig. 7-22 we can approximate the length l of the collector junction depletion region in the n material from Eq. (5-23b) with V_0 replaced by $V_0 - V_{CB}$ and V_{CB} taken to be large and negative:

$$l = \left(\frac{2\epsilon V_{BC}}{qN_d}\right)^{1/2} \tag{7-56}$$

If the reverse bias on the collector junction is increased far enough, it is possible to decrease W_b to the extent that the collector depletion region essentially fills the entire base. In this punch-through condition holes are swept directly from the emitter region to the collector, and transistor action is lost. Punch-through is a breakdown effect that is generally avoided in circuit design. In most cases, however, avalanche breakdown of the collector junction occurs before punch-through is reached. We shall discuss the effects of avalanche multiplication in the following section.

In devices with graded base doping, base narrowing is of less importance. For example, if the donor concentration in the base region of a p-n-p increases with position from the collector to the emitter, the intrusion of the collector space charge region into the base becomes less important with increased bias as more donors are available to accommodate the space charge.

7.7.3   Avalanche Breakdown

Before punch-through occurs in most transistors, avalanche multiplication at the collector junction becomes important (see Section 5.4.2). As Fig. 7-23 indicates, the collector current increases sharply at a well-defined breakdown

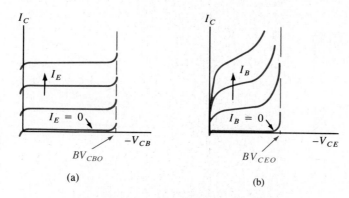

Figure 7-23
Avalanche
breakdown in a
transistor:
(a) common-base
configuration;
(b) common-emitter
configuration.

voltage BV_{CBO} for the common-base configuration. For the common-emitter case, however, there is a strong influence of carrier multiplication over a fairly broad range of collector voltage. Furthermore, the breakdown voltage in the common-emitter case BV_{CEO} is significantly smaller than BV_{CBO}. We can understand these effects by considering breakdown for the condition $I_E = 0$ in the common-base case and for $I_B = 0$ in the common-emitter case. These conditions are implied by the O in the subscripts of BV_{CEO} and BV_{CBO}. In each case the terminal current I_C is the current entering the collector depletion region multiplied by the factor M. Including multiplication due to impact ionization, Eq. (7-37b) becomes

$$I_C = (\alpha_N I_E + I_{CO})M = (\alpha_N I_E + I_{CO}) \frac{1}{1 - (V_{BC}/BV_{CBO})^n} \qquad (7\text{-}57)$$

where for M we have used the empirical expression given in Eq. (5-44).

For the limiting common-base case of $I_E = 0$ (the lowest curve in Fig. 7-23a), I_C is simply MI_{CO}, and the breakdown voltage is well defined, as in an isolated junction. The term BV_{CBO} signifies the collector junction breakdown voltage in common-base with the emitter open. In the common-emitter case the situation is somewhat more complicated. Setting $I_B = 0$, and therefore $I_C = I_E$ in Eq. (7-57), we have

$$I_C = \frac{MI_{CO}}{1 - M\alpha_N} \qquad (7\text{-}58)$$

We notice that in this case the collector current increases indefinitely when $M\alpha_N$ approaches unity. By contrast, M must approach infinity in the common-base case before BV_{CBO} is reached. Since α_N is close to unity in most transistors, M need be only slightly larger than unity for Eq. (7-58) to approach breakdown. Avalanche multiplication thus dominates the current in a common-emitter transistor well below the breakdown voltage of the isolated collector junction. The sustaining voltage for avalanching in the common-emitter case BV_{CEO} is therefore smaller than BV_{CBO}.

We can understand physically why multiplication is so important in the common-emitter case by considering the effect of M on the base current. When an ionizing collision occurs in the collector junction depletion region, a secondary hole and electron are created. The primary and secondary holes are swept into the collector in a p-n-p, but the electron is swept into the base by the junction field. Therefore, the supply of electrons to the base is increased, and from our charge control analysis we conclude that hole injection at the emitter must increase to maintain space charge neutrality. This is a regenerative process, in which an increased injection of holes from the emitter causes an increased multiplication current at the collector junction; this in turn increases the rate at which secondary electrons are swept into the base, calling for more hole injection. Because of this regenerative effect, it is easy to understand why the multiplication factor M need be only slightly greater than unity to start the avalanching process.

7.7.4   Injection Level; Thermal Effects

In discussions of transistor characteristics we have assumed that α and β are independent of carrier injection level. Actually, the parameters of a practical transistor may vary considerably with injection level, which is determined by the magnitude of I_E or I_C. For very low injection, the assumption of negligible recombination in the junction depletion regions is invalid (see Section 5.6.2). This is particularly important in the case of recombination in the emitter junction, where any recombination tends to degrade the emitter injection efficiency γ. Thus we expect that α and β should decrease for low values of I_C, causing the curves of the collector characteristics to be spaced more closely for low currents than for higher currents.

As I_C is increased beyond the low injection level range, α and β increase but fall off again at very high injection. The primary cause of this fall-off is the increase of majority carriers at high injection levels (see Section 5.6.1). For example, as the concentration of excess holes injected into the base becomes large, the matching excess electron concentration can become greater than the background n_n. This conductivity modulation effect results in a decrease in γ as more electrons are injected across the emitter junction into the emitter region.

Large values of I_C may be accompanied by significant power dissipation in the transistor and therefore heating of the device. In particular, the product of I_C and the collector voltage V_{BC} is a measure of the power dissipated at the collector junction. This dissipation is due to the fact that carriers swept through the collector junction depletion region are given increased kinetic energy, which in turn is given up to the lattice in scattering collisions. It is very important that the transistor be operated in a range such that $I_C V_{BC}$ does not exceed the maximum power rating of the device. In devices designed for high power capability, the transistor is mounted on an efficient heat sink, so that thermal energy can be transferred away from the junction.

If the temperature of the device is allowed to increase due to power dissipation or thermal environment, the transistor parameters change. The most important parameters dependent on temperature are the carrier lifetimes and diffusion coefficients. In Si or Ge devices the lifetime τ_p increases with temperature for most cases, due to thermal reexcitation from recombination centers. This increase in τ_p tends to increase β for the transistor. On the other hand, the mobility decreases with increasing temperature in the lattice-scattering range, varying approximately as $T^{-3/2}$ (see Fig. 3-22). Thus from the Einstein relation, we expect D_p to decrease as the temperature increases, thereby causing a drop in β due to an increasing transit time τ_t. Of these competing processes, the effect of increasing lifetime with temperature usually dominates, and β becomes larger as the device is heated. It is clear from this effect that *thermal runaway* can occur if the circuit is not designed to prevent it. For example, a large power dissipation in the device can cause an increase in T; this results in a larger β and therefore a larger I_C for a given base current; the larger I_C causes more collector dissipation and the cycle continues. This type of runaway of the collector current can result in overheating and destruction of the device.

7.7.5 Base Resistance and Emitter Crowding

A number of structural effects are important in determining the operation of a transistor. For example, the emitter and collector areas are considerably different in the diffused transistor of Fig. 7-24a. This and most other structural effects can be accounted for by differences in α_N, α_I, and the other parameters in

Figure 7-24
Effects of base resistance: (a) cross section of a diffused transistor; (b) and (c) top view, showing emitter and base areas and metallized contacts; (d) illustration of base resistance; (e) expanded view of distributed resistance in the active part of the base region.

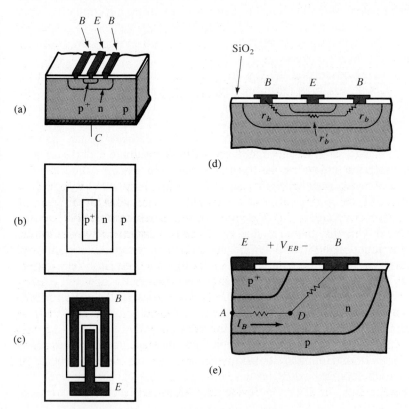

the Ebers–Moll model. Several effects caused by the structural arrangement of real transistors deserve special attention, however. One of the most important of these effects is the fact that base current must pass from the active part of the base region to the base contacts B. Thus, to be accurate, we should include a resistance r_b in equivalent models for the transistor to account for voltage drops which may occur between B and the active part of the base. Because of r_b, it is common to contact the base with the metallization pattern on both sides of the emitter, as in Fig. 7-24c.

If the transistor is designed so that the n-type regions leading from the base to the contacts are large in cross-sectional area, the base resistance r_b may be negligible. On the other hand, the distributed resistance r_b' along the thin base region is almost always important.[†] Since the width of the base between emitter and collector is very narrow, this distributed resistance is usually quite high. Therefore, as base current flows from points within the base region toward each end, a voltage drop occurs along r_b'. In this case the forward bias across the emitter-base junction is not uniform, but instead varies with position according to the voltage drop in the distributed base resistance. In particular, the forward bias of the emitter junction is largest at the corner of the emitter region near the base contact. We can see that this is the case by considering the simplified example of Fig. 7-24e. Neglecting variations in the base current along the path from point A to the contact B, the forward bias of the emitter junction above point A is approximately

$$V_{EA} = V_{EB} - I_B(R_{AD} + R_{DB}) \tag{7-59}$$

Actually, the base current is not uniform along the active part of the base region, and the distributed resistance of the base is more complicated than we have indicated. But this example does illustrate the point of nonuniform injection. Whereas the forward bias at A is approximately described by Eq. (7-59), the emitter bias voltage at point D is

$$V_{ED} = V_{EB} - I_B R_{DB} \tag{7-60}$$

which can be significantly closer to the applied voltage V_{EB}.

Since the forward bias is largest at the edge of the emitter, it follows that the injection of holes is also greatest there. This effect is called *emitter crowding,* and it can strongly affect the behavior of the device. The most important result of emitter crowding is that high-injection effects described in the previous section can become dominant locally at the corners of the emitter before the overall emitter current is very large. In transistors designed to handle appreciable current, this is a problem which must be dealt with by proper structural design. The most effective approach to the problem of emitter crowding is to distribute the emitter current along a relatively large emitter edge, thereby reducing the current density at any one point. Clearly, what is needed is an emitter region with a large perimeter compared with its area. A likely geometry to accomplish this is a long thin stripe for the emitter, with base contacts on each side (Fig. 7-24b and c). With this geometry the total emitter current I_E is

---

[†]The distributed resistance r_b' is often called the *base spreading resistance.*

spread out along a rather long edge on each side of the stripe. An even better geometry is several emitter stripes, connected electrically by the metallization and separated by interspersing base contacts (Figs. 7-25 and 7-26). Many such thin emitter and base contact "fingers" can be interlaced to provide for handling large current in a power transistor. This is often called very descriptively an *interdigitated* geometry.

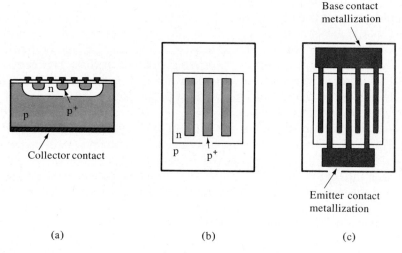

(a)   (b)   (c)

Figure 7-25 An interdigitated geometry to compensate for the effects of emitter crowding in a power transistor: (a) cross section; (b) top view of diffused regions; (c) top view with metallized contacts. The metal interconnections are isolated from the device by an oxide layer except where they contact the appropriate base and emitter regions at "windows" in the oxide.

Figure 7-26 Interdigitated power transistors: (a) a Si power transistor mounted on a header designed for efficient heat dissipation. The collector is alloyed to a metallized region on a ceramic substrate, and the base and emitter contact "fingers" are connected to two other metallized regions by Al wires. (Photograph courtesy of Texas Instruments, Inc.)

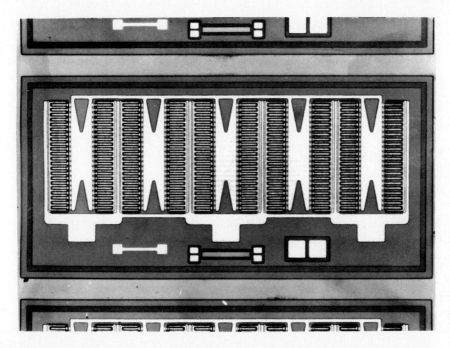

Figure 7-26 (cont.) (b) A Si transistor with many interlaced base and emitter regions. This device is rated at 25 W and 175 MHz. (Photograph courtesy of Motorola Semiconductor Products, Inc.)

Calculate the base spreading resistance r_b' for the transistor shown, in which the base is contacted on one side and all base current must flow along the base region to this contact. Assume low injection, so that emitter crowding is negligible (I_E is uniform across the emitter).   **EXAMPLE 7-5**

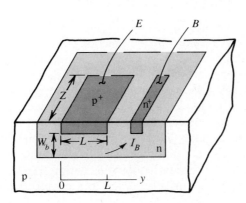

We must find an average value $\langle V_B \rangle$ for the lateral voltage drop in the base region. The base current $I_B(y) = I_B y/L$ if I_E is uniform. The differential voltage drop is   **SOLUTION**

$$dV_B(y) = \frac{\rho_b}{ZW_b} \, dy \, I_B(y) = \frac{\rho_B I_B}{ZLW_b} y \, dy$$

where ρ_b is the resistivity of the base. Integrating from 0 to y, we have

$$V_B(y) = \frac{\rho_b y^2 I_B}{2ZW_b L}$$

Thus

$$\langle V_B \rangle = \frac{1}{L} \int_0^L V_B(y)\,dy = \frac{\rho_B I_B L}{6ZW_b}$$

Then the base spreading resistance is

$$r_b' = \frac{\langle V_B \rangle}{I_B} = \frac{\rho_b L}{6ZW_b}$$

If the base were contacted on both sides, r_b' would be one-half this value.

■

**7.8
FREQUENCY
LIMITATIONS
OF TRANSISTORS**

In this section we discuss the properties of bipolar transistors under high-frequency operation. Some of the frequency limitations are junction capacitance, charging times required when excess carrier distributions are altered, and transit time of carriers across the base region. Our aim here is not to attempt a complete analysis of high-frequency operation, but rather to consider the physical basis of the most important effects. Therefore, we shall include the dominant capacitances and charging times and discuss the effects of the transit time on high-frequency devices.

7.8.1 Capacitance and Charging Times

The most obvious frequency limitation of transistors is the presence of junction capacitance at the emitter and collector junctions. We have considered this type of capacitance in Chapter 5, and we can include junction capacitors C_{je} and C_{jc} in circuit models for the transistor (Fig. 7-27a). If there is some equivalent resistance r_b between the base contact and the active part of the base region, we can also include it in the model, along with r_c to account for a series collector resistance.[†] Clearly, the combinations of r_b with C_{je} and r_c with C_{jc} can introduce important time constants into a-c circuit applications of the device.

From Section 5.5.3 we recall that capacitive effects can arise from the requirements of altering the carrier distributions during time-varying injection. In a-c circuits the transistor is usually biased to a certain steady state operating

---

[†]Since elements such as r_b and r_c in Fig. 7-27 are added to the basic transistor model we have previously analyzed, it is most convenient to refer here to the terminal voltages and currents as v_{be}', i_c', and so on. In this way we can use previously derived expressions involving the internal quantities [i_b and v_{eb} in Eq. (7-65), for example]. In most circuits texts the primes are instead used for the internal quantities, just the opposite method from that used here.

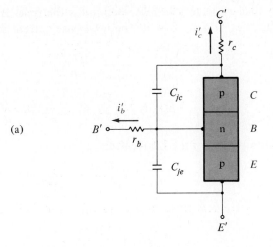

(a)

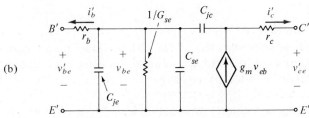

(b)

Figure 7-27
Models for a-c operation:
(a) inclusion of base and collector resistances and junction capacitances;
(b) hybrid-pi model synthesizing Eqs. (7-65) and (7-66).

point characterized by the d-c quantities V_{BE}, V_{CE}, I_C, I_B, and I_E; then a-c signals are superimposed upon these steady state values. We shall call the a-c terms v_{be}, v_{ce}, i_c, i_b, and i_e. Total (a-c + d-c) quantities will be lowercase with capitalized subscripts.

If a small a-c signal is applied to the emitter p-n junction along with a d-c level, we can show (Prob. 7-22) that

$$\Delta p_E(t) \simeq \Delta p_E(\text{d-c}) \left(1 + \frac{q v_{eb}}{kT} \right) \qquad (7\text{-}61)$$

We can relate this time-varying excess hole concentration to the stored charge in the base region, and then use Eq. (7-43) to determine the resulting currents. For simplicity we shall assume the device is biased in the normal active mode and use only $Q_N(t)$. Assuming an essentially triangular excess hole distribution in the base, Eq. (7-23) gives

$$Q_N(t) = \tfrac{1}{2} q A W_b \Delta p_E(t) = \tfrac{1}{2} q A W_b \Delta p_E(\text{d-c}) \left[1 + \frac{q v_{eb}}{kT} \right] \qquad (7\text{-}62)$$

The terms outside the brackets constitute the d-c stored charge $I_B \tau_p$:

$$Q_N(t) = I_B \tau_p \left(1 + \frac{q v_{eb}}{kT} \right) \qquad (7\text{-}63)$$

Now that we have a simple relation for the time-dependent stored charge, we can use Eq. (7-43c) to write the total base current as

$$i_B(t) = \frac{Q_N(t)}{\tau_p} + \frac{dQ_N(t)}{dt}$$

$$= I_B + \frac{q}{kT} I_B v_{eb} + \frac{q}{kT} I_B \tau_p \frac{dv_{eb}}{dt}$$

$$(7\text{-}64)$$

The a-c component of the base current is

$$i_b = G_{se} v_{eb} + C_{se} \frac{dv_{eb}(t)}{dt} \tag{7-65}$$

where

$$G_{se} \equiv \frac{q}{kT} I_B \quad \text{and} \quad C_{se} \equiv \frac{q}{kT} I_B \tau_p = G_{se} \tau_p$$

Thus, as in the case of the simple diode, an a-c conductance and capacitance are associated with the emitter–base junction due to charge storage effects. From Eq. (7-43b) we have

$$i_C(t) = \frac{Q_N(t)}{\tau_t} = \beta I_B + \frac{q}{kT} \beta I_B v_{eb}$$

$$i_c = g_m v_{eb} \quad \text{where} \quad g_m \equiv \frac{q}{kT} \beta I_B = \frac{C_{se}}{\tau_t} \tag{7-66}$$

The quantity g_m is an a-c *transconductance*, which is evaluated at the steady-state value of collector current $I_C = \beta I_B$. We can synthesize Eqs. (7-65) and (7-66) in an equivalent a-c circuit as in Fig. 7-27b. In this equivalent circuit the voltage v_{be} used in the calculations appears "inside" the device, so that a new applied voltage v'_{be} must be used external to r_b to refer to the voltage applied between the contacts, and similarly for v'_{ce}. This equivalent model is discussed in detail in most electronic circuits texts; it is often called a *hybrid-pi* model.

From Fig. 7-27b it is clear that several charging times are important in the a-c operation of a transistor; the most important are the time required to charge the emitter and collector depletion regions and the delay time in altering the charge distribution in the base region. Other delay times included in a complete analysis of high-frequency transistors are the transit time through the collector depletion region and the charge storage time in the collector region. If all of these are included in a single delay time τ_d, we can estimate the upper frequency limit of the device. This is usually defined as the *cutoff frequency* for the transistor $f_T \equiv (2\pi\tau_d)^{-1}$. It is possible to show that f_T represents the frequency at which the a-c amplification for the device $[\beta(\text{a-c}) \equiv h_{fe} = \partial i'_c / \partial i'_b]$ drops to unity.

7.8.2 Transit Time Effects

In high-frequency transistors the ultimate limitation is often the transit time across the base. For example, in a p-n-p device the time τ_t required for holes to diffuse from emitter to collector can determine the maximum frequency of operation for the device. We can calculate τ_t for a transistor with normal biasing and $\gamma = 1$ from Eq. (7-20) and the relation $\beta \simeq \tau_p/\tau_t$:

$$\beta \simeq \frac{\mathrm{csch}\ W_b/L_p}{\tanh\ W_b/2L_p} = \frac{2L_p^2}{W_b^2} = \frac{2D_p\tau_p}{W_b^2} = \frac{\tau_p}{\tau_t}$$

$$\tau_t = \frac{W_b^2}{2D_p} \tag{7-67}$$

Another instructive way of calculating τ_t is to consider that the diffusing holes *seem* to have an average velocity $\langle v(x_n) \rangle$ (actually the individual hole motion is completely random, as discussed in Section 4.4.1). The hole current $i_p(x_n)$ is then given by

$$i_p(x_n) = qAp(x_n)\langle v(x_n) \rangle \tag{7-68}$$

The transit time is

$$\tau_t = \int_0^{W_b} \frac{dx_n}{\langle v(x_n) \rangle} = \int_0^{W_b} \frac{qAp(x_n)}{i_p(x_n)} dx_n \tag{7-69}$$

For a triangular distribution as in Fig. 7-10b, the diffusion current is almost constant at $i_p = qAD_p \Delta p_E/W_b$, and τ_t becomes

$$\tau_t = \frac{qA \Delta p_E W_b/2}{qAD_p \Delta p_E/W_b} = \frac{W_b^2}{2D_p} \tag{7-70}$$

as before. The average velocity concept should not be pushed too far in the case of diffusion, but it does serve to illustrate the point that a delay time exists between the injection and collection of holes.

We can estimate the transit time for a typical device by choosing a value of W_b, say 1 μm (10^{-4} cm). For Si, a typical number for D_p is about 10 cm^2/s; then for this transistor $\tau_t = 0.5 \times 10^{-9}$ s. Approximating the upper frequency limit as $(2\pi\tau_t)^{-1}$, we can use the transistor to about 320 MHz. Actually, this estimate is too optimistic because of other delay times. A factor of four improvement can be obtained in τ_t by reducing W_b to $\frac{1}{2}$ μm. By careful diffusion or elaborate fabrication processing such as ion implantation, it is possible to reduce W_b to about 0.1 μm. The transit time can also be reduced by making use of field-driven currents in the base. For the diffused transistor of Fig. 7-21, the holes drift in the built-in field from emitter to collector over most of the base region. By increasing the doping gradient in the base, we can reduce the transit time and thereby increase the maximum frequency of the transistor.

7.8.3 High-Frequency Transistors

The most obvious generality we can make about the fabrication of high-frequency transistors is that the physical size of the device must be kept small.

The base width must be narrow to reduce the transit time, and the emitter and collector areas must be small to reduce junction capacitance. Unfortunately, the requirement of small size generally works against the requirements of power rating for the device. Since we usually require a trade-off between frequency and power, the dimensions and other design features of the transistor must be tailored to the specific circuit requirements. On the other hand, many of the fabrication techniques useful for power devices can be adapted to increase the frequency range. For example, the method of interdigitation (Fig. 7-25) provides a means of increasing the useful emitter edge length while keeping the overall emitter area to a minimum. Therefore, some form of interdigitation is generally used in transistors designed for high frequency and reasonable power requirements (Fig. 7-28).

Another set of parameters that must be considered in the design of a high-frequency device is the effective resistance associated with each region of the transistor. Since the emitter, base, and collector resistances affect the various RC charging times, it is important to keep them to a minimum. Therefore, the metallization patterns contacting the emitter and base regions must not present significant series resistance. Furthermore, the semiconductor regions themselves must be designed to reduce resistance. For example, the series base resistance r_b of an n-p-n device can be reduced greatly by performing a p^+ diffusion between the contact area on the surface and the active part of the base region. Further reduction of base resistance by heavy doping of the base requires the use of a heterojunction (Section 7.9) to maintain γ at an acceptable value.

In Si, n-p-n transistors are usually preferred, since the electron mobility and diffusion coefficient are higher than for holes. It is common to fabricate n-p-n transistors in n-type epitaxial material grown on an n^+ substrate. The heav-

Figure 7-28
A 4-GHz Si bipolar transistor. The five emitter stripes are each 0.5 μm wide and are separated by 5 μm. (Photograph courtesy of Hewlett-Packard Company.)

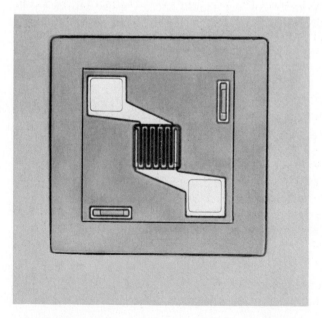

ily doped substrate provides a low-resistance contact to the collector region, while maintaining low doping in the epitaxial collector material to ensure a high breakdown voltage of the collector junction. It is important, however, to keep the collector depletion region as small as possible to reduce the transit time of carriers drifting through the collector junction. This can be accomplished by making the lightly doped collector region narrow so that the depletion region under bias extends to the n^+ substrate.

In addition to the various parameters of the device itself, the transistor must be packaged properly to avoid parasitic resistance, inductance, or capacitance at high frequencies. We shall not attempt to describe the many techniques for mounting and packaging transistors here, since methods vary greatly among manufacturers. We should mention, however, that the beam-lead method to be described in Chapter 9 offers many advantages for high-frequency discrete transistors.

In Section 7.4.4 we saw that the emitter injection efficiency of a bipolar transistor is limited by the fact that carriers can flow from the base into the emitter region, over the emitter junction barrier, which is reduced by the forward bias. According to Eq. (7-25) it is necessary to use lightly doped material for the base region and heavily doped material for the emitter to maintain a high value of γ and, therefore, α and β. Unfortunately, the requirement of light base doping results in undesirably high base resistance. This resistance is particularly noticeable in transistors with very narrow base regions, and represents a major limitation in high frequency applications of BJTs. Furthermore, heavy doping of the emitter to improve its injection efficiency tends to increase the emitter junction capacitance. Therefore, a more suitable BJT for high frequency would have a heavily doped base and a lightly doped emitter. This is just the opposite of the traditional BJT discussed thus far in this chapter. To accomplish such a radically different transistor design, we need some other mechanism instead of doping to control the relative amount of injection of electrons and holes across the emitter junction.

If transistors are made in materials that allow heterojunctions to be used, the emitter injection efficiency can be increased without strict requirements on doping. In Fig. (7-29) an n-p-n transistor made in a single material (*homojunction*) is contrasted with a *heterojunction bipolar transistor (HBT)*, in which the emitter is a wider band gap semiconductor. It is possible in such a structure for the barrier for electron injection (qV_n) to be smaller than the hole barrier (qV_p). Since carrier injection varies exponentially with the barrier height, even a small difference in these two barriers can make a very large difference in the transport of electrons and holes across the emitter junction. Neglecting differences in carrier mobilities and other effects, we can approximate the dependence of carrier injection across the emitter as

$$\frac{I_n}{I_p} \propto \frac{N_d^E}{N_a^B} e^{\Delta E_g / kT} \tag{7-71}$$

**7.9
HETEROJUNCTION
BIPOLAR
TRANSISTORS**

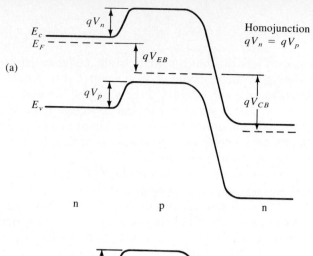

(a)

Homojunction
$qV_n = qV_p$

qV_n

E_c
E_F

qV_{EB}

qV_p

E_v

qV_{CB}

n   p   n

Figure 7-29
Contrast of carrier injection at the emitter of (a) a homojunction BJT and (b) a heterojunction bipolar transistor (HBT). In the forward-biased homojunction emitter, the electron barrier qV_n and the hole barrier qV_p are the same. In the HBT with a wide band gap emitter, the electron barrier is smaller than the hole barrier, resulting in preferential injection of electrons across the emitter junction.

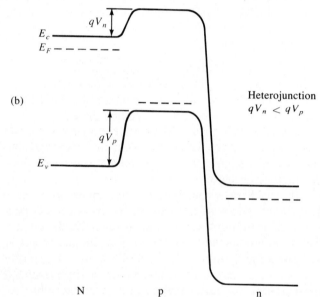

(b)

qV_n

E_c
E_F

Heterojunction
$qV_n < qV_p$

qV_p

E_v

N   p   n

In this expression, the ratio of electron current I_n to hole current I_p crossing the emitter junction is proportional to the ratio of the doping in the emitter N_d^E and the base N_a^B. In the homojunction BJT this doping ratio is all we have to work with in designing a useful emitter junction. However, in the HBT there is an additional factor in which the band gap difference ΔE_g between the wide band gap emitter and the narrow band gap base appears in an exponential factor. As a result, a relatively small value of ΔE_g in the exponential term can dominate Eq. (7-71). This allows us to choose the doping terms for lower base resistance and emitter junction capacitance. In particular, we can choose a heavily doped base to reduce the base resistance and a lightly doped emitter to reduce junction capacitance.

The heterojunction shown in Fig. 7-29 has a smooth barrier, without the spike and notch commonly observed for heterojunctions (see Fig. 5-37). It is possible to smooth out such discontinuities in the bands by grading the composition of the ternary or quaternary alloy between the two materials (Fig. 7-30).

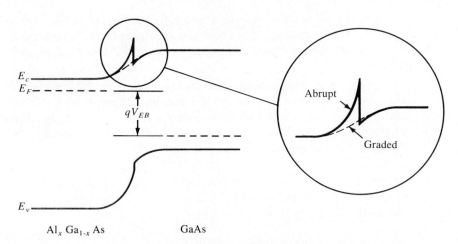

E_c

E_F

qV_{EB}

Abrupt

Graded

E_v

Al$_x$ Ga$_{1-x}$ As   GaAs

Removal of the conduction band spike by grading the alloy composition (x) in the heterojunction. In this example the junction is graded from the composition used in the AlGaAs emitter to x = 0 at the GaAs base. This grading typically takes place over a distance of 100 Å or less.

Figure 7-30

Clearly, grading out the conduction band spike improves the electron injection by reducing the barrier that electrons must overcome. There are some HBT designs, however, that make use of the spike as a "launching ramp" to inject hot electrons into the base.

Materials commonly used in HBTs obviously include the AlGaAs/GaAs system because of its wide range of lattice-matched compositions. In addition, the InGaAsP system (including In$_{0.53}$Ga$_{0.47}$As) grown on InP has become popular in HBT design. InGaAs has much lower surface recombination than GaAs, and the Γ-L and Γ-X intervalley band separations are much larger than in GaAs (see Fig. 3-6). The lower rate of surface recombination reduces loss of injected carriers at the surface of the device. This is a particularly important effect in small-geometry devices, in which the base perimeter-to-area ratio, and thus the intersection of the base with the surface, is large. The larger intervalley band separation in InGaAs helps ensure that the electrons remain in the low-mass (high mobility) Γ valley during field-enhanced transport through the base region.

It is also possible to make HBTs using elemental semiconductor heterostructures such as Si/Si$_{1-x}$Ge$_x$. In this material system, the band gap difference ΔE_g between the Si emitter and the narrower band gap Si$_{1-x}$Ge$_x$ base occurs primarily in the valence band. As a result, a rather small addition of Ge to the base results in higher electron injection efficiency than is possible in homojunction silicon bipolar transistors.

PROBLEMS

 7.1. (a) Sketch the energy band diagrams for a p-n-p and an n-p-n transistor at equilibrium and under normal bias conditions.

(b) Use the n-p-n band diagrams and a figure similar to Fig. 7-5 to discuss the injection and collection of electrons in this structure; what are the important components of base current in the n-p-n?

7.2. Sketch the ideal collector characteristics (i_C, $-v_{CE}$) for the transistor of Fig. 7-6; let i_B vary from zero to 0.2 mA in increments of 0.02 mA, and let $-v_{CE}$ vary from 0 to 10 V. Draw a load line on the resulting characteristics for the circuit of Fig. 7-6, and find the steady state value of $-V_{CE}$ graphically for $I_B = 0.1$ mA.

7.3. Sketch the masks required for fabrication of the double-diffused transistor of Fig. 7-7. Refer to Fig. 5-4 as an aid. For simplicity, sketch the masks for a single transistor rather than for an array.

7.4. Given the data of Prob. 5.22, plot the doping profiles $N_a(x)$ and $N_d(x)$ for the following double-diffused transistor: The starting wafer is n-type Si with $N_d = 5 \times 10^{16}$ cm^{-3}; $N_s = 5.82 \times 10^{13}$ cm^{-2} boron atoms are deposited on the surface, and these atoms are diffused into the wafer at 1100°C for 1 hr ($D = 3 \times 10^{-13}$ cm^2/sec for B in Si at 1100°C); then the wafer is placed in a phosphorus diffusion furnace at 1000°C for 15 min ($D = 3 \times 10^{-14}$ cm^2/sec for P in Si at 1000°C); during the emitter diffusion the surface concentration is held constant at 5×10^{20} cm^{-3}. You may assume the base doping profile does not change appreciably during the emitter diffusion, which takes place at a lower temperature and for a shorter time. Find the width of the base region from plots of $N_a(x)$ and $N_d(x)$. *Hint:* Use five-cycle semilog paper and let x vary from zero to about 1.5 μm in steps that are chosen to be simple multiples of $2\sqrt{Dt}$.

7.5. Calculate and plot the excess hole distribution $\delta p(x_n)$ in the base of a p-n-p transistor from Eq. (7-14), assuming $W_b/L_p = 0.5$. The calculations are simplified if the vertical scale is measured in units of $\delta p/\Delta p_E$ and the horizontal scale in units of x_n/L_p. In good transistors, W_b/L_p is much smaller than 0.5; however, $\delta p(x_n)$ is quite linear even for this rather large base width.

7.6. Derive Eq. (7-19) from the charge control approach by integrating Eq. (7-11) across the base region and applying Eq. (7-13).

7.7. Derive the emitter injection efficiency for a p-n-p transistor. Use superscripts to identify the material as shown in Eq. (7-25).

7.8. Modify Eqs. (7-25) and (7-26) for an n-p-n transistor, and work Example 7-3 for the n-p-n.

7.9. A symmetrical p^+-n-p^+ Si bipolar transistor has the following properties:

	Emitter	Base
$A = 10^{-4}$ cm^2	$N_a = 10^{17}$	$N_d = 10^{15}$ cm^{-3}
$W_b = 1\ \mu m$	$\tau_n = 0.1\ \mu s$	$\tau_p = 10\ \mu s$
	$\mu_p = 200$	$\mu_n = 1300$ cm^2/V-s
	$\mu_n = 700$	$\mu_p = 450$

(a) Calculate the saturation current $I_{ES} = I_{CS}$.

(b) With $V_{EB} = 0.3$ V and $V_{CB} = -40$ V, calculate the base current I_B, assuming perfect emitter injection efficiency.

7.10. For the transistor described in Prob. 7.9, calculate the emitter injection efficiency γ and the amplification factor β, assuming the emitter region is long compared to L_n.

7.11. (a) How much charge (in coulombs) due to excess holes is stored in the base of the transistor shown in Fig. 7-6 at the d-c bias given?

(b) Why is the base transport factor B different in the normal and inverted modes for the transistor shown in Fig. 7-7?

7.12. (a) How is it possible that the average time an injected hole spends in transit across the base τ_t is shorter than the hole lifetime in the base τ_p?

(b) Explain why the turn-on transient of a BJT is faster when the device is driven into oversaturation.

7.13. Using appropriate diagrams, explain how you would *measure* I_{ES} and I_{CO} for a p-n-p transistor.

7.14. The symmetrical p^+-n-p^+ transistor of Fig. P7-14 is connected as a diode in the four configurations shown. Assume that $V \gg kT/q$. Sketch $\delta p(x_n)$ in the base region for each case. Which connection seems most appropriate for use as a diode? Why?

Figure P7-14

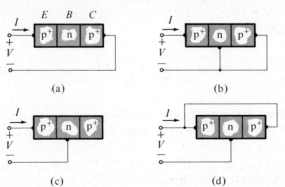

7.15. For the transistor connection in Fig. P7-14a, (a) show that $V_{EB} = (kT/q)$ ln 2; (b) find the expression for I when $V \gg kT/q$ and sketch I vs. V.

7.16. (a) Find the expression for the current I for the transistor connection of Fig. P7-14b; compare the result with the narrow base diode problem (Prob. 6.5).
(b) How does the current I divide between the base lead and the collector lead?

7.17. Suppose that V is negative in Fig. P7-14c.
(a) Find I from the Ebers–Moll equations.
(b) Find the expression for V_{CB}.
(c) Sketch $\delta p(x_n)$ in the base.

7.18. For the transistor connection of Fig. P7-14d, (a) find the expression for $\delta p(x_n)$ in the base region; (b) find the current I.

7.19. It is obvious from Eqs. (7-35) and (7-36) that I_{EO} and I_{CO} are the saturation currents of the emitter and collector junctions, respectively, with the opposite junction open circuited.
(a) Show that this is true from Eq. (7-32).
(b) Find expressions for the following excess concentrations: Δp_C with the emitter junction forward biased and the collector open; Δp_E with the collector junction forward biased and the emitter open.
(c) Sketch $\delta p(x_n)$ in the base for the two cases of part (b).

7.20. (a) Show that the definitions of Eq. (7-40) are correct; what does q_N represent?
(b) Show that Eqs. (7-39) correspond to Eqs. (7-34), using the definitions of Eqs. (7-40).

7.21. Show that the stored charge $Q_b(t)$ decays according to Eq. (7-51) when the base current is driven from I_B to $-I_B$ at $t = 0$.

7.22. Derive Eq. (7-61) for $\Delta p_E(t)$ assuming that the emitter has an applied voltage

$$v_{EB}(t) = V_{EB} + v_{eb}(t)$$

where $V_{EB} \gg kT/q$. For $v_{eb} \ll kT/q$, the approximation $e^x \simeq 1 + x$ can be employed.

7.23. If the graded-base transistor of Fig. 7-21 were redrawn for an n-p-n, would the electric field in the base assist or oppose the electron flow? Explain.

7.24. Use Example 5-6 to design an N-p-n heterojunction bipolar transistor with reasonable γ and base resistance.

*7.25. The current amplification factor β of a BJT is very sensitive to the base width as well as to the ratio of the base doping to the emitter doping. Calculate and plot β for a p-n-p BJT with $L_p^n = L_n^p$, for:
  (a) $n_n = p_p$, $W_b/L_p^n = 0.01$ to 1;
  (b) $W_b = L_p^n$, $n_n/p_p = 0.01$ to 1.
Neglect mobility variations ($\mu_n^p = \mu_p^n$).

*7.26. Calculate and plot the common-base characteristics at 300 K of a symmetrical Si p-n-p BJT with a base area of 10^2 cm^{-2}, a base width of 500 Å, base doping of 10^{14} cm^{-3}, emitter and collector doping of 10^{17} cm^{-3}, and carrier lifetimes $\tau_n = \tau_p = 10^{-6}$ s, for emitter currents of 0, 0.04, 0.08, and 0.12 A.

READING LIST

Becker, J. A. and J. N. Shive. "The Transistor—A New Semiconductor Amplifier." *Proceedings of the IEEE* 72, no. 12 (December 1984): 1696–1703.

Eastman, L. F. "Compound-Semiconductor Transistors." *Physics Today* 39, no. 10 (October 1986): 77–83.

Eden, R. C., A. R. Livingston, and B. M. Welch. "Integrated Circuits: The Case for Gallium Arsenide." *IEEE Spectrum* 20, no. 12 (December 1983): 30–37.

Ferry, D. K., ed. *Gallium Arsenide Technology.* Indianapolis: Howard W. Sams & Co., 1985.

Grove, A. S. *Physics and Technology of Semiconductor Devices.* New York: John Wiley, 1967.

Moll, J. L. "Junction Transistor Electronics." *Proceedings of the IRE* 43 (December 1955): 1807–19.

Moll, J. L. *Physics of Semiconductors.* New York: McGraw-Hill, 1964.

Muller, R. S. and T. I. Kamins. *Device Electronics for Integrated Circuits.* 2d ed. New York: John Wiley, 1986.

Neudeck, G. W. *The Modular Series on Solid State Devices. Vol. III: The Bipolar Junction Transistor,* 2nd edition. Reading, Mass.: Addison-Wesley, 1989.

Pulfrey, D. L. and N. G. Tarr. *Introduction to Microelectronic Devices.* Englewood Cliffs, NJ: Prentice Hall, 1989.

Shur, M. *GaAs Devices and Circuits.* New York: Plenum Press, 1987.

Sze, S. M. *Physics of Semiconductor Devices.* New York: John Wiley, 1981.

Sze, S. M. *Semiconductor Devices.* New York: John Wiley, 1985.

chapter 8

FIELD-EFFECT TRANSISTORS

In this chapter we continue the study of transistors by considering the *field-effect* transistor (*FET*). Like its bipolar counterpart, the FET is a three-terminal device in which the current through two terminals is controlled at the third. Unlike the BJT, however, field-effect devices are controlled by a voltage at the third terminal rather than by a current. Another difference is that the FET is a *unipolar* device; that is, the current involves only majority carriers.

The field-effect transistor comes in several forms. In a *junction* FET (called a *JFET*) the control (*gate*) voltage varies the depletion width of a reverse-biased p-n junction. A similar device results if the junction is replaced by a Schottky barrier (*metal-semiconductor* FET, called a *MESFET*). Alternatively, the metal gate electrode may be separated from the semiconductor by an insulator (metal–insulator–semiconductor FET, called a *MISFET*). A common special case of this type uses an oxide layer as the insulator (*MOSFET*).

The various types of FET are characterized by a high *input impedance,* since the control voltage is applied to a reverse-biased junction or Schottky barrier, or across an insulator. These devices are particularly well suited for controlled switching between a conducting state and a nonconducting state, and are therefore useful in digital circuits. They are also suitable for integration of many devices on a single chip, as we shall see in Chapter 9. In fact, millions of MOS transistors are commonly used together in semiconductor memory devices.

8.1
THE JUNCTION FET

In a *junction FET* (*JFET*) the voltage-variable depletion region width of a junction is used to control the effective cross-sectional area of a conducting *channel*. In the device of Fig. 8-1, the current I_D flows through an n-type channel between two p^+ regions. A reverse bias between these p^+ regions and the channel causes the depletion regions to intrude into the n material, and therefore the effective width of the channel can be restricted. Since the resistivity of the channel region is fixed by its doping, the channel resistance varies with changes in the effective cross-sectional area. By analogy, the variable depletion regions serve as the two doors of a gate, which open and close on the conducting channel.

In Fig. 8-1 electrons in the n-type channel drift from right to left, opposite to current flow. The end of the channel from which electrons flow is called the *source,* and the end toward which they flow is called the *drain*. The p^+ regions are called *gates*. If the channel were p-type, holes would flow from the source to the drain, in the same direction as the current flow, and the gate regions would be n^+.[†] It is common practice to connect the two gate regions electrically; therefore, the voltage V_G refers to the potential from each gate region G to the source S. Since the conductivity of the heavily doped p^+ regions is high, we can assume that the potential is uniform throughout each gate. In the lightly doped channel material, however, the potential varies with position (Fig. 8-1b). If the channel of Fig. 8-1 is considered as a distributed resistor carrying a current I_D it is clear that the voltage from the drain end of the channel D to the

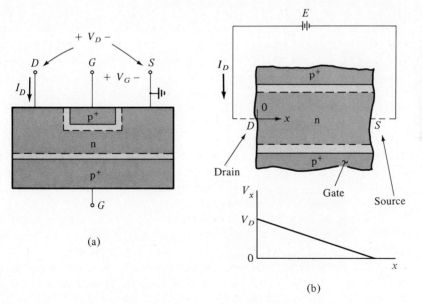

(a)

(b)

Figure 8-1
Simplified cross-sectional view of a junction FET: (a) transistor geometry; (b) detail of the channel and voltage variation along the channel with $V_G = 0$ and small I_D.

[†]In early treatments of field-effect transistors p-channel devices were common, and it was most convenient to draw the source on the left and drain on the right, with hole current flowing from left to right on the diagram. Since n-channel devices are now more common, we choose in this chapter to draw FET devices with the current flowing left to right, from drain to source. In comparing figures in this chapter with other books, the student will find that many use the convention of placing the source on the left.

source electrode S must be greater than the voltage from a point near the source end to S. For low values of current we can assume a linear variation of voltage V_x in the channel, varying from V_D at the drain end to zero at the source end (Fig. 8-1b).

8.1.1 Pinch-off and Saturation

In Figure 8-2 we consider the channel in a simplified way by neglecting voltage drops between the source and drain electrodes and the respective ends of the channel. For example, we assume that the potential at the drain end of the channel is the same as the potential at the electrode D. This is a good approximation if the source and drain regions are relatively large, so that there is little resistance between the ends of the channel and the electrodes. In Fig. 8-2 the gates are short circuited to the source ($V_G = 0$), such that the potential at $x = L$ is the same as the potential everywhere in the gate regions. For very small currents, the widths of the depletion regions are close to the equilibrium values (Fig. 8-2a). As the current I_D is increased, however, it becomes important that V_x is large near the drain end and small near the source end of the channel. Since the reverse bias across each point in the gate-to-channel junction is simply V_x when V_G is zero, we can estimate the shape of the depletion regions as in Fig. 8-2b. The reverse bias is relatively large near the drain ($V_{GD} = -V_D$) and decreases toward zero near the source. As a result, the de-

Figure 8-2
Depletion regions in the channel of a JFET with zero gate bias for several values of V_D: (a) linear range; (b) near pinch-off; (c) beyond pinch-off.

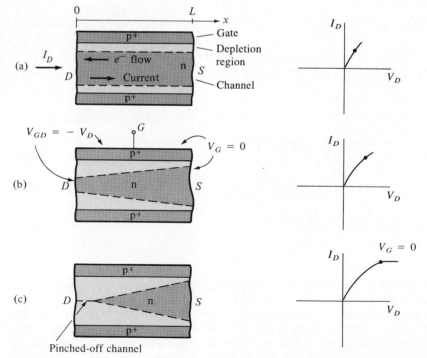

Pinched-off channel

pletion region intrudes into the channel near the drain, and the effective channel area is constricted.

Since the resistance of the constricted channel is higher, the *I–V* plot for the channel begins to depart from the straight line that was valid at low current levels. As the voltage V_D and current I_D are increased still further, the channel region near the drain becomes more constricted by the depletion regions and the channel resistance continues to increase. As V_D is increased, there must be some bias voltage at which the depletion regions meet near the drain and essentially *pinch off* the channel (Fig. 8-2c). When this happens, the current I_D cannot increase significantly with further increase in V_D. Beyond pinch-off the current is *saturated* approximately at its value at pinch-off.[†] Once electrons from the channel enter the electric field of the depletion region, they are swept through and ultimately flow to the positive drain contact. After the current saturates beyond pinch-off, the differential channel resistance dV_D/dI_D becomes very high. To a good approximation, we can calculate the current at the critical pinch-off voltage and assume there is no further increase in current as V_D is increased.

8.1.2   Gate Control

The effect of a negative gate bias $-V_G$ is to increase the resistance of the channel and induce pinch-off at a lower value of current (Fig. 8-3). Since the depletion regions are larger with V_G negative, the effective channel width is smaller and its resistance is higher in the low-current range of the characteristic. Therefore, the slopes of the I_D vs. V_D curves below pinch-off become smaller as the gate voltage is made more negative (Fig. 8-3b). The pinch-off condition is reached at a lower drain-to-source voltage, and the saturation current is lower than for the case of zero gate bias. As V_G is varied, a family of curves is obtained for the *I–V* characteristic of the channel, as in Fig. 8-3b.

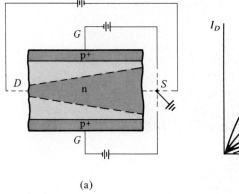

(a)

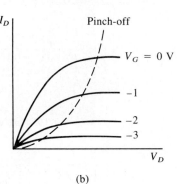

(b)

Figure 8-3
Effects of a negative gate bias: (a) increase of depletion region widths with V_G negative; (b) family of current–voltage curves for the channel as V_G is varied.

[†]*Saturation* is used by device engineers in more different contexts than any other word. We have discussed velocity saturation, reverse saturation current of a junction, saturation of a BJT, and now the saturation of FET characteristics. The student has probably also reached saturation by now in trying to absorb these various meanings.

Beyond the pinch-off voltage the drain current I_D is controlled by V_G. By varying the gate bias we can obtain amplification of an a-c signal. Since the input control voltage V_G appears across the reverse-biased gate junctions, the input impedance of the device is high.

We can calculate the pinch-off voltage rather simply by representing the channel in the approximate form of Fig. 8-4. If the channel is symmetrical and the effects of the gates are the same in each half of the channel region, we can restrict our attention to the channel half-width $h(x)$, measured from the center line ($y = 0$). The metallurgical half-width of the channel (i.e., neglecting the depletion region) is a. We can find the pinch-off voltage by calculating the reverse bias between the n channel and the p^+ gate at the drain end of the channel ($x = 0$). For simplicity we shall assume that the channel width at the drain decreases uniformly as the reverse bias increases to pinch-off. If the reverse bias between the gate and the drain is $-V_{GD}$, the width of the depletion region at $x = 0$ can be found from Eq. (5-57):

$$W(x = 0) = \left[\frac{2\epsilon(-V_{GD})}{qN_d} \right]^{1/2} \quad (V_{GD} \text{ negative}) \qquad (8\text{-}1)$$

In this expression we assume the equilibrium contact potential V_0 is negligible compared with V_{GD} and the depletion region extends primarily into the channel for the p^+-n junction.

Pinch-off occurs at the drain end of the channel when

$$h(x = 0) = a - W(x = 0) = 0 \qquad (8\text{-}2)$$

Figure 8-4
Simplified diagram of the channel with definitions of dimensions and differential volume for calculations.

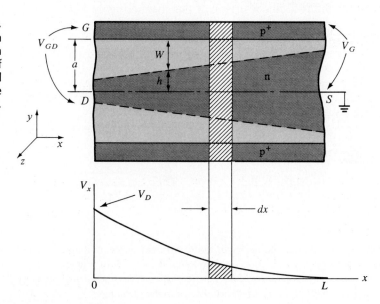

that is, when $W(x = 0) = a$. If we define the value of $-V_{GD}$ at pinch-off as V_P, we have

$$\left[\frac{2\epsilon V_P}{qN_d}\right]^{1/2} = a$$

$$V_P = \frac{qa^2N_d}{2\epsilon} \qquad (8\text{-}3)$$

The pinch-off voltage V_P is a positive number; its relation to V_D and V_G is

$$V_P = -V_{GD}(\text{pinch-off}) = -V_G + V_D \qquad (8\text{-}4)$$

where V_G is zero or negative for proper device operation. A forward bias on the gate would cause hole injection from the p^+ regions into the channel, eliminating the field-effect control of the device. From Eq. (8-4) it is clear that pinch-off results from a combination of gate-to-source voltage and drain-to-source voltage. Pinch-off is reached at a lower value of V_D (and therefore a lower I_D) when a negative gate bias is applied, in agreement with Fig. 8-3b.

8.1.3   Current–Voltage Characteristics

Calculation of the exact channel current is complicated, although the mathematics is relatively straightforward below pinch-off. The approach we shall take is to find the expression for I_D just at pinch-off, and then assume the saturation current beyond pinch-off remains fairly constant at this value.

In the coordinate system defined in Fig. 8-4, the center of the channel at the drain end is taken as the origin. The length of the channel in the x-direction is L, and the depth of channel in the z-direction is Z. We shall call the resistivity of the n-type channel material ρ (valid only in the neutral n material, outside the depletion regions). If we consider the differential volume of neutral channel material $Z2h(x)\,dx$, the resistance of the volume element is $\rho dx/Z2h(x)$ [see Eq. (3-44)]. Since the current does not change with distance along the channel, I_D is related to the differential voltage drop in the element $-dV_x$ by the conductance of the element:

$$I_D = -\frac{Z2h(x)}{\rho}\frac{dV_x}{dx} \qquad (8\text{-}5)$$

The minus sign associated with dV_x simply indicates that V_x decreases as x increases along the channel. The term $2h(x)$ is the channel width at x.

The half-width of the channel at point x depends on the local reverse bias between gate and channel $-V_{Gx}$:

$$h(x) = a - W(x) = a - \left[\frac{2\epsilon(-V_{Gx})}{qN_d}\right]^{1/2} = a\left[1 - \left(\frac{V_x - V_G}{V_P}\right)^{1/2}\right]$$

$$\qquad (8\text{-}6)$$

since $V_{Gx} = V_G - V_x$ and $V_P = qa^2N_d/2\epsilon$. Implicit in Eq. (8-6) is the assumption that the expression for $W(x)$ can be obtained by a simple extension of Eq. (8-1) to point x in the channel. This is called the *gradual approximation;* it is valid if $h(x)$ does not vary abruptly in any element dx.

The voltage V_{Gx} will be negative since the gate voltage V_G is chosen zero or negative for proper operation. Substituting Eq. (8-6) into Eq. (8-5), we have

$$\frac{2Za}{\rho}\left[1 - \left(\frac{V_x - V_G}{V_P}\right)^{1/2}\right]dV_x = -I_D\,dx \tag{8-7}$$

We can solve this equation (Prob. 8.2) to obtain

$$I_D = G_0V_P\left[\frac{V_D}{V_P} + \frac{2}{3}\left(-\frac{V_G}{V_P}\right)^{3/2} - \frac{2}{3}\left(\frac{V_D - V_G}{V_P}\right)^{3/2}\right] \tag{8-8}$$

where V_G is negative and $G_0 \equiv 2aZ/\rho L$ is the conductance of the channel for negligible $W(x)$, i.e., with no gate voltage and low values of I_D. This equation is valid only up to pinch-off, where $V_D - V_G = V_P$. If we assume the saturation current remains essentially constant at its value at pinch-off, we have

$$\begin{aligned}I_D(\text{sat.}) &= G_0V_P\left[\frac{V_D}{V_P} + \frac{2}{3}\left(-\frac{V_G}{V_P}\right)^{3/2} - \frac{2}{3}\right]\\ &= G_0V_P\left[\frac{V_G}{V_P} + \frac{2}{3}\left(-\frac{V_G}{V_P}\right)^{3/2} + \frac{1}{3}\right]\end{aligned} \tag{8-9}$$

where

$$\frac{V_D}{V_P} = 1 + \frac{V_G}{V_P}$$

The resulting family of *I–V* curves for the channel agrees with the results we predicted qualitatively (Fig. 8-3b). The saturation current is greatest when V_G is zero and becomes smaller as V_G is made negative.

We can represent the device biased in the saturation region by an equivalent circuit where changes in drain current are related to gate voltage changes by

$$g_m(\text{sat.}) = \frac{\partial I_D(\text{sat.})}{\partial V_G} = G_0\left[1 - \left(-\frac{V_G}{V_P}\right)^{1/2}\right] \tag{8-10}$$

The quantity g_m is called the *mutual transconductance,* with units (A/V) of siemens (S), sometimes called mhos. As a figure of merit for FET devices it is common to describe the transconductance per unit channel width Z. This quantity g_m/Z is usually given in units of millisiemens per millimeter.

It is found experimentally that a square-law characteristic closely approximates the drain current in saturation:

$$I_D(\text{sat.}) \simeq I_{DSS}\left(1 + \frac{V_G}{V_P}\right)^2, \qquad (V_G\ \textit{negative}) \tag{8-11}$$

where I_{DSS} is the saturated drain current with $V_G = 0$.

The appearance of a constant value of channel resistivity (in the G_0 term) in Eqs. (8-8)–(8-10) implies that the electron mobility is constant. As mentioned in Sec. 3.4.4, electron velocity saturation at high fields may make this assumption invalid. This is particularly likely for very short channels, where even moderate drain voltage can result in a high field along the channel. Another departure from the ideal model results from the fact that the effective channel length decreases as the drain voltage is increased beyond pinch-off, as Fig. 8-2(c) suggests. In short-channel devices this effect can cause I_D to increase beyond pinch-off, since L appears in the denominator of Eq. (8-9), in G_0. Therefore, the assumption of constant saturation current is not valid for very short-channel devices.

The depletion of the channel discussed above for a JFET can be accomplished by the use of a reverse-biased Schottky barrier instead of a p-n junction. The resulting device is called a MESFET, indicating that a metal–semiconductor junction is used. This device is useful in high-speed digital or microwave circuits, where the simplicity of Schottky barriers allows fabrication to close geometrical tolerances. There are particular speed advantages for MESFET devices in III–V compounds such as GaAs or InP, which have higher mobilities and carrier drift velocities than Si.

**8.2
THE METAL–
SEMICONDUCTOR
FET**

8.2.1   The GaAs MESFET

Figure 8-5 shows schematically a simple MESFET in GaAs. The substrate is undoped or doped with chromium, which has an energy level near the center of the GaAs band gap. In either case the Fermi level is near the center of the gap, resulting in very high resistivity material ($\sim 10^8$ Ω-cm), generally called *semi-insulating* GaAs. On this nonconducting substrate a thin layer of lightly-doped n-type GaAs is grown epitaxially, to form the channel region of the FET.[†] The

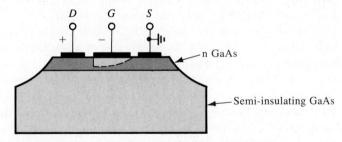

GaAs MESFET formed on an n-type GaAs layer grown epitaxially on a semi-insulating substrate. Common metals for the Schottky gate in GaAs are Al or alloys of Ti, W, and Au. The ohmic source and drain contacts may be an alloy of Au and Ge. In this example the device is isolated from others on the same chip by etching through the n region to the semi-insulating substrate.

Figure 8-5

[†]In many cases a high resistivity GaAs epitaxial layer (called a *buffer layer*) is grown between the two layers shown in Fig. 8-5.

photolithographic processing consists of defining patterns in the metal layers for source and drain ohmic contacts (e.g., Au–Ge) and for the Schottky barrier gate (e.g., Al). By reverse biasing the Schottky gate, the channel can be depleted to the semi-insulating substrate, and the resulting I–V characteristics are similar to the JFET device.

By using GaAs instead of Si, a higher electron mobility is available (see Appendix III), and furthermore GaAs can be operated at higher temperatures (and therefore higher power levels). Since no diffusions are involved in Fig. 8-5, close geometrical tolerances can be achieved and the MESFET can be made very small. Gate lengths $L \lesssim 1$ μm are common in these devices. This is important at high frequencies, since drift time and capacitances must be kept to a minimum.

It is possible to avoid the epitaxial growth of the n-type layer and the etched isolation in Fig. 8-5 by using ion implantation. Starting with a semi-insulating GaAs substrate, a thin n-type layer at the surface of each transistor region can be formed by implanting Si or a column VI donor impurity such as Se. This implantation requires an anneal to remove the radiation damage, but the epitaxial growth step is eliminated. In either the fully implanted device or the epitaxial device of Fig. 8-5, the source and drain contacts may be improved by further n^+ implantation in these regions. Because of the relative simplicity of implanted GaAs MESFETs and the isolation between devices provided by the semi-insulating substrate, these structures are commonly used in GaAs integrated circuits.

8.2.2   The High Electron Mobility Transistor (HEMT)

Since the metal–semiconductor field effect transistor (MESFET) is compatible with the use of III–V compounds, it is possible to exploit the band gap engineering available with heterojunctions in these materials. In order to maintain high transconductance in a MESFET, the channel conductivity must be as high as possible. Obviously, the conductivity can be increased by increasing the doping in the channel and thus the carrier concentration. However, increased doping also causes increased scattering by the ionized impurities, which leads to a degradation of mobility (see Fig. 3-23). What is needed is a way of creating a high electron concentration in the channel of a MESFET by some means other than doping. A clever approach to this requirement is to grow a thin undoped well (e.g., GaAs) bounded by wider band gap, doped barriers (e.g., AlGaAs). This configuration, called *modulation doping*, results in conductive GaAs when electrons from the doped AlGaAs barriers fall into the well and become trapped there, as shown in Fig. 8-6(a). Since the donors are in the AlGaAs rather than the GaAs, there is no impurity scattering of electrons in the well. If a MESFET is constructed with the channel along the GaAs well (perpendicular to the page in Fig. 8-6), we can take advantage of this reduced scattering and resulting higher mobility. The effect is especially strong at low temperature where lattice (phonon) scattering is also low. This device is called a *modulation doped field-effect transistor (MODFET)* and is also called a *high electron mobility transistor (HEMT)*.

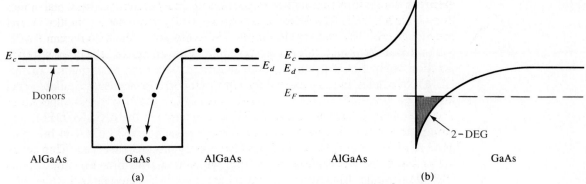

(a) Simplified view of modulation doping, showing only the conduction band. **Figure 8-6**
Electrons in the donor-doped AlGaAs fall into the GaAs potential well and
become trapped. As a result, the undoped GaAs becomes n-type, without the
scattering by ionized donors which is typical of bulk n-type material. (b) Use of
a single AlGaAs/GaAs heterojunction to trap electrons in the undoped GaAs.
The thin sheet of charge due to free electrons at the interface forms a
two-dimensional electron gas (2-DEG), which can be exploited in HEMT
devices.

In Fig. 8-6(a) we have left out the band-bending expected at the AlGaAs/GaAs
interfaces. Based upon the discussion in Section 5.8, we expect the electrons
to accumulate at the corners of the well due to band-bending at the heterojunction. In fact, only one heterojunction is required to trap electrons, as shown in
Fig. 8-6(b). Generally, the donors in the AlGaAs layer are purposely separated
from the interface by ~ 100 Å. Using this configuration, we can achieve a high
electron concentration in the channel while retaining high mobility, since the
GaAs channel region is spatially separated from the ionized impurities which
provide the free carriers.

In Fig. 8-6(b), mobile electrons generated by the donors in the AlGaAs
diffuse into the small band gap GaAs layer, and they are prevented from
returning to the AlGaAs by the potential barrier at the AlGaAs/GaAs interface.
The electrons in the (almost) triangular well form a two-dimensional electron
gas (sometimes abbreviated 2-DEG). Sheet carrier densities as high as
$10^{12} \mathrm{cm}^{-2}$ can be obtained at a single interface such as that shown in Fig. 8-6(b).
Ionized impurity scattering is greatly reduced simply by separating the electrons
from the donors. Also, screening effects due to the extremely high density of
the two-dimensional electron gas can reduce ionized impurity scattering further. In properly designed structures, the electron transport approaches that of
bulk GaAs with no impurities, so that mobility is limited by lattice scattering.
As a result, mobilities above 250,000 cm^2/V-s at 77 K and 2,000,000 cm^2/V-s
at 4 K can be achieved.

The advantages of a HEMT are its ability to locate a large electron density
($\sim 10^{12}$ cm^{-2}) in a very thin layer (<100 Å thick) very close to the gate while
simultaneously eliminating ionized impurity scattering. The AlGaAs layer in
a HEMT is fully depleted under normal operating conditions, and since the

electrons are confined to the heterojunction, device behavior closely resembles that of a MOSFET. The advantages of the HEMT over the Si MOSFET are the higher mobility and maximum electron velocity in GaAs compared with Si, and the smoother interfaces possible with an AlGaAs/GaAs heterojunction compared with the Si/SiO$_2$ interface. The high performance of the HEMT translates into an extremely high cutoff frequency, and devices with fast access times.

Although we have discussed the HEMT in terms of the AlGaAs/GaAs heterojunction, other materials are also promising, such as the InGaAsP/InP system. A motivation for avoiding Al$_x$Ga$_{1-x}$As is the presence of a deep-level defect called the DX center for $x > 0.2$, which traps electrons and impairs the HEMT operation. Since very thin layers are involved, materials with slight lattice mismatch can be grown to form *pseudomorphic* HEMTs. An example of such a system is the use of a thin layer of InGaAs grown pseudomorphically on GaAs, followed by AlGaAs. An advantage of this system is that a useful band discontinuity can be achieved using AlGaAs of low enough Al composition to avoid the DX center problem.

The HEMT, or MODFET, is also referred to as a *two-dimensional electron gas FET (2-DEG FET, or TEGFET)* to emphasize the fact that conduction along the channel occurs in a thin sheet of charge. The device has also been called a *separately doped FET (SEDFET),* to emphasize the fact that the doping occurs in a separate region from the channel.

8.2.3 Short Channel Effects

As mentioned in Section 8.1.3, a variety of modifications to the simple theory of JFET and MESFET operation must be made when the channel length is small (typically <1 μm). Until recently these short-channel effects would be considered unusual, but now it is common to encounter FET devices in which these effects dominate the I–V characteristics. For example, high-field effects occur when 1 V appears across a channel length of 1 μm (10^{-4} cm), giving an electric field of 10 kV/cm.

A simple piecewise-linear approximation to the velocity-field curve assumes a constant mobility (linear) dependence up to some critical field $\mathscr{E}_c$ and a constant saturation velocity v_s for higher fields. For Si a better approximation is

$$v_d = \frac{\mu \mathscr{E}}{1 + \mu \mathscr{E}/v_s} \qquad (8\text{-}12)$$

where μ is the low-field mobility. These two approximations are shown in Fig. 8-7 (a). If we assume that the electrons passing through the channel drift with a constant saturation velocity v_s, the current takes a simple form

$$I_D = qnv_s A = qN_d v_s Zh \qquad (8\text{-}13)$$

where h is a slow function of V_G. In this case the saturated current follows the velocity saturation, and does not require a true pinch-off in the sense of depletion regions meeting at some point in the channel. In the saturated velocity case, the transconductance g_m is essentially constant, in contrast with the constant

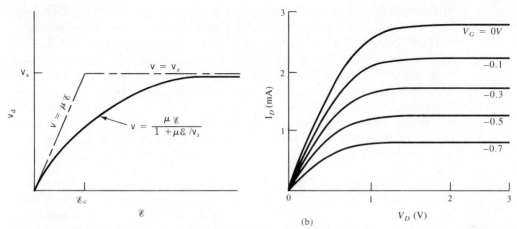

Effects of electron velocity saturation at high electric fields: (a) approximations to the saturation of drift velocity with increasing field; (b) drain current–voltage characteristics for the saturated velocity case, showing almost equally spaced curves with increasing gate voltage.

Figure 8-7

mobility case described by Eq. (8-10). As shown in Fig. 8-7(b), the I_d–V_d curves are more evenly spaced if constant saturation velocity dominates, compared with the V_G-dependent spacing shown in Fig. 8-3(b) for the long-channel constant-mobility case.

Most devices operate with characteristics intermediate between the constant mobility and the constant velocity regimes. Depending on the details of the field distribution, it is possible to divide up the channel into regions dominated by the two extreme cases, or to use an approximation such as Eq. (8-12).

Another important short-channel effect, described in Section 8.1.3, is the reduction in effective channel length after pinch-off as the drain voltage is increased. This effect is not significant in long-channel devices, since the change in L due to intrusion of the depletion region is a minor fraction of the total channel length. In short-channel devices, however, the effective channel length can be substantially shortened, leading to a slope in the saturated I–V characteristic that is analogous to the Early (base-width narrowing) effect in bipolar transistors (Fig. 7-22).

One of the most widely used electronic devices, particularly in digital integrated circuits, is the *metal–insulator–semiconductor (MIS) transistor*. In this device the channel current is controlled by a voltage applied at a gate electrode that is isolated from the channel by an insulator. The resulting device may be referred to generically as an insulated-gate field effect transistor (IGFET). However, since most such devices are made using silicon for the semiconductor, SiO_2 for the insulator, and Al or other metal for the gate electrode, the term *MOS transistor* (MOST) is commonly used.

**8.3
THE METAL–
INSULATOR–
SEMICONDUCTOR
FET**

8.3.1 Basic Operation

The basic MOS transistor is illustrated in Fig. 8-8 for the case of an n-type channel formed on a p-type Si substrate. The n⁺ source and drain regions are diffused or implanted into a relatively lightly doped p-type substrate, and a thin oxide layer separates the Al metal gate from the Si surface. No current flows from drain to source without a conducting n channel between them, since the drain–substrate–source combination includes oppositely directed p-n junctions in series.

When a positive voltage is applied to the gate relative to the substrate (which is connected to the source in this case), positive charges are in effect

Figure 8-8
An enhancement-type n-channel MOS transistor: (a) cross section of the device; (b) schematic illustration of the induced n-channel and the depletion region near pinch-off; (c) drain current–voltage characteristics as a function of gate voltage.

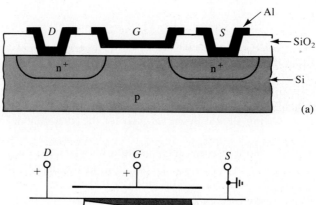

(a)

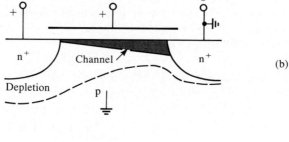

(b)

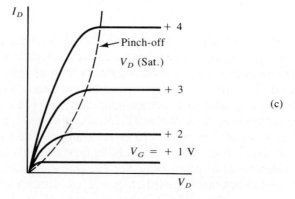

(c)

deposited on the gate metal. In response, negative charges are induced in the underlying Si, by the formation of a depletion region and a thin surface region containing mobile electrons. These induced electrons form the channel of the FET, and allow current to flow from drain to source. As Fig. 8-8c suggests, the effect of the gate voltage is to vary the conductance of this induced channel for low drain-to-source voltage, analogous to the JFET case. For a given value of V_G there will be some drain voltage V_D for which the current becomes saturated, after which it remains essentially constant.

An important parameter in MOS transistors is the *threshold voltage V_T*, which is the minimum gate voltage required to induce the channel. In general, the positive gate voltage of an n-channel device (such as that shown in Fig. 8-8) must be larger than some value V_T before a conducting channel is induced. Similarly, a p-channel device (made on an n-type substrate with p-type source and drain implants or diffusions) requires a gate voltage more negative than some threshold value to induce the required positive charge (mobile holes) in the channel. There are exceptions to this general rule, however, as we shall see. For example, some n-channel devices have a channel already with zero gate voltage, and in fact a negative gate voltage is required to turn the device off. Such a "normally on" device is called a *depletion-mode* transistor, since gate voltage is used to deplete a channel which exists at equilibrium. The more common MOS transistor is "normally off" with zero gate voltage, and operates in the *enhancement mode* by applying a gate voltage large enough to induce a conducting channel.

The MOS transistor is particularly useful in digital circuits, in which it is switched from the "off" state (no conducting channel) to the "on" state. The control of drain current is obtained at a gate electrode which is insulated from the source and drain by the oxide. Thus the d-c input impedance of an MOS circuit can be very large.

Both n-channel and p-channel MOS transistors are in common usage. The n-channel type illustrated in Fig. 8-8 is generally preferred because it takes advantage of the fact that the electron mobility in Si is larger than the mobility of holes. In much of the discussion to follow we will use the n-channel (p-type substrate) example, although the p-channel case will be kept in mind also.

8.3.2 The Ideal MOS Capacitor

The surface effects that arise in an apparently simple MOS structure are actually quite complicated. Although many of these effects are beyond the scope of this discussion, we will be able to identify those which control typical MOS transistor operation. We begin by considering an uncomplicated idealized case, and then include effects encountered in real surfaces in the next section.

Some important definitions are made in the energy band diagram of Fig. 8-9. The work function characteristic of the metal (see Section 2.2.1) can be defined in terms of the energy required to move an electron from the Fermi level to outside the metal. In MOS work it is more convenient to use a *modified work function $q\Phi_m$* for the metal–oxide interface. The energy $q\Phi_m$ is measured from

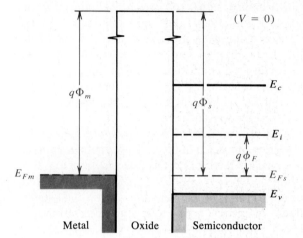

Figure 8-9
Band diagram for
the ideal MOS
structure at
equilibrium.

the metal Fermi level to the conduction band of the oxide.[†] Similarly, $q\Phi_s$ is the modified work function at the semiconductor–oxide interface. In this idealized case we assume that $\Phi_m = \Phi_s$, so there is no difference in the two work functions. Another quantity that will be useful in later discussions is $q\phi_F$, which measures the position of the Fermi level below the intrinsic level E_i for the semiconductor. This quantity indicates how strongly p-type the semiconductor is [see Eq. (3-25)].

The MOS structure of Fig. 8-9 is essentially a capacitor in which one plate is a semiconductor. If we apply a negative voltage between the metal and the semiconductor (Fig. 8-10a), we effectively deposit a negative charge on the metal. In response, we expect an equal net positive charge to accumulate at the surface of the semiconductor. In the case of a p-type substrate this occurs by *hole accumulation* at the semiconductor–oxide interface.

Since the applied negative voltage *depresses* the electrostatic potential of the metal relative to the semiconductor, the electron energies are *raised* in the metal relative to the semiconductor.[‡] As a result, the Fermi level for the metal E_{Fm} lies above its equilibrium position by qV, where V is the applied voltage.

Since Φ_m and Φ_s do not change with applied voltage, moving E_{Fm} up in energy relative to E_{Fs} causes a tilt in the oxide conduction band. We expect such a tilt since an electric field causes a gradient in E_i (and similarly in E_v and E_c) as described in Section 4.4.2:

$$\mathscr{E}(x) = \frac{1}{q}\frac{dE_i}{dx} \qquad \text{(see 4-26)}$$

[†]On the MOS band diagrams of this section we show a break in the electron energy scale leading to the insulator conduction band, since the band gap of SiO_2 (or other typical insulators) is much greater than that of the Si.

[‡]Recall that an electrostatic potential diagram is drawn for positive test charges, in contrast with an electron energy diagram which is drawn for negative charges.

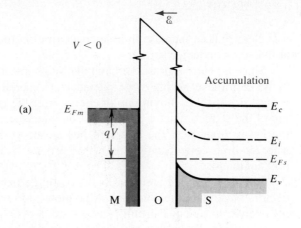

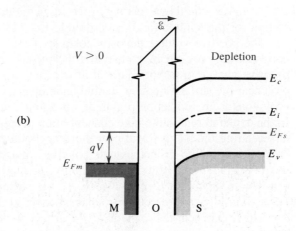

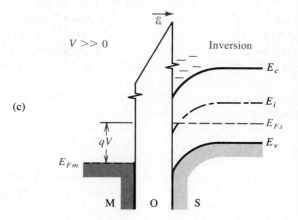

Figure 8-10
Effects of applied voltage on the ideal MOS capacitor: (a) negative voltage causes hole accumulation in the p-type semiconductor; (b) positive voltage depletes holes from the semiconductor surface; (c) a larger positive voltage causes inversion—an "n-type" layer at the semiconductor surface.

The energy bands of the semiconductor bend near the interface to accommodate the accumulation of holes. Since

$$p = n_i e^{(E_i - E_F)/kT}$$

(see 3-25)

it is clear that an increase in hole concentration implies an increase in $E_i - E_F$ at the semiconductor surface.

Since no current passes through the MOS structure, there can be no variation in the Fermi level within the semiconductor. Therefore, if $E_i - E_F$ is to increase, it must occur by E_i moving up in energy near the surface. The result is a bending of the semiconductor bands near the interface. We notice in Fig. 8-10a that the Fermi level near the interface lies closer to the valence band, indicating a larger hole concentration than that arising from the doping of the p-type semiconductor.

In Fig. 8-10b we apply a positive voltage from the metal to the semiconductor. This raises the potential of the metal, lowering the metal Fermi level by qV relative to its equilibrium position. As a result, the oxide conduction band is again tilted. We notice that the slope of this band, obtained by simply moving the metal side down relative to the semiconductor side, is in the proper direction for the applied field, according to Eq. (4-26).

The positive voltage deposits positive charge on the metal and calls for a corresponding net negative charge at the surface of the semiconductor. Such a negative charge in p-type material arises from *depletion* of holes from the region near the surface, leaving behind uncompensated ionized acceptors. This is analogous to the depletion region at a p-n junction discussed in Section 5.2.3. In the depleted region the hole concentration decreases, moving E_i closer to E_F, and bending the bands down near the semiconductor surface.

If we continue to increase the positive voltage, the bands at the semiconductor surface bend down more strongly. In fact, a sufficiently large voltage can bend E_i *below* E_F (Fig. 8-10c). This is a particularly interesting case, since $E_F \gg E_i$ implies a large electron concentration in the conduction band.

The region near the semiconductor surface in this case has conduction properties typical of n-type material, with an electron concentration given by Eq. (3-25a). This n-type surface layer is formed not by doping, but instead by *inversion* of the originally p-type semiconductor due to the applied voltage. This inverted layer, separated from the underlying p-type material by a depletion region, is the key to MOS transistor operation.

We should take a closer look at the inversion region, since it becomes the conducting channel in the FET. In Fig. 8-11 we define a potential ϕ at any point x, measured relative to the equilibrium position of E_i. The energy $q\phi$ tells us the extent of band bending at x, and $q\phi_s$ represents the band bending at the surface. We notice that $\phi_s = 0$ is the *flat band* condition for this ideal MOS case (i.e., the bands look like Fig. 8-9). When $\phi_s < 0$, the bands bend up at the surface, and we have hole accumulation (Fig. 8-10a). Similarly, when $\phi_s > 0$, we have depletion (Fig. 8-10b). Finally, when ϕ_s is positive and larger than ϕ_F, the bands at the surface are bent down such that $E_i(x = 0)$ lies below E_F, and inversion is obtained.

While it is true that the surface is inverted whenever ϕ_s is larger than ϕ_F, a practical criterion is needed to tell us whether a true n-type conducting channel exists at the surface. The best criterion for *strong inversion* is that the surface should be as strongly n-type as the substrate is p-type. That is, E_i should lie as

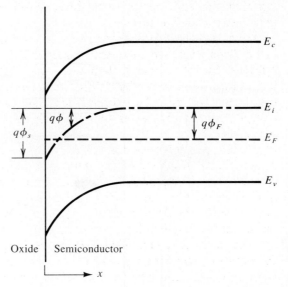

E_c

$q\phi$

$q\phi_s$

E_i

$q\phi_F$

E_F

E_v

Oxide | Semiconductor

x

Figure 8-11
Bending of the semiconductor bands at the onset of strong inversion: the surface potential ϕ_s is twice the value of ϕ_F in the neutral p material.

far below E_F at the surface as it is above E_F far from the surface. This condition occurs when

$$\phi_s(\text{inv.}) = 2\phi_F = 2\frac{kT}{q} \ln \frac{N_a}{n_i} \qquad (8\text{-}14)$$

A surface potential of ϕ_F is required to bend the bands down to the intrinsic condition at the surface ($E_i = E_F$), and E_i must then be depressed another $q\phi_F$ at the surface to obtain the condition we call strong inversion.

The electron and hole concentrations are related to the potential $\phi(x)$ defined in Fig. 8-11. Since the equilibrium electron concentration is

$$n_0 = n_i e^{(E_F - E_i)/kT} = n_i e^{-q\phi_F/kT} \qquad (8\text{-}15)$$

we can easily relate the electron concentration at any x to this value:

$$n = n_i e^{-q(\phi_F - \phi)/kT} = n_0 e^{q\phi/kT} \qquad (8\text{-}16)$$

and similarly for holes:

$$p = p_0 e^{-q\phi/kT} \qquad (8\text{-}17)$$

at any x. We could combine these equations with Poisson's equation (8-18) and the usual charge density expression (8-19) to solve for $\phi(x)$:

$$\frac{\partial^2 \phi}{\partial x^2} = -\frac{\rho(x)}{\epsilon_s} \qquad (8\text{-}18)$$

$$\rho(x) = q(N_d^+ - N_a^- + p - n) \qquad (8\text{-}19)$$

The charge distribution, electric field, and electrostatic potential for the inverted surface are sketched in Fig. 8-12. For simplicity we use the depletion

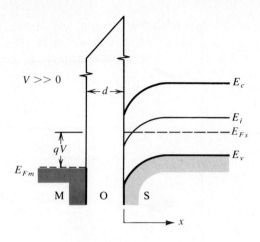

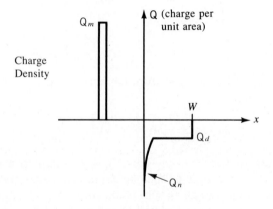

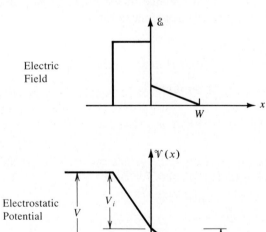

Figure 8-12
Approximate
distributions of
charge, electric
field, and
electrostatic
potential in the
ideal MOS
capacitor in
inversion. The
relative width of
the inverted region
is exaggerated for
illustrative
purposes, but is
neglected in the
field and potential
diagrams.

approximation of Chapter 5 in this figure, assuming complete depletion for $0 < x < W$, and neutral material for $x > W$. In this approximation the charge per unit area due to uncompensated acceptors in the depletion region is $-qN_aW$.[†] The positive charge Q_m on the metal is balanced by the negative charge Q_s in the semiconductor, which is the depletion layer charge plus the charge due to the inversion region Q_n:

$$Q_m = -Q_s = qN_aW - Q_n \qquad (8\text{-}20)$$

The width of the inversion region is exaggerated in Fig. 8-12 for illustrative purposes. Actually, the width of this region is generally less than 100 Å. Thus we have neglected it in sketching the electric field and potential distribution. In the potential distribution diagram we see that an applied voltage V appears partially across the insulator (V_i) and partially across the depletion region of the semiconductor (ϕ_s):

$$V = V_i + \phi_s \qquad (8\text{-}21)$$

The voltage across the insulator is obviously related to the charge on either side, divided by the capacitance:

$$V_i = \frac{-Q_s d}{\epsilon_i} = \frac{-Q_s}{C_i} \qquad (8\text{-}22)$$

where ϵ_i is permittivity of the insulator and C_i is the insulator capacitance per unit area. The charge Q_s will be negative for n-channel, giving a positive V_i.

Using the depletion approximation, we can solve for W as a function of ϕ_s (Prob. 8.6). The result is the same as would be obtained for an n^+-p junction in Chapter 5, for which the depletion region extends almost entirely into the p region:

$$W = \left[\frac{2\epsilon_s\phi_s}{qN_a}\right]^{1/2} \qquad (8\text{-}23)$$

This depletion region grows with increased voltage across the capacitor until strong inversion is reached. After that, further increases in voltage result in stronger inversion rather than in more depletion. Thus the maximum value of the depletion width is

$$W_m = \left[\frac{2\epsilon_s\phi_s(\text{inv.})}{qN_a}\right]^{1/2} = 2\left[\frac{\epsilon_s kT \ln(N_a/n_i)}{q^2 N_a}\right]^{1/2} \qquad (8\text{-}24)$$

using Eq. (8-14). We know the quantities in this expression, so W_m can be calculated.

[†]In this chapter we will use charge per unit area Q and capacitance per unit area C to avoid carrying A throughout the discussion.

EXAMPLE 8-1   Find the maximum width of the depletion region for an ideal MOS capacitor on p-type Si with $N_a = 10^{15}$ cm^{-3}.

SOLUTION   The relative dielectric constant of Si is 11.8 from Appendix III. We get ϕ_F from Eq. (8-14):

$$\phi_F = \frac{kT}{q} \ln \frac{N_a}{n_i} = 0.0259 \ln \frac{10^{15}}{1.5 \times 10^{10}} = 0.288 \text{ V}$$

Thus

$$W_m = 2\sqrt{\frac{\epsilon_s \phi_F}{qN_a}} = 2\left[\frac{(11.8)(8.85 \times 10^{-14})(0.288)}{(1.6 \times 10^{-19})(10^{15})}\right]^{1/2}$$

$$= 8.67 \times 10^{-5} \text{ cm} = 0.867 \ \mu\text{m}$$

∎

The charge per unit area in the depletion region Q_d at strong inversion is[†]

$$Q_d = -qN_a W_m = -2(\epsilon_s qN_a \phi_F)^{1/2} \tag{8-25}$$

The applied voltage must be large enough to create this depletion charge plus the surface potential ϕ_s(inv.). Therefore, the *threshold* voltage required for strong inversion in this ideal MOS capacitor is

$$V_T = -\frac{Q_d}{C_i} + 2\phi_F \qquad \text{(ideal case)} \tag{8-26}$$

This assumes the negative charge at the semiconductor surface Q_s at inversion is mostly due to the depletion charge Q_d. The threshold voltage represents the minimum voltage required to achieve strong inversion, and is an extremely important quantity for MOS transistors. We will see in the next section that other terms must be added to this expression for real MOS structures.

The capacitance–voltage characteristics of this ideal MOS structure (Fig. 8-13) vary depending on whether the semiconductor surface is in accumulation, depletion, or inversion. For negative voltage, holes are accumulated at the surface (Fig. 8-10a). As a result, the MOS structure appears almost like a parallel-plate capacitor, dominated by the insulator properties $C_i = \epsilon_i/d$. As the voltage becomes positive, the semiconductor surface is depleted. Thus a depletion-layer capacitance C_d is added in series with C_i:

$$C_d = \frac{\epsilon_s}{W} \tag{8-27}$$

where ϵ_s is the semiconductor permittivity and W is the width of the depletion layer from Eq. (8-23). The total capacitance is

$$C = \frac{C_i C_d}{C_i + C_d} \tag{8-28}$$

The capacitance decreases with positive voltage as W grows, until finally

---

[†]In the p-channel (n-type substrate) case, for which ϕ_F is negative, we use $Q_d = +qN_d W_m = 2(\epsilon_s qN_d |\phi_F|)^{1/2}$.

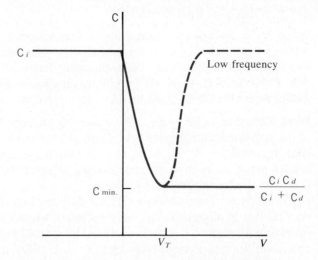

Figure 8-13
Capacitance–voltage
relation for the
ideal n-channel
(p-substrate) MOS
capacitor. The
dashed curve for
$V > V_T$ is observed
only at very low
measurement
frequencies.

inversion is reached at V_T. With inversion there is no further change in C_d, since the depletion width has reached its maximum, W_m. Actually, if the capacitance measurement is made using a very low frequency sampling voltage (e.g., ~ 10 Hz), the recombination–generation kinetics of the electrons in the inversion region can vary in response to the voltage variations. Therefore, the a-c measurement samples small variations in the inversion region rather than in the depletion region. Because of this effect, for measurements at very low frequency the MOS capacitor in inversion resembles the parallel-plate capacitor C_i again (dashed line in Fig. 8-13). This effect can be induced at higher measurement frequencies by speeding up the carrier generation rate (e.g., by shining light on the sample).

EXAMPLE 8-2

Using the conditions of Example 8-1 and a 1000-Å-thick SiO_2 layer, we can calculate major points on the C–V curve of Fig. 8-13. The relative dielectric constant of SiO_2 is 3.9.

$$C_i = \frac{\epsilon_i}{d} = \frac{(3.9)(8.85 \times 10^{-14})}{10^{-5}} = 3.45 \times 10^{-8} \text{ F/cm}^2$$

$$Q_d = -qN_aW_m = -(1.6 \times 10^{-19})(10^{15})(0.867 \times 10^{-4})$$
$$= -1.39 \times 10^{-8} \text{ C/cm}^2$$

$$V_T = -\frac{Q_d}{C_i} + 2\phi_F = \frac{1.39 \times 10^{-8}}{3.45 \times 10^{-8}} + 2(0.288) = 0.98 \text{ V}$$

At V_T:

$$C_d = \frac{\epsilon_s}{W_m} = \frac{(11.8)(8.85 \times 10^{-14})}{0.867 \times 10^{-4}} = 1.2 \times 10^{-8} \text{ F/cm}^2$$

$$C_{min} = \frac{C_iC_d}{C_i + C_d} = \frac{3.45 \times 1.2}{3.45 + 1.2}10^{-8} = 0.89 \times 10^{-8} \text{ F/cm}^2$$

8.3.3 Effects of Real Surfaces

When MOS devices are made using typical materials (e.g., Al–SiO$_2$–Si), departures from the ideal case described in the previous section can strongly affect V_T and other properties. First, the work function of Al is not the same as that of Si. Second, there are inevitably charges at the Si–SiO$_2$ interface and within the oxide which must be taken into account.

Work Function Difference. We expect Φ_s to vary depending on the doping of the semiconductor. Figure 8-14 illustrates the work function potential difference $\Phi_{ms} = \Phi_m - \Phi_s$ for Al on Si as the doping is varied. We note that Φ_{ms} is always negative for this case, and is most negative for heavily doped p-type Si (i.e., for E_F close to the valence band).

If we try to construct an equilibrium diagram with Φ_{ms} negative (Fig. 8-15a), we find that in aligning E_F we must include a tilt in the oxide conduction band (implying an electric field). Thus the metal is positively charged and the semiconductor surface is negatively charged at equilibrium, to accommodate the work function difference. As a result, the bands bend down near the semiconductor surface. In fact, if Φ_{ms} is sufficiently negative, an inversion region can exist with no external voltage applied. To obtain the *flat band* condition pictured in Fig. 8-15b, we must apply a negative voltage to the metal ($V_{FB} = \Phi_{ms}$).

Interface Charge. In addition to the work function difference, the equilibrium MOS structure is affected by charges in the insulator and at the semiconductor–oxide interface (Fig. 8-16). For example, alkali metal ions (particularly Na$^+$) can be incorporated inadvertently in the oxide during growth or subsequent processing steps. Since sodium is a common contaminant, it is necessary to use extremely clean chemicals, water, gases, and processing environment to minimize its effect on dielectric layers. Sodium ions introduce positive charges (Q_m) in the oxide, which in turn induce negative charges in the semiconductor.

Figure 8-14
Variation of the metal–semiconductor work function potential difference Φ_{ms} with substrate doping concentration, for Al–Si.

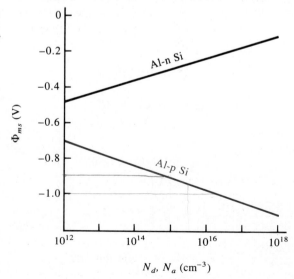

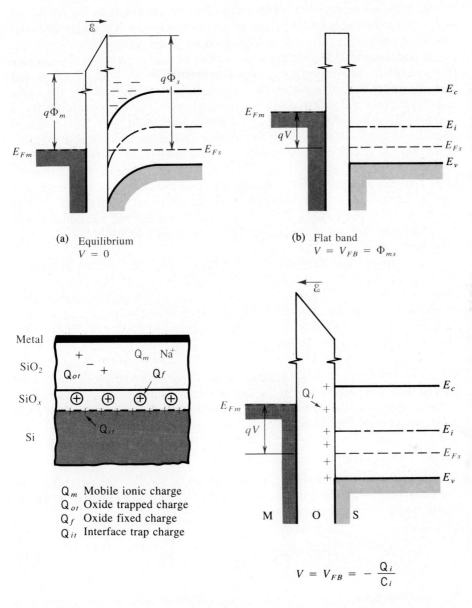

Figure 8-15
Effect of a negative work function difference ($\Phi_{ms} < 0$): (a) band bending and formation of negative charge at the semiconductor surface; (b) achievement of the flat band condition by application of a negative voltage.

(a) Equilibrium
$V = 0$

(b) Flat band
$V = V_{FB} = \Phi_{ms}$

Figure 8-16
Effects of charges in the oxide and at the interface: (a) definitions of charge densities (C/cm^2) due to various sources; (b) representing these charges as an equivalent sheet of positive charge Q_i at the oxide–semiconductor interface. This positive charge induces an equivalent negative charge in the semiconductor, which requires a negative gate voltage to achieve the flat band condition.

Q_m Mobile ionic charge
Q_{ot} Oxide trapped charge
Q_f Oxide fixed charge
Q_{it} Interface trap charge

$$V = V_{FB} = -\frac{Q_i}{C_i}$$

The effect of such positive ionic charges in the oxide depends upon the number of ions involved and their distance from the semiconductor surface (Prob. 8.11). The negative charge induced in the semiconductor is greater if the Na^+ ions are near the interface than if they are farther away. The effect of this ionic charge on threshold voltage is complicated by the fact that Na^+ ions are relatively mobile in SiO_2, particularly at elevated temperatures, and can thus drift in an applied electric field. Obviously, a device with V_T dependent on its past history of voltage bias is unacceptable. Fortunately, Na contamination of the oxide can be reduced to tolerable levels by proper care in processing. The oxide also contains trapped charges (Q_{ot}) due to imperfections in the SiO_2.

In addition to oxide charges, a set of positive charges arises from *interface states* at the Si–SiO$_2$ interface. These charges, which we will call Q_{it}, result from the sudden termination of the semiconductor crystal lattice at the oxide interface. Near the interface is a transition layer (SiO_x) containing fixed charges (Q_f). As oxidation takes place in forming the SiO_2 layer, Si is removed from the surface and reacts with the oxygen. When the oxidation is stopped, some ionic Si is left near the interface. These ions, along with uncompleted Si bonds at the surface, result in a sheet of positive charge Q_f near the interface. This charge depends on oxidation rate and subsequent heat treatment, and also on crystal orientation. For carefully treated Si–SiO$_2$ interfaces, typical charge densities due to Q_{it} and Q_f are about 10^{10} charges/cm^2 for samples with {100} surfaces. The interface charge density is about a factor of ten higher on {111} surfaces.

For simplicity we will include the various oxide and interface charges in an *effective* positive charge at the interface Q_i (C/cm^2). The effect of this charge is to induce an equivalent negative charge in the semiconductor. Thus an additional component must be added to the flat band voltage:

$$V_{FB} = \Phi_{ms} - \frac{Q_i}{C_i} \qquad (8\text{-}29)$$

Since the difference in work function and the positive interface charge both tend to bend the bands down at the semiconductor surface, a negative voltage must be applied to the metal relative to the semiconductor to achieve the flat band condition of Fig. 8-16b.

8.3.4 Threshold Voltage

The voltage required to achieve flat band should be added to the threshold voltage equation (8-26) obtained for the ideal MOS structure (for which we assumed a zero flat band voltage)

$$V_T = \Phi_{ms} - \frac{Q_i}{C_i} - \frac{Q_d}{C_i} + 2\phi_F \qquad (8\text{-}30)$$

Thus the voltage required to create strong inversion must be large enough to first achieve the flat band condition (Φ_{ms} and Q_i/C_i terms), then accommodate the charge in the depletion region (Q_d/C_i), and finally to induce the inverted region ($2\phi_F$). This equation accounts for the dominant threshold voltage effects in typical MOS devices. It can be used for both n-type and p-type substrates[†] if appropriate signs are included for each term (Fig. 8-17). Typically Φ_{ms} is nega-

---

[†]It is important to remember that n-channel devices are made on p-type substrates, and p-channel devices have n-type substrates.

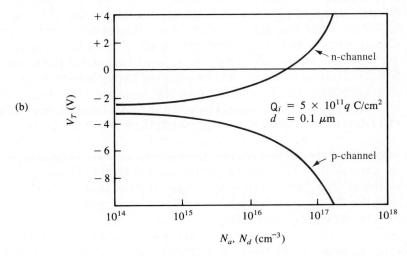

$$V_T = \quad \Phi_{ms} \quad\Bigg|\quad -\frac{Q_i}{C_i} \quad\Bigg|\quad -\frac{Q_d}{C_i} \quad\Bigg|\quad +2\phi_F$$

(a)

Φ_{ms} (−)

$-\dfrac{Q_i}{C_i}$ (−)

$-\dfrac{Q_d}{C_i}$ (+) n channel (−) p channel

$+2\phi_F$ (+) n channel (−) p channel

(b)

$Q_i = 5 \times 10^{11} q$ C/cm^2
$d = 0.1$ μm

n-channel

p-channel

V_T (V) axis: +4, +2, 0, −2, −4, −6, −8

N_a, N_d (cm^{-3}) axis: 10^{14}, 10^{15}, 10^{16}, 10^{17}, 10^{18}

Figure 8-17
Influence of materials parameters on threshold voltage: (a) the threshold voltage equation, indicating signs of the various contributions; (b) variation of V_T with substrate doping for n-channel and p-channel Al–SiO$_2$–Si devices. A rather large value of equivalent interface charge Q_i is chosen, to illustrate the effect of negative V_T for the n-channel case. With proper processing this effect is less pronounced. The oxide thickness is also large in this example.

tive, although its value varies as in Fig. 8-14. The interface charge is positive, so the contribution of the $-Q_i/C_i$ term is negative for either substrate type. On the other hand, the charge in the depletion region is negative for ionized acceptors (p-type substrate, n-channel device) and is positive for ionized donors (n-type substrate, p channel). Also, the term ϕ_F, which is defined as $(E_i - E_F)/q$ in the neutral substrate, can be positive or negative, depending on the conductivity type of the substrate. Considering the signs in Fig. 8-17, we see that all four terms give negative contributions in the p-channel case. Thus we expect negative threshold voltages for typical p-channel devices. On the other hand, n-channel devices may have either positive or negative threshold voltages, depending on the relative values of terms in Eq. (8-30).

All terms in Eq. (8-30) except Q_i/C_i depend on the doping in the substrate. The terms Φ_{ms} and ϕ_F have relatively small variations as E_F is moved up or down by the doping. Larger changes can occur in Q_d, which varies with the square root of the doping impurity concentration as in Eq. (8-25). We illustrate the variation of threshold voltage with substrate doping in Fig. 8-17. As expected from Eq. (8-30), V_T is always negative for the p-channel case. In the n-channel case, the negative flat band voltage terms can dominate for lightly doped p-type substrates, resulting in a negative threshold voltage. However, for more heavily doped substrates, the increasing contribution of N_a to the Q_d term dominates, and V_T becomes positive.

We should pause here and consider what positive or negative V_T means for the two cases. In a p-channel device we expect to apply a negative voltage from metal to semiconductor in order to induce the positive charges in the channel. In this case a negative threshold voltage means that the negative voltage we apply must be larger than V_T in order to achieve strong inversion. In the n-channel case we expect to apply a positive voltage to the metal to induce the channel. Thus a positive value for V_T means the applied voltage must be larger than this threshold value to obtain strong inversion and a conducting n channel. On the other hand, a negative V_T in this case means that a channel exists at $V = 0$ due to the Φ_{ms} and Q_i effects (Figs. 8-15 and 8-16), and we must apply a negative voltage V_T to turn the device off. Since lightly doped substrates are desirable to maintain a high breakdown voltage for the drain junction, Fig. 8-17 suggests that V_T will be negative for n-channel devices made by standard processing. This tendency for the formation of depletion mode (normally on) n-channel transistors is a problem which must be dealt with by special fabrication methods to be described in Section 8.3.6.

EXAMPLE 8-3   We can calculate V_T for the MOS structure described in Examples 8-1 and 8-2, including the effects of flat band voltage. If Al is used for the gate, Fig. 8-14 indicates $\Phi_{ms} = -0.9$ V for $N_a = 10^{15}$ cm^{-3}. Assuming an interface charge of 5×10^{11} q (C/cm^2), we obtain

$$V_T = \Phi_{ms} + 2\phi_F - \frac{1}{C_i}(Q_i + Q_d)$$

$$= -0.9 + 0.576 - \frac{(5 \times 10^{11} \times 1.6 \times 10^{-19}) - 1.39 \times 10^{-8}}{3.45 \times 10^{-8}}$$

$$= -2.24 \text{ V}$$

This value corresponds to the $N_a = 10^{15}$-cm^{-3} point in Fig. 8-17 for the n-channel case.

∎

8.3.5 The MOS Field-Effect Transistor

The MOS transistor is also called a surface field-effect transistor, since it depends on control of current through a thin channel at the surface of the semiconductor (Fig. 8-8). When an inversion region is formed under the gate, current can flow from drain to source (for an n-channel device). In this section we analyze the conductance of this channel and find the $I_D - V_D$ characteristics as a function of gate voltage V_G. As in the JFET case, we will find these characteristics below saturation and then assume I_D remains essentially constant above saturation.

The applied gate voltage V_G is accounted for by Eq. (8-21) plus the voltage required to achieve flat band:

$$V_G = V_{FB} - \frac{Q_s}{C_i} + \phi_s \qquad (8\text{-}31))$$

The induced charge Q_s in the semiconductor is composed of mobile charge Q_n and fixed charge in the depletion region Q_d. Substituting $Q_n + Q_d$ for Q_s, we can solve for the mobile charge:

$$Q_n = -C_i\left[V_G - \left(V_{FB} + \phi_s - \frac{Q_d}{C_i}\right)\right] \qquad (8\text{-}32)$$

At threshold the term in brackets can be written $V_G - V_T$ from Eq. (8-30).

With a voltage V_D applied, there is a voltage drop V_x from each point x in the channel to the source. Thus the potential $\phi_s(x)$ is that required to achieve strong inversion ($2\phi_F$) plus the voltage V_x:

$$Q_n = -C_i\left[V_G - V_{FB} - 2\phi_F - V_x - \frac{1}{C_i}\sqrt{2q\epsilon_s N_a(2\phi_F + V_x)}\right]$$
$$(8\text{-}33)$$

If we neglect the variation of $Q_d(x)$ with bias V_x, Eq. (8-33) can be simplified to

$$Q_n(x) = -C_i(V_G - V_T - V_x) \qquad (8\text{-}34)$$

This equation describes the mobile charge in the channel at point x (Fig. 8-18). The conductance of the differential element dx is $\overline{\mu}_n Q_n(x)Z/dx$, where Z is the depth of the channel and $\overline{\mu}_n$ is *a surface* electron mobility (indicating the mobil-

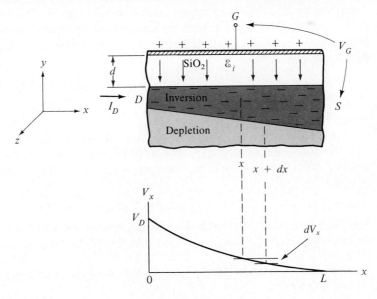

Figure 8-18
Schematic view of the n-channel region of a MOS transistor under bias below pinch-off, and the variation of voltage V_x along the conducting channel.

ity in a thin region near the surface is not the same as in the bulk material). At point x we have

$$I_D \, dx = \overline{\mu}_n Z Q_n(x) \, dV_x \tag{8-35}$$

Integrating from drain to source,

$$\int_0^L I_D \, dx = -\overline{\mu}_n Z C_i \int_{V_D}^0 (V_G - V_T - V_x) \, dV_x$$

$$I_D = \frac{\overline{\mu}_n Z C_i}{L} [(V_G - V_T)V_D - \tfrac{1}{2}V_D^2] \tag{8-36}$$

In this analysis the depletion charge Q_d in the threshold voltage V_T is simply the value with no drain current. This is an approximation, since $Q_d(x)$ varies considerably when V_D is applied, to reflect the variation in V_x (see Fig. 8-8b). However, Eq. (8-36) is a fairly accurate description of drain current for low values of V_D, and is often used in approximate design calculations because of its simplicity. A more accurate and general expression is obtained by including the variation of $Q_d(x)$. Performing the integration of Eq. (8-35) using Eq. (8-33) for $Q_n(x)$, one obtains (Prob. 8.12)

$$I_D = \frac{\overline{\mu}_n Z C_i}{L}$$

$$\times \left\{ (V_G - V_{FB} - 2\phi_F - \tfrac{1}{2}V_D)V_D - \frac{2}{3} \frac{\sqrt{2\epsilon_s q N_a}}{C_i} [(V_D + 2\phi_F)^{3/2} - (2\phi_F)^{3/2}] \right\} \tag{8-37}$$

The drain characteristics that result from these questions are shown in Fig. 8-8c. If the gate voltage is above threshold ($V_G > V_T$), the drain current is described by Eq. (8-37) or approximately by Eq. (8-36) for low V_D. Initially the channel appears as an essentially linear resistor, dependent on V_G. The conductance of the channel in this linear region can be obtained from Eq. (8-36) with $V_D \ll (V_G - V_T)$:

$$g = \frac{\partial I_D}{\partial V_D} \simeq \frac{Z}{L} \overline{\mu}_n C_i (V_G - V_T) \tag{8-38}$$

where $V_G > V_T$ for a channel to exist.

As the drain voltage is increased, the voltage across the oxide decreases near the drain, and Q_n becomes smaller there. As a result the channel becomes pinched off at the drain end, and the current saturates. The saturation condition is approximately given by

$$V_D(\text{sat.}) \simeq V_G - V_T \tag{8-39}$$

The drain current at saturation remains essentially constant for larger values of drain voltage. Substituting Eq. (8-39) into Eq. (8-36), we obtain

$$I_D(\text{sat.}) \simeq \tfrac{1}{2}\overline{\mu}_n C_i \frac{Z}{L} (V_G - V_T)^2 = \frac{Z}{2L} \overline{\mu}_n C_i V_D^2(\text{sat.}) \tag{8-40}$$

for the approximate value of drain current at saturation.

The transconductance in the saturation range can be obtained approximately by differentiating Eq. (8-40) with respect to the gate voltage:

$$g_m(\text{sat.}) = \frac{\partial I_D(\text{sat.})}{\partial V_G} \simeq \frac{Z}{L}\bar{\mu}_n C_i (V_G - V_T) \qquad (8\text{-}41)$$

The similarity of this expression with Eq. (8-38) is due to the approximations used. A more careful analysis reveals a difference in the subtractive term V_T.

The derivations presented here are based on the n-channel device. For the p-channel enhancement transistor the voltages V_D, V_G, and V_T are negative, and current flows from source to drain (Fig. 8-19).

8.3.6   Control of Threshold Voltage

Since the threshold voltage determines the requirements for turning the MOS transistor on or off, it is very important to be able to adjust V_T in designing the device. For example, if the transistor is to be used in a circuit driven by a 3-V battery, it is clear that a 4-V threshold voltage is unacceptable. Some applications require not only a low value of V_T, but also a precisely controlled value to match other devices in the circuit.

All of the terms in Eq. (8-30) can be controlled to some extent. The work function potential difference Φ_{ms} is determined by choice of the gate conductor material; ϕ_F depends on the substrate doping; Q_i can be reduced by proper oxidation methods and by using Si grown in the {100} orientation; Q_d can be ad-

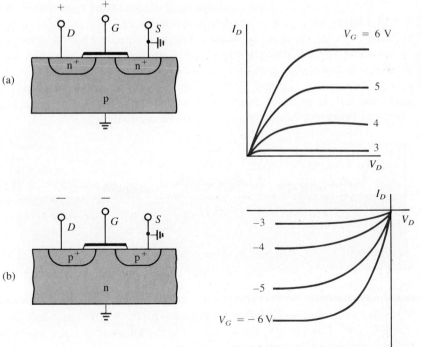

Figure 8-19
Drain current–voltage characteristics for enhancement transistors: (a) for n-channel V_D, V_G, V_T, and I_D are positive; (b) for p-channel all these quantities are negative.

justed by doping of the substrate; and C_i depends on the thickness and dielectric constant of the insulator. We shall discuss here several methods of controlling these quantities in device fabrication.

Silicon Gate Technology. A straightforward method for reducing Φ_{ms} is to deposit Si for the gate electrode instead of Al. Silicon can be deposited in a silane reactor (Section 1.3.4), and heavily doped deposited Si can approximate the desired properties of a metal electrode. Since the deposition occurs onto the insulator layer, the resulting Si layer is polycrystalline.

The use of polycrystalline Si as a gate conductor provides a close match between Φ_m and Φ_s (depending on the Fermi level positions in the two materials). There are additional advantages of this method that make it attractive for MOS integrated-circuit applications. Unlike Al, a Si gate layer can be raised to high temperature. This allows considerable flexibility in device processing, as we shall see in Section 8.3.8.

Control of C_i. Since a low value of V_T is usually desired, a thin oxide layer is used in the gate region to increase $C_i = \epsilon_i/d$ in Eq. (8-30). From Fig. 8-17 we see that increasing C_i makes V_T less negative for p-channel devices and less positive for n-channel with $-Q_d > Q_i$. For practical considerations, the gate oxide thickness is generally 100–1000 Å (0.01–0.1 μm). To achieve a clean oxide of this thickness, the oxide used as a diffusion mask in forming the source and drain is etched away in the gate region, and a fresh thin gate oxide is regrown.

Although a low threshold voltage is desirable in the gate region of a transistor, a large value of V_T is needed between devices. For example, if a number of transistors are interconnected on a single Si chip, we do not want inversion layers to be formed inadvertently between devices (generally called the *field*). One way to avoid such parasitic channels is to increase V_T in the field by using a very thick oxide. Figure 8-20 illustrates a transistor with a gate oxide 0.1 μm thick and a field oxide of 1 μm. Such a thick oxide layer can be deposited by chemical vapor deposition (e.g., the oxidation of silane).

Figure 8-20
Thin oxide in the gate region and thick oxide in the field between transistors for V_T control.

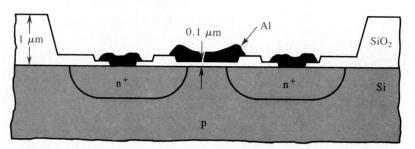

EXAMPLE 8-4 Consider an Al–SiO$_2$–Si p-channel device with $N_d = 10^{15}$ cm^{-3} and $Q_i = 5 \times 10^{11}q$ C/cm^2. Calculate V_T for a gate oxide thickness of 0.1 μm and repeat for a field oxide thickness of 1 μm.

Values of ϕ_F, Q_i, and Q_d can be obtained from Examples 8-2 and 8-3 if we use appropriate signs as in Fig. 8-17a. The value of C_i for the thin oxide case is the same as in Example 8-2. From Fig. 8-14, $\Phi_{ms} = -0.3$ V.

$$V_T = -0.3 - 0.576 - \frac{8 \times 10^{-8} + 1.39 \times 10^{-8}}{3.45 \times 10^{-8}} = -3.6 \text{ V}$$

This value corresponds to that expected from Fig. 8-17b. In the field region where $d = 1$ μm,

$$V_T = -0.876 - \frac{9.39 \times 10^{-8}}{3.45 \times 10^{-9}} = -28 \text{ V}$$

The value of C_i can also be controlled by varying ϵ_i. For example, Si_3N_4 has a relative dielectric constant of about 7, compared with 3.9 for SiO_2. Sandwich structures of SiO_2 covered with Si_3N_4 provide good $Si-SiO_2$ interface properties along with higher ϵ_i. Problems arise, however, with charges at the oxide-nitride interface. We shall discuss control of these charges in Section 9.5 as a means of programming V_T values in integrated circuits.

Threshold Adjustment by Ion Implantation.   The most valuable tool for controlling threshold voltage is ion implantation (Section 5.1.4). Since very precise quantities of impurity can be introduced by this method, it is possible to maintain close control of V_T. For example, Fig. 8-21 illustrates a boron implantation through the gate oxide of a p-channel device such that the implanted peak occurs just below the Si surface. The negatively charged boron acceptors serve to reduce the effects of the positive depletion charge Q_d. As a result, V_T

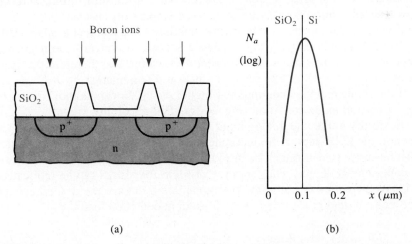

(a)   (b)

Adjustment of V_T in a p-channel transistor by boron implantation: (a) boron ions are implanted through the thin gate oxide but are absorbed within the thick oxide regions; (b) variation of implanted boron concentration in the gate region—here the peak of the boron distribution lies just below the Si surface.

Figure 8-21

becomes less negative. Similarly, a shallow boron implant into the p-type substrate of an n-channel transistor can make V_T positive, as required for an enhancement device.

EXAMPLE 8-5   For the p-channel transistor of Example 8-4, calculate the boron ion dose F_B (B^+ ions/cm^2) required to reduce V_T from -3.6 V to -1.0 V. Assume that the implanted acceptors form a sheet of negative charge just below the Si surface.

SOLUTION

$$-1 = -3.6 + \frac{qF_B}{C_i}$$

$$F_B = \frac{3.45 \times 10^{-8}}{1.6 \times 10^{-19}}(2.6) = 5.6 \times 10^{11} \text{ cm}^{-2}$$

For a beam current of 1 μA scanned over a 100-cm^2 target area,

$$\frac{10^{-6}(\text{C/s})}{100 \text{ cm}^2}t(\text{s}) = 5.6 \times 10^{11}(\text{ions/cm}^2) \times 1.6 \times 10^{-19}(\text{C/ion})$$

The implant time is $t = 9$ s.

∎

If the implantation is performed at higher energy, or into the bare Si instead of through an oxide layer, the impurity distribution lies deeper below the surface. In such cases the essentially gaussian impurity concentration profile cannot be approximated by a spike at the Si surface. Therefore, effects of distributed charge on the Q_d term of Eq. (8-30) must be considered. Calculations of the effects on V_T in this case are more complicated, and the shift of threshold voltage with implantation dose is often obtained empirically instead.

The implantation energy required for shallow V_T adjustment implants is low (50–100 keV), and relatively low doses are needed. A typical V_T adjustment requires only about 10 s of implantation for each wafer, and therefore this procedure is compatible with large-scale production requirements.

If the implantation is continued to higher doses, V_T can be moved past zero to the *depletion-mode* condition (Fig. 8-22). This capability provides considerable flexibility to the integrated-circuit designer, by allowing enhancement- and depletion-mode devices to be incorporated on the same chip. For example, a depletion-mode transistor can be used instead of a resistor as a load element for the enhancement device. Thus an array of MOS transistors can be fabricated in an IC layout, with some adjusted by implantation to have the desired enhancement mode V_T and others implanted to become depletion loads.

8.3.7   Substrate Bias Effects

In the derivation of Eq. (8-36) for current along the channel, we assumed that the source S was connected to the substrate B (Fig. 8-19). In fact, it is possible to apply a voltage between S and B (Fig. 8-23). With a reverse bias between

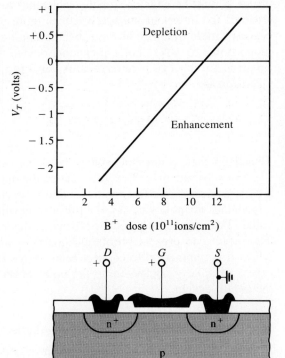

Figure 8-22
Typical variation of V_T for a p-channel device with increased implanted boron dose. The originally enhancement p-channel transistor becomes a depletion-mode device ($V_T > 0$) by sufficient B implantation.

Figure 8-23

Substrate bias resulting from application of a voltage V_B from the substrate (i.e., bulk) to the source. For n channel, V_B must be zero or negative to avoid forward bias of the source junction. For p channel, V_B must be zero or positive.

the substrate and the source (V_B negative for an n-channel device), the depletion region is widened and the threshold gate voltage required to achieve inversion must be increased to accommodate the larger Q_d. A simplified view of the result is that W is widened uniformly along the channel, so that Eq. (8-25) should be changed to

$$Q_d' = -[2\epsilon_s q N_a(2\phi_F - V_B)]^{1/2} \tag{8-42}$$

The change in threshold voltage due to the substrate bias is

$$\Delta V_T = \frac{\sqrt{2\epsilon_s q N_a}}{C_i}[(2\phi_F - V_B)^{1/2} - (2\phi_F)^{1/2}] \tag{8-43}$$

If the substrate bias V_B is much larger than $2\phi_F$ (typically ~0.6 V), the threshold voltage is dominated by V_B and

$$\Delta V_T \simeq \frac{\sqrt{2\epsilon_s q N_a}}{C_i}(-V_B)^{1/2} \qquad \text{(n channel)} \tag{8-44}$$

where V_B will be negative for the n-channel case. As the substrate bias is increased, the threshold voltage becomes more positive. The effect of this bias becomes more dramatic as the substrate doping is increased, since ΔV_T is also proportional to $\sqrt{N_a}$. For a p-channel device the bulk-to-source voltage V_B is positive to achieve a reverse bias, and the approximate change ΔV_T for $V_B \gg 2\phi_F$ is

$$\Delta V_T \simeq -\frac{\sqrt{2\epsilon_s q N_d}}{C_i} V_B^{1/2} \qquad \text{(p channel)} \qquad (8\text{-}45)$$

Thus the p-channel threshold voltage becomes more negative with substrate bias.

The substrate bias effect (also called the *body effect*) increases V_T for either type of device. This effect can be used to raise the threshold voltage of a marginally enhancement device ($V_T \simeq 0$) to a somewhat larger and more manageable value. This can be an asset for n-channel devices particularly (see Fig. 8-17). The effect can present problems, however, in MOS integrated circuits for which it is impractical to connect each source region to the substrate. In these cases, possible V_T shifts due to the body effect must be taken into account in the circuit design.

8.3.8   Capacitance Effects and Self-Aligned Transistors

The high-frequency operation of MOS transistors depends upon the capacitances associated with various parts of the device. The intrinsic MOS capacitance due to the gate region itself is essentially that of the MOS capacitor of Fig. 8-13, taking into account the effects of threshold voltage. Since the inversion region is contacted by the source and drain in the transistor, the supply of carriers at high frequency is not limited by thermal generation as in the MOS capacitor. Thus the capacitance after inversion is C_i (Fig. 8-24). For a 100-Å gate oxide thickness, this capacitance is 34.5×10^{-8} F/cm^2. Assuming a gate region with $L = 1.0$ μm and $Z = 5$ μm, the gate capacitance is 0.017 pF.

Figure 8-24
Gate
capacitance–voltage
characteristic of a
p-channel MOS
transistor with
$V_B = 0$.

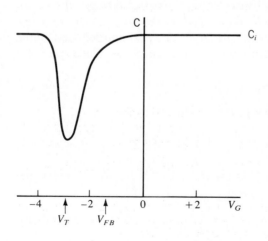

In addition to the intrinsic MOS capacitance, there are several parasitic capacitances which limit the high-frequency operation of MOS transistors. The junction capacitance associated with the junctions between source and drain and the substrate is described by Eq. (5-61). This capacitance is often comparable to the MOS gate capacitance, and varies with junction voltage. Therefore, the junction capacitances are significant in high-frequency or fast-switching operations. Another parasitic capacitance arises from metal interconnections over the field oxide. Since the field oxide is generally thick, this stray capacitance is most important in circuits with considerable metallization.

A particularly troublesome stray capacitance results from the overlap of the gate with the source and drain regions (Fig. 8-25a). If the gate does not extend to the source and drain diffusions, an incomplete channel will be formed and the device will not operate. To avoid this possibility, standard Al gate processing calls for some overlap of the gate past the source and drain edges. This overlap must be large enough to allow for mask registration tolerances and variation in lateral diffusion in the source and drain diffusion step. Because of this overlap, a parasitic capacitance develops between the gate and source and between gate and drain. Since the gate is usually the input and the drain is the output in typical circuit configurations, this overlap capacitance introduces an undesirable feedback effect between input and output at high frequencies (sometimes called the *Miller effect*). To avoid this problem, several methods of achieving *self-aligned gate* regions have been developed.

One method for avoiding the gate overlap capacitance is to use ion implantation to align the source and drain with the edges of the gate (Fig. 8-25b). In this method the gate metal is made narrower than the distance between the gate and source diffusions. An ion implantation is then performed, using the metal gate as a mask, to extend the source and drain regions to the edges of the gate. This procedure takes advantage of the fact that the ion beam arrives essentially perpendicular to the surface. The primary disadvantage of this self-aligned gate procedure is the requirement of a relatively high dose to achieve sufficient conduction in the implanted source and drain extensions, and the need for subsequent annealing of the radiation damage. Since the Al metal gate cannot be raised to high temperature, annealing must be done in the range of 500°C. Al-

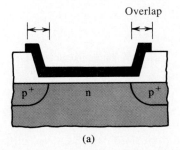

(a)

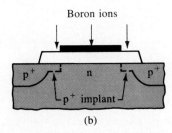

(b)

Reduction of parasitic capacitance due to gate overlap: (a) standard processing with overlap allowed for mask registration tolerance; (b) extension of source and drain to the edges of the Al gate by boron implantation.

Figure 8-25

though reasonable annealing can result, the residual damage does reduce the conductivity of the implanted regions.

The silicon gate method described in Section 8.3.6 has the additional advantage of fairly good gate alignment. In this procedure the polycrystalline Si gate material is deposited over the gate oxide, and windows for the source and drain diffusion are opened in the Si film and the underlying oxide (Fig. 8-26). Since the polycrystalline Si can be raised to high temperature, the source and drain diffusion can be done with the gate in place. Two important advantages result from this method — the gate oxide is covered immediately after growth by the deposited Si, thereby protecting it during subsequent processing; in addition the edges of the gate serve to define the windows for the source and drain diffusion. As a result, the overlap is only that which results from lateral diffusion. The mask registration tolerance is eliminated.

An additional advantage of the Si-gate procedure is that the deposited Si can also be used for interconnections to adjacent devices (Fig. 8-26c). After covering the polycrystalline Si with oxide, another layer of metallization (e.g., Al) can be added. Such multilayer interconnection is of great value in the layout of complex integrated circuits. The reduction of gate–substrate work function difference, the self-aligned gate feature, protection of the gate oxide during most processing steps, and the ease of multilayer interconnection all provide attractive advantages for polycrystalline Si gate technology.

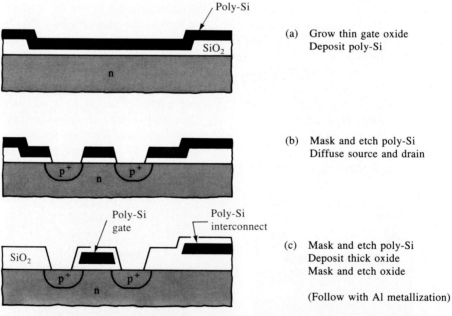

(a) Grow thin gate oxide
     Deposit poly-Si

(b) Mask and etch poly-Si
     Diffuse source and drain

(c) Mask and etch poly-Si
     Deposit thick oxide
     Mask and etch oxide

     (Follow with Al metallization)

Figure 8-26 Silicon gate technology: (a) polycrystalline Si deposited over SiO_2; (b) diffusion of source and drain using the Si gate as a mask to define the channel length; (c) after further processing, the poly-Si gate and interconnection regions are covered with SiO_2. The source and drain regions can now be contacted by Al metal, and Al interconnections can pass over the buried poly-Si interconnections.

8.3.9   Short-Channel Effects

When MOS transistors are made very small, as required in many high-speed devices and integrated circuits, the size of the depletion region near the drain in saturation can be comparable to the channel length (Fig. 8-27). When this occurs, the saturation current does not remain constant with increased V_D. In deriving Eq. (8-40) we assumed that the channel length L was not changed by the formation of the depletion region at the drain end beyond pinch-off. For short channels the spreading of this depletion region with increased V_D beyond pinch-off causes the effective channel length L' to decrease. Since I_D (sat.) is inversely proportional to channel length in Eq. (8-40), the drain current increases as the drain depletion region spreads into the channel. As a result, I_D (sat.) increases somewhat with V_D, rather than remaining constant. Velocity saturation effects (Section 8.2.3) are also likely to occur in short-channel devices.

A variety of other effects occur as the MOS device dimensions are reduced. For example, as Fig. 8-28 illustrates, the source and drain depletion regions can intrude into the channel even without bias, as these junctions are brought closer together in short-channeled devices. This effect is called *charge*

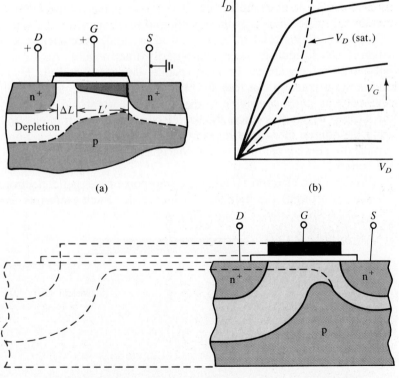

Figure 8-27
Short-channel effects:
(a) reduction of the effective channel length L' in saturation;
(b) increase of I_D beyond pinch-off due to the reduction of L'.

Intrusion of the source and drain depletion regions into the channel as the device dimensions are reduced from the long-channel case (dashed lines). The effect of these depletion regions in reducing the usable channel length is called *charge sharing,* and *drain-induced barrier lowering* can result in loss of gate control.

Figure 8-28

sharing, but it should probably be called charge hogging, because the source and drain in effect take part of the channel charge, which would otherwise be controlled by the gate. As the drain depletion region continues to increase with bias, it can actually interact with the source-to-channel junction and lower the potential barrier. This problem is known as *drain-induced barrier lowering* (DIBL). When the source junction barrier is reduced, electrons are easily injected into the channel and the gate voltage no longer has control of the drain current. Under extreme conditions of encroaching source and drain depletion regions, the two can meet. This punch-through effect results in a continuous depleted region from drain to source. Carriers injected at the source are driven to the drain by the high field between the two electrodes. The current is called *space-charge-limited* current and is proportional to V_D^2/L^3.

An effect that is exacerbated by short-channel designs is the *subthreshold current,* which arises from the fact that some electrons are induced in the channel even before strong inversion is established. For the low electron concentrations typical of the subthreshold regime, we expect diffusion currents (proportional to carrier gradients) to dominate over drift currents (proportional to carrier concentrations). For very short channel lengths, such carrier diffusion from source to drain can make it impossible to turn off the device below threshold. The subthreshold current is made worse by the DIBL effect just mentioned, which increases the injection of electrons from the source.

Electric fields tend to be increased at small geometries, since device voltages are difficult to scale to arbitrarily small values. As a result, various hot carrier effects appear in short-channel devices. The field in the reverse-biased drain junction can lead to impact ionization and carrier multiplication. The resulting holes contribute to the substrate current and some may move to the source, where they lower the source barrier and result in electrons injected from the source into the p-region. In fact, n-p-n transistor action can result within the source–channel–drain configuration and prevent gate control of the current.

Another hot electron effect is the transport of energetic electrons over (or tunneling through) the barrier into the oxide. Such electrons can become

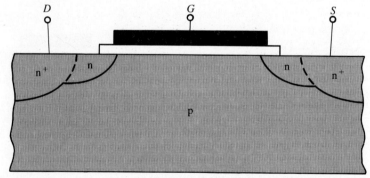

Figure 8-29 The lightly doped drain (LDD) structure, in which the source and drain regions are doped more lightly near the channel to reduce the electric fields and therefore reduce hot electron effects.

trapped in the oxide, where they change the threshold voltage and the I-V characteristics of the device. Hot electron effects can be reduced by reducing the doping in the source and drain regions, so that the junction fields are smaller. However, lightly doped source and drain regions are incompatible with small geometry devices because of contact resistance and other problems. A compromise design called the *lightly doped drain* (LDD) uses two doping levels, with heavy doping over most of the source and drain areas but with light doping in a region adjacent to the channel (Fig. 8-29). The LDD structure decreases the field between the drain and channel regions, thereby reducing injection into the oxide, impact ionization, and other hot electron effects.

PROBLEMS

8.1. Calculate the incremental channel resistance $\partial V_D / \partial I_D$ for a JFET below saturation. What does the simple theory imply about this resistance beyond pinch-off?

8.2. Show that Eq. (8-8) results from integration of Eq. (8-7).

8.3. Modify Eqs. (8-8) and (8-9) to include effects of the contact potential V_0 at the p^+-n junction on either side of the channel.

8.4. The current I_D varies almost linearly with V_D in a JFET for low values of V_D.
   (a) Use the binomial expansion with $V_D/(-V_G) < 1$ to rewrite Eq. (8-8) as an approximation to this case.
   (b) Show that the expression for the channel conductance I_D/V_D in the linear range is the same as g_m(sat.) given by Eq. (8-10).
   (c) What value of gate voltage V_G turns the device off such that the channel conductance goes to zero?

8.5. Redraw Figs. 8-10, 8-11, and 8-12 for the p-channel (n-type substrate) case.

8.6. Show that the width of the depletion region in Fig. 8-12 is given by Eq. (8-23). Assume the carriers are completely swept out within W, as was done in Section 5.2.3.

8.7. An Al-gate n-channel MOS transistor is made on a p-type Si substrate with $N_a = 5 \times 10^{15}$ cm^{-3}. The SiO$_2$ thickness is 500 Å in the gate region, and the effective interface charge Q_i is $4 \times 10^{10} q$ C/cm^2.
   (a) Find W_m, V_{FB}, and V_T.
   (b) Sketch the C–V curve for this device and give important numbers for the scale.

8.8. An Al-gate p-channel MOS transistor is made on an n-type Si substrate with $N_d = 5 \times 10^{16}$ cm^{-3}. The SiO$_2$ thickness is 1000 Å in the gate region, and the effective interface charge Q_i is $2 \times 10^{11}q$ C/cm^2. Find W_m, V_{FB}, and V_T. Sketch the C–V curve for this device and give important numbers for the scale.

8.9. Find the threshold voltage for a Si n-channel MOS transistor with $N_a = 10^{17}$ cm^{-3}, $\Phi_{ms} = -0.95$ V, $Q_i = 10^{11}q$ C/cm^2, and an SiO$_2$ thickness $d = 800$ Å. Repeat for a p-channel device ($N_d = 10^{17}$ cm^{-3}) with the same parameters (except for Φ_{ms}, which can be calculated from the change in E_F).

8.10. Calculate sufficient points to construct a figure such as Fig. 8-17b for $Q_i = 10^{11}q$ C/cm^2 and $d = 0.05$ μm.

8.11. (a) Find the voltage V_{FB} required to reduce to zero the negative charge induced at the semiconductor surface by a sheet of positive charge Q_{ox} located x' below the metal. (b) In the case of an arbitrary distribution of charge $\rho(x')$ in the oxide, show that

$$V_{FB} = -\frac{1}{C_i} \int_0^d \frac{x'}{d} \rho(x') \, dx'$$

8.12. Obtain Eq. (8-37) by integration of Eq. (8-35), using Eq. (8-33) for Q_n.

8.13. Find the dose of boron (ions/cm^2) required to form (a) an n-channel enhancement transistor with $V_T = +2$ V, and (b) a p-channel depletion transistor with $V_T = +2$ V, for substrate doping of 5×10^{16} cm^{-3} in each case and the parameters of Fig. 8-17b. Assume that the implanted boron resides just below the Si surface and all impurities are ionized.

8.14. When an MOS transistor is biased with $V_D > V_D$(sat.), the effective channel length is reduced by ΔL and the current I_D' is larger than I_D(sat.), as shown in Fig. 8-27. Assuming that the depleted region ΔL is described by an expression similar to Eq. (8-23) with $V_D - V_D$(sat.) for the voltage across ΔL, show that the conductance beyond saturation is

$$g_D' = \frac{\partial I_D'}{\partial V_D} = I_D(\text{sat.}) \frac{\partial}{\partial V_D} \left(\frac{L}{L - \Delta L} \right)$$

and find the expression for g_D' in terms of V_D.

8.15. Show that the conductance of the differential element in Fig. 8-18 is $\bar{\mu}_n Q_n(x) Z / dx$.

8.16. Construct a high-frequency $C-V$ curve for an MOS capacitor with parameters of the device discussed in Example 8-3.

8.17. The flat band voltage is shifted to $-4V$ for an Al-SiO$_2$-Si capacitor with parameters discussed in Example 8-2. Redraw Fig. 8-13 for this case and find the value of interface charge Q_i required to cause this shift in V_{FB}, with Φ_{ms} given by Fig. 8-14.

8.18. Plot I_D vs. V_D with several values of V_G for the thin-oxide p-channel transistor described in Example 8-4. Use the p-channel version of Eq. (8-36), and assume that I_D(sat.) remains constant beyond pinch-off. Assume that $\overline{\mu}_p = 200$ cm^2/V-s, and $Z = 10L$.

8.19. A typical figure of merit for high-frequency operation of MOS transistors is the cutoff frequency $f_c = g_m/2\pi C_G LZ$, where the gate capacitance C_G is essentially C_i over most of the voltage range. Express f_c above pinch-off in terms of materials parameters and device dimensions, and calculate f_c for the transistor of Prob. 8.18, with $L = 1$ μm.

*8.20. Use Eqs. (8-8) and (8-9) to calculate and plot $I_D(V_D, V_G)$ at 300 K for a Si JFET with $a = 1000$ Å, $N_d = 7 \times 10^{17}$ cm^{-3}, $Z = 100$ μm, and $L = 5$ μm. Allow V_D to range from 0 to 5 V and allow V_G to take on values of 0, -1, -2, -3, -4, and -5 V.

*8.21. Use Eq. (8-37) to calculate and plot $I_D(V_D, V_G)$ at 300 K for an n-channel Si MOSFET with an oxide thickness $d = 200$ Å, a channel mobility $\overline{\mu}_n = 1000$ cm^2/V-s, $Z = 100$ μm, $L = 5$ μm, and N_a of 10^{14}, 10^{15}, 10^{16}, and 10^{17} cm^{-3}. Allow V_D to range from 0 to 5 V and allow V_G to take on values of 0, 1, 2, 3, 4, and 5 V. Assume that $Q_i = 5 \times 10^{11}q$ C/cm^2.

*8.22 Use the field-dependent mobility expression to calculate and plot $I_D(V_D, V_G)$ for the Si JFET described in Prob. 8.20 for gate lengths of 0.25, 0.50, 1.0, 2.0, and 5.0 μm and a gate voltage of 0 V. The field dependent mobility model of Eq. (8-12) takes the form

$$\mu(\mathscr{E}) = \frac{v_d}{\mathscr{E}} = \frac{\mu_0}{1 + (\mu_0\mathscr{E}/v_s)}$$

where μ_0 is the low field mobility (as given in Fig. 3-23) and v_s is the saturation velocity (10^7 cm/s). Assume that $\mathscr{E} = V_D/L$ and assume that the drain current saturates when it reaches a maximum. Discuss the advantages of having a short gate length device.

READING LIST

Bell, T. E., J. M. Poate, R. C. Dynes, L. F. Eastman, M. I. Nathan, and M. Heiblum. "The Quest for Ballistic Action." *IEEE Spectrum* 23, no. 2 (February 1986): 36–47.

DiLorenzo, J. V., and D. D. Khandelwal, eds. *GaAs FET Principles and Technology.* Dedham, Mass.: Artech House, 1982.

Drummond, T. J., W. T. Masselink, and H. Morkoç. "Modulation-Doped GaAs/(Al, Ga) As Heterojunction Field-Effect Transistors: MODFETs." *Proceedings of the IEEE* 74 (June 1986): 773–822.

Eastman, L. F. "Compound-Semiconductor Transistors." *Physics Today* (October 1986): 77–83.

Eden, R. C., A. R. Livingston, and B. M. Welch. "Integrated Circuits: The Case for Gallium Arsenide." *IEEE Spectrum* 20, no. 12 (December 1983): 30–37.

D. K. Ferry, ed. *Gallium Arsenide Technology.* Indianapolis: Howard W. Sams & Co., 1985.

Frensly, W. R. "Gallium Arsenide Transistors." *Scientific American* 257, no. 2 (August 1987): 80–87.

Morkoç, H., and P. M. Solomon. "The HEMT: A Superfast Transistor." *IEEE Spectrum* 21, no. 2 (February 1984): 28–35.

Nicollian, E. H. and J. R. Brews. *MOS (Metal Oxide Semiconductor) Physics and Technology.* New York: John Wiley, 1982.

Pierret, R. F. *The Modular Series on Solid State Devices. Vol. IV: Field Effect Devices.* Reading, Mass.: Addison-Wesley, 1983.

Pulfrey, D. L., and N. G., Tarr. *Introduction to Microelectronic Devices.* Englewood Cliffs, NJ: Prentice Hall, 1989.

Sah, C. T. "Evolution of the MOS Transistor—From Conception to VLSI," *Proceedings of the IEEE* 76 (October 1988): 1280–1326.

Schroder, D. K. *The Modular Series on Solid State Devices. Vol. VII: Advanced MOS Devices.* Reading, MA: Addison-Wesley, 1987.

Shur, M. *GaAs Devices and Circuits.* New York: Plenum Press, 1987.

Sze, S. M. *Physics of Semiconductor Devices.* New York: John Wiley, 1981.

Sze, S. M. *Semiconductor Devices.* New York: John Wiley, 1985.

chapter 9

INTEGRATED CIRCUITS

Just as the transistor revolutionized electronics by offering more flexibility, convenience, and reliability than the vacuum tube, the integrated circuit enables new applications for electronics that were not possible with discrete devices. Integration allows complex circuits consisting of many thousands of transistors, diodes, resistors, and capacitors to be included in a chip of semiconductor. This means that sophisticated circuitry can be miniaturized for use in space vehicles, in large-scale computers, and in other applications where a large collection of discrete components would be impractical. In addition to offering the advantages of miniaturization, the simultaneous fabrication of hundreds of ICs on a single Si wafer greatly reduces the cost and increases the reliability of each of the finished circuits. Certainly discrete components have played an important role in the development of electronic circuits; however, most circuits are now fabricated on the Si chip rather than with a collection of individual components. Therefore, the traditional distinctions between the roles of circuit and system designers do not apply to IC development.

In this chapter we shall discuss various types of ICs and the fabrication steps used in their production. We shall investigate techniques for building large numbers of transistors, diodes, and resistors on a single chip of Si, as well as the interconnection, contacting, and packaging of these circuits in usable form. All the processing techniques discussed here are very basic and general. There would be no purpose in attempting a comprehensive review of all the subtleties of device fabrication in a book of this type. In fact, the only way to keep up with such an expanding field is to study the current literature. Many good reviews are suggested in the reading list at the end of this chapter; more important, current issues of those periodicals cited can be consulted for up-to-date information regarding IC technology. Having the background of this chapter, one should be able to read the current literature and thereby keep abreast of the present trends in this very important field of electronics.

9.1
BACKGROUND

In this section we provide an overview of the nature of integrated circuits and the motivation for using them. It is important to realize the reasons, both technical and economic, for the dramatic rise of ICs to their present role in electronics. We shall discuss several main types of ICs and point out some of the applications of each. More specific fabrication techniques will be presented in later sections.

9.1.1 Advantages of Integration

It might appear that building complicated circuits, involving many interconnected components on a single Si substrate, would be risky both technically and economically. In fact, however, modern techniques allow this to be done reliably and relatively inexpensively; in most cases an entire circuit on a Si chip can be produced much more inexpensively and with greater reliability than a similar circuit built up from individual components. The basic reason is that hundreds of identical circuits can be built simultaneously on a single Si wafer (Fig. 9-1); this process is called *batch fabrication*. Although the processing steps for the wafer are complex and expensive, the large number of resulting integrated circuits makes the ultimate cost of each fairly low. Furthermore, the processing steps are essentially the same for a circuit containing 1,000,000 transistors as for a simpler circuit. This drives the IC industry to build increasingly complex circuits and systems on each chip, and use larger Si wafers

Figure 9-1
A wafer of integrated circuits after scribing and breaking into individual chips. (Photograph courtesy of American Microsystems, Inc.)

(e.g., 8 in. in diameter). As a result, the number of components in each circuit increases without a proportional increase in the ultimate cost of the system. The implications of this principle are tremendous for circuit designers; it greatly increases the flexibility of design criteria. Unlike circuits with individual transistors and other components wired together or placed on a circuit board, ICs allow many "extra" components to be included economically. Thus redundancy and "back-up" circuitry can be included without greatly raising the cost of the final product. Reliability is also improved since all devices and interconnections are made on a single rigid substrate, greatly minimizing failures due to the soldered interconnections of discrete component circuits.

The advantages of ICs in terms of miniaturization are obvious. Since many circuit functions can be packed into a small space, complex electronic equipment can be employed in many applications where weight and space are critical, such as in aircraft or space vehicles. In large-scale computers it is now possible not only to reduce the size of the overall unit but also to facilitate maintenance by allowing for the replacement of entire circuits quickly and easily. Applications of ICs are pervasive in such consumer products as watches, calculators, automobiles, telephones, television, and appliances. Miniaturization and the cost reduction provided by ICs mean that we all have increasingly more sophisticated electronics at our disposal.

Some of the most important advantages of miniaturization pertain to response time and the speed of signal transfer between circuits. For example, in high-frequency circuits it is necessary to keep the separation of various components small to reduce time delay of signals. Similarly, in very high speed computers it is important that the various logic and information storage circuits be placed close together. Since electrical signals are ultimately limited by the speed of light (about 1 ft/ns), physical separation of the circuits can be an important limitation. As we shall see in Section 9.5, *large-scale integration* of many circuits on a Si chip have led to major reductions in computer size, thereby tremendously increasing speed and function density. In addition to decreasing the signal transfer time, integration can reduce parasitic capacitance and inductance between circuits. Reduction of these parasitics can provide significant improvement in the operating speed of the system.

We have discussed several advantages of reducing the size of each unit in the batch fabrication process, such as miniaturization, high-frequency and switching speed improvements, and cost reduction due to the large number of circuits fabricated on a single wafer. Another important advantage has to do with the percentage of usable devices (often called the *yield*) which results from batch fabrication. Faulty devices usually occur because of some defect in the Si wafer or in the fabrication steps. Defects in the Si can occur because of lattice imperfections and strains introduced in the crystal growth, cutting, and handling of the wafers. Usually such defects are extremely small, but their presence can ruin devices built on or around them. Reducing the size of each device greatly increases the chance for a given device to be free of such defects. The same is true for fabrication defects, such as the presence of a dust particle on a photolithographic mask. For example, a lattice defect or dust particle $\frac{1}{2}$ μm in diameter can easily ruin a circuit which includes the damaged area. If a fairly large circuit is built around the defect it will be faulty; however, if the device size is reduced so that four circuits occupy the same area on the wafer, chances are good that only the one containing the defect will be faulty and the other three will be good. Therefore, the percentage yield of usable circuits increases over a certain range of decreasing chip area. There is an optimum area for each circuit, above which defects are needlessly included and below which the elements are spaced too closely for reliable fabrication.

9.1.2 Types of Integrated Circuits

There are several ways of categorizing ICs as to their use and method of fabrication. The most common categories are *linear* or *digital* according to application, and *monolithic* or *hybrid* according to fabrication.

A linear IC is one that performs an amplification or other essentially linear operations on signals. Examples of linear circuits are simple amplifiers, operational amplifiers, and analog communications circuits. Digital circuits involve logic and memory, for applications in computers, calculators, microprocessors, and the like. By far the greatest volume of ICs has been in the digital field, since large numbers of such circuits are required. Since digital circuits generally require only "on-off" operation of transistors, the design requirements for integrated digital circuits are often less stringent than for linear circuits. Although transistors can be fabricated as easily in integrated form as in discrete form, passive elements (resistors and capacitors) are usually more difficult to produce to close tolerances in ICs.

9.1.3 Monolithic and Hybrid Circuits

Integrated circuits that are included entirely on a single chip of semiconductor (usually Si) are called *monolithic* circuits (Fig. 9-1). The word monolithic literally means "one stone" and implies that the entire circuit is contained in a single piece of semiconductor. Any additions to the semiconductor sample, such as insulating layers and metallization patterns, are intimately bonded to the surface of the chip. A *hybrid* circuit may contain one or more monolithic circuits or individual transistors bonded to an insulating substrate with resistors, capacitors, or other circuit elements, with appropriate interconnections (Fig. 9-2). Monolithic circuits have the advantage that all components are contained in a single rigid structure which can be batch fabricated; that is, hundreds of identical circuits can be built simultaneously on a Si wafer. On the other hand, hybrid circuits offer excellent isolation between components and allow the use of more precise resistors and capacitors. Furthermore, hybrid circuits are often less expensive to build in small numbers.

When resistors and capacitors are made external to the monolithic Si chip, basically two types of technology are used; the passive elements are fabricated and interconnected by *thick-film* or *thin-film* processes. Although the dividing line between thin and thick films is not precise, they are fairly well separated

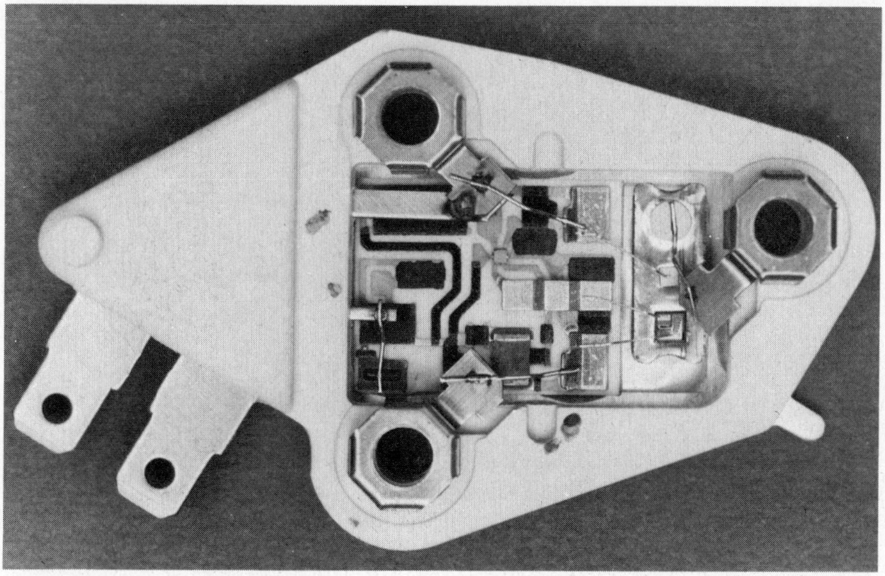

Figure 9-2
A hybrid circuit employing thick-film printing. This automobile voltage regulator circuit contains thick-film resistors and thermistors, a ceramic chip capacitor, and monolithic Si chips mounted on an insulating substrate. (Photograph courtesy of Delco Electronics Corp.)

in application to ICs: "thin" films are typically 0.1 to 0.5 μm, and "thick" films are about 25 μm.

The processing steps for the two hybrid techniques are quite different. In thick-film circuits the resistors and interconnection patterns are "printed" on a ceramic substrate (Fig. 9-2) by silk-screen or similar processes. Conductive and resistive pastes consisting of metal powders in organic binders are printed on the substrate and cured in an oven. One advantage of this process is that resistors can be made below the rated values and then trimmed by abrasion, or by selective evaporation using a pulsed laser. These corrections can be made quickly with automated procedures while the resistance values are under test. Small ceramic chip capacitors can be bonded into place in the interconnection pattern, along with monolithic circuits or individual transistors.

Thin-film technology allows for greater precision and miniaturization, and is generally preferred when space is an important limitation. Thin-film interconnection patterns and resistors can be vacuum deposited on a glass or glazed ceramic substrate (Fig. 9-3). The resistive films are usually made of tantalum or other resistive metal, and the conductors are often aluminum or gold. In general, the resistive materials must be deposited by sputtering. Pattern definition for the resistors and conductor paths can be achieved by depositing the films through metal shields which contain appropriate apertures. Better definition is obtained by metallizing the entire substrate, or large parts of it, and using photolithographic methods to remove the metal except in the desired pattern. This technique is similar to that used in forming metal contact patterns on monolithic circuits. Capacitors can be fabricated by thin-film techniques by depositing an insulating layer between two metal films or by oxidizing the sur-

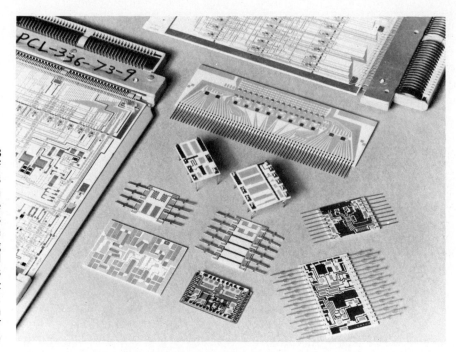

Figure 9-3
Examples of thin-film circuits employing tantalum films to achieve precision resistors and capacitors on insulating substrates. Several of these circuits include monolithic Si chips. (Photograph courtesy of AT&T Bell Laboratories.)

face of one film and then depositing a second film on top. Whether thin-film or thick-film techniques are used, the object is to fabricate passive components of greater precision than could be obtained on the Si substrate.

An important passive component that is missing from the discussion thus far is the inductor. Inductors are difficult to integrate into small substrates and are usually avoided in the circuit design or placed external to the IC. This limitation is serious but seldom disastrous, since the function of an inductor can often be obtained by proper design of an active circuit. Since active devices can be made easily in ICs, it is usually possible to avoid the use of inductors altogether.

**9.2
FABRICATION
OF MONOLITHIC
CIRCUITS**

The most important element of IC technology is the monolithic chip, which may contain several thousands or millions of individual transistors, all properly interconnected. Such monolithic circuits are batch fabricated on a Si wafer 6 or 8 inches in diameter; each wafer may contain hundreds of the monolithic circuits. These circuits are then separated by selective etching, sawing, or by scribing the wafer with a diamond stylus or laser and breaking it apart into small squares or rectangles containing individual circuits (Fig. 9-1). Each circuit is then mounted on an appropriate substrate, contacted, and packaged.

The fabrication process for making the wafer of monolithic circuits is the subject of this section. We shall review the processes of masking and selective doping and shall point out some of the steps in obtaining the various monolithic components. Many of the techniques discussed here are similar to those described earlier for discrete components, but there are also many other methods which are peculiar to monolithic circuits. We shall discuss the problem of isolation between components, special techniques for fabricating monolithic transistors, and the various ways of obtaining diodes, resistors, and capacitors on a monolithic chip.

9.2.1 Masking and Selective Doping

Although the process of selective doping has been discussed in previous chapters, it is worthwhile to review it here and investigate the extensions to monolithic ICs. As with discrete components, the object of the fabrication process is to selectively dope certain regions of the semiconductor and to properly interconnect the resulting components with a metallization pattern. Including oxidation steps, the number of operations involved in fabricating a monolithic circuit can be quite large. As an example, let us review the simple diffused transistor of Fig. 7-7. The basic steps in the process are as follows:

1. Grow the first oxide layer.
2. Open a window in the SiO_2 for the base diffusion.
3. Perform a boron diffusion.
4. Grow a second oxide layer.
5. Open a window for the emitter diffusion.
6. Perform a phosphorus diffusion.
7. Grow a third oxide layer.
8. Open windows for the base and emitter contacts.
9. Evaporate Al on the surface.
10. Remove the Al except in the desired metallization pattern.

In this simple example there are three oxidation steps, two diffusions, and one metallization. The number of masks required is four: two for diffusions, one for the contacting windows, and one for the metallization definition. Many more steps and therefore more masks are required for monolithic circuits. The important point, however, is that many identical circuits are made simultaneously on the wafer, thus making the process economically feasible. Therefore, it is important to reduce the size of each circuit and use large wafers, to multiply the number of usable devices resulting from the batch fabrication. This means that the various masks must be extremely accurate and well aligned during each photolithographic step.

Generation of the masks required for IC fabrication is one of the most time-consuming and expensive steps in the process. In the early days of IC technology the complexity of circuits was such that original artwork could be made by a draftsman. For most present-day complex circuits this is no longer possible, and computer-controlled mask generation must be employed. In general, a precise copy of the desired pattern for one masking step is prepared for one member of the array of circuits. This original copy is photographed and reduced in size. A *step-and-repeat* camera is then used to photograph the reduced pattern, perform the final reduction, and repeat the process for every rectangle in the final array, which may contain hundreds of identical patterns. The final pattern array is printed on a glass mask (Fig. 9-4) to be placed over the Si wafer in the photolithographic process.

Contact printing, in which the mask touches the photoresist layer on the wafer during the exposure, requires a 1X image and has the disadvantage of contamination of the mask after several runs. To avoid this problem, *projection printing* is used. In this method the light passes through the mask and is focused onto the sample surface by a precision lens system. Since the mask does not touch the sample, it can be used repeatedly without degradation. In a projection system the mask may contain an array of individual circuit masks such as in Fig. 9-4, or the mask may be for only a portion of the array. In the latter case the mask pattern is reduced and focused onto the wafer, building up the array on the wafer by a precision step-and-repeat process. Such a step-and-repeat system allows focusing of the image on different parts of the wafer to accomodate slight problems with surface flatness.

For very complex circuits it is imperative to use automated mask generation equipment. One type of system utilizes computer-controlled light flashes to build up the pattern on a photographic film by a series of line or block exposures. The resulting film is then reduced and handled in a step-and-repeat system to create the production mask. Alternatively, the master mask can be generated by an electron-beam exposure system, again controlled by computer. We will discuss this use of electron beams, as well as direct exposure of photoresist on the wafer by electron beams in the following section.

Figure 9-4
A photolithographic mask used in the fabrication of an array of integrated circuits on a Si wafer (shown below the mask). By shining ultraviolet light through the mask, a layer of photoresist on the wafer is exposed in the desired pattern for one step in the fabrication process. (Photograph courtesy of American Microsystems, Inc.)

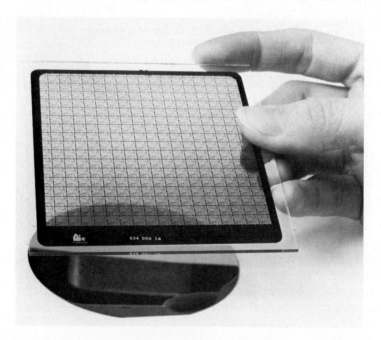

9.2.2   Fine-Line Lithography

The effort to pack ever-increasing functional density on a chip of Si provides a strong motivation for making device elements (e.g., transistors) smaller. Speed and power consumption requirements also drive the device designer to use smaller dimensions. One soon reaches limits in this size reduction process, however, imposed by the photolithography being used. If ultraviolet light is used to expose photoresist through a mask, minimum linewidths are ultimately limited to a few wavelengths because of diffraction effects. For a uv wavelength of 0.35 μm, for example, one might not expect linewidths smaller than about 1 μm. In fact, after allowing for mask registration and other tolerances in a device requiring several masking steps, a minimum linewidth of several microns is more reasonable with traditional optical lithography. Clearly, it is necessary to expose the photoresist at shorter wavelengths for submicron geometries. Therefore, optical systems using deep ultraviolet wavelengths have been developed for submicron lithography.

The de Broglie theorem states that the wavelength of a particle varies inversely with its momentum:

$$\lambda = \frac{h}{p} \tag{9-1}$$

Thus more massive particles or energetic photons should be considered to achieve shorter wavelengths. Viable candidates for this application are electrons, ions, or X rays. For example, electron beams are easily generated, focused, and deflected. The basic technology for this process has been developed over many years in scanning electron microscopy. Since a 10-keV electron has a wavelength of about 0.1 Å, the linewidth limits become the size of the focused beam and its interaction with the photoresist layer. It is possible to achieve linewidths of a fraction of a micrometer by direct electron beam writing on the wafer photoresist. Furthermore, the computer-controlled electron beam exposure requires no masks. This capability allows extremely dense packing of circuit elements on the chip, but direct writing of complex patterns is slow.

Because of the time required for electron beam wafer exposure, it is in many cases advantageous to use electron beam technology for mask generation, but to expose the wafer photoresist through the mask using photons. For example, if a heavy metal is used in the mask, soft X rays ($\lambda \sim 10$ Å) can be used to expose the wafer with submicron resolution. A particularly high flux of X rays can be obtained from the synchrotron radiation emitted by electrons accelerated in a storage ring or synchrotron.

9.2.3   Isolation

An important step in the fabrication of monolithic circuits is the provision of electrical isolation between components. If bipolar transistors were made on a monolithic chip by the technique of Fig. 7-7, all collector regions would be in common. It is therefore necessary to isolate most components and then interconnect them with the metallization pattern.

For n-p-n bipolar transistors, a convenient method of isolation involves diffusing a pattern of p-type "moats" into an n-type epitaxial layer on a p-type substrate (Fig. 9-5). The p-type substrate provides mechanical support for the structure and, in conjunction with the diffused p-type pattern, defines isolated regions of n-type material. The epitaxial layer is less than 10 μm thick in most cases, which is thick enough to accommodate subsequent diffusion of n-p-n transistors and other elements. Since each component can be contained in an island of n-type material, good isolation results if the substrate p material is held at the most negative potential in the circuit. In effect, each component is surrounded by a reverse-biased p-n junction. The room temperature saturation current for Si junctions is typically in the nanoampere range. Thus the isolation between circuit elements is good, except for capacitance effects at high frequencies.

A disadvantage of the junction isolation method is the capacitance inherent at the isolating p-n junctions. The capacitance associated with the sidewalls between the n region and the diffused junctions in Fig. 9-5 can be eliminated by the use of various oxide-isolated configurations. For example, Fig. 9-6 il-

Figure 9-5
Junction isolation, in which an n-type epitaxial region is surrounded by p-type diffused junctions extending to the p-type substrate.

Figure 9-6
Local oxidation of silicon using a nitride mask: (a) isolation of n-type regions; (b) expanded view showing intrusion of oxide under the edge of the nitride, causing a bird's beak shape; (c) use of local oxidation for the field oxide in an MOS transistor.

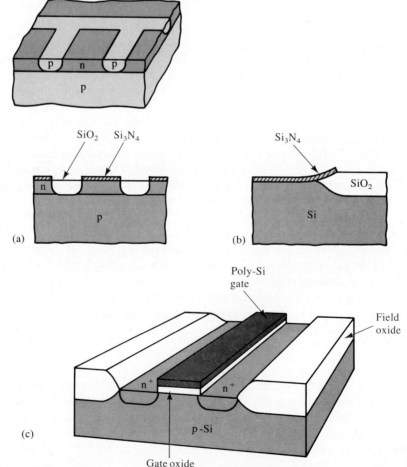

lustrates a method of oxide isolation which takes advantage of the fact that silicon nitride can be used to mask the underlying Si against oxidation. Using a nitride mask, it is thus possible to oxidize the Si in selected regions. This *local oxidation* method is used in both bipolar and MOS devices. Figure 9-6a illustrates an n-epitaxial region isolated by such a local oxidation. The area that will contain the transistor is covered with Si_3N_4, and the unprotected areas are then etched about halfway through the epitaxial layer. The reason for this etching step is that the thickness of a thermally grown SiO_2 layer is about twice the thickness of the Si consumed in the oxidation. During the oxidation, SiO_2 grows in the areas not protected by the nitride. When the oxidation takes place near the edge of the nitride mask, there is some intrusion of the oxidation laterally under the nitride. As a result, the nitride is pushed up slightly, forming an irregular surface (Fig. 9-6b). The shape of the oxide, resembling a bird's beak, can be improved by using careful patterning and oxidation techniques. The local oxidation process is widely used in MOS devices to provide a thick field oxide between transistors (Fig. 9-6c).

An isolation scheme that is particularly well suited for high density circuits involves the formation of relatively deep trenches, backfilled with polysilicon (Fig. 9-7). In this process a nitride layer is patterned and used as an etch mask for a deep, anisotropic etch of the silicon to form the trench. Using reactive ion etching, a deep, narrow trench can be formed with very straight sidewalls. Oxidation inside the trench forms an insulating layer, and the trench is then filled with polysilicon by chemical vapor deposition (CVD). Since the top of the polysilicon can be etched flat and oxidized, a planar surface results over the trench (Fig. 9-7c).

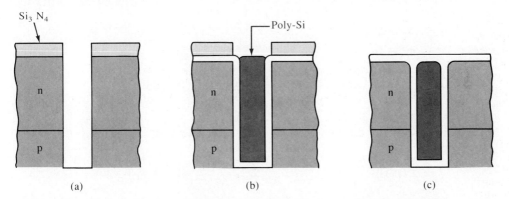

Trench isolation: (a) using a nitride mask and reactive ion etching, a narrow trench with straight sidewalls is etched in the Si; (b) the sidewalls are oxidized and the trench is filled with polysilicon by CVD; (c) the top surface is etched flat and oxidized to form a planar region over the trench.

Figure 9-7

Now we shall consider the various elements that make up an integrated circuit, and some of steps in their fabrication. The basic elements are fairly easy to name—transistors, resistors, capacitors, and some form of interconnection.

9.3 MONOLITHIC DEVICE ELEMENTS

There are some elements in integrated circuits, however, which do not have simple counterparts in discrete devices. We shall consider one of these, the merged transistor (or integrated injection) logic approach, and another example will be discussed in Section 9.4, charge transfer devices. Discussion of fabrication technology is difficult in a book of this type, since device fabrication engineers seem to make changes faster than typesetters do! Since this important and fascinating field is changing so rapidly, the reader should obtain a basic understanding of device design and processing from this discussion and then search out new innovations in the current literature.

9.3.1 Bipolar Transistors

Early bipolar integrated circuits used junction isolation and diffusion for selective doping (Fig. 9-8). One important difference between integrated transistors and the discrete transistors we have discussed earlier is that in the monolithic case all three terminals must be available on the top surface of the chip. For example, interconnection of several isolated bipolar transistors requires that collector contacts be made at the surface. As a result, collector current must pass along a high-resistance path in the lightly doped n-epitaxial material while flowing from the active part of the collector to the contact. The resulting series collector resistance can be detrimental to the transistor properties and should be avoided if possible. One common method of decreasing the collector resistance is to include a heavily doped n layer just below the collector. Usually, this n^+ *buried layer* is included by diffusion of n^+ regions into the p-type substrate before the epitaxial layer is grown (Fig. 9-8a). After the

Figure 9-8
An early type of diffused, junction-isolated bipolar transistor for integrated circuits: (a) buried layer diffusion; (b) growth of epitaxial layer; (c) isolation diffusion; (d) double-diffused epitaxial transistor; (e) improved collector contact by deep n^+ diffusion; (f) lateral p-n-p transistor. The oxide layers in (c)–(f) and metallizations in (d)–(f) are omitted for clarity.

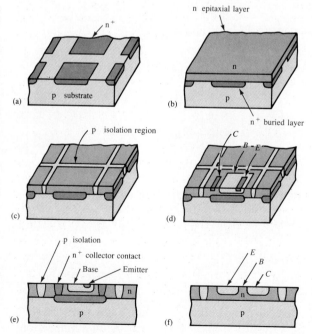

n-epitaxial layer is grown and the p-type isolation diffusion is made, the wafer has a pattern typified by Fig. 9-8c. During the isolation diffusion and in subsequent diffusions, there is some movement of the n^+ buried layer into the epitaxial region. To minimize this spreading of the buried layer, it is usually doped with Sb, which diffuses rather slowly in Si (Prob. 9.3).

After the isolation pattern has been diffused through the epitaxial layer, the wafer is masked for the base diffusion (boron) and then again for the emitter diffusion (phosphorus). During the shallow n^+ emitter diffusion, it is common to diffuse n^+ stripes above the collector areas to improve the collector contact. The resulting transistor structure is shown in Fig. 9-8d (the oxide and metallization layers are omitted for clarity). If very low collector resistance is required, deep n^+ diffusion can be made from the surface to the buried layer (Fig. 9-8e). This step is done before the base diffusion. After all diffusions have been completed, the surface is passivated and the interconnection pattern is included, as discussed below.

We have presented the processes for fabrication of n-p-n transistors, since this is the most common type of bipolar device used in ICs. If p-n-p transistors are to be included as well, special techniques must be used. One way to make a p-n-p device in the n-type island of Fig. 9-8c would be to employ a triple diffusion: a p-layer, then an n-layer, and finally another p-layer. This method is undesirable because of the extra diffusion step required and because the ultimate emitter region has been compensated three times. When p-n-p transistors are required in an otherwise n-p-n circuit, they are usually made by the method shown in Fig. 9-8f. This is called a *lateral transistor,* since the current flows laterally from the emitter to the base to the collector. The base width and other parameters are more difficult to control in the lateral transistor than in *planar* transistors, in which the emitter and collector junctions are parallel. In fact, the β of lateral transistors is often near unity; therefore, the lateral p-n-p transistor is used most often in conjunction with a planar n-p-n such that the n-p-n boosts the gain of the overall unit (Prob. 9.5).

In modern device fabrication, the diffusion steps of Fig. 9-8 are largely replaced by ion implantation. Also, the local oxidation method of isolation described in Section 9.2.3 provides an excellent opportunity for size reduction of transistors. In Fig. 9-9 local oxidation is used not only between devices, but also between the base and the collector contact region of the transistor. As a result, the mask for the base implant or diffusion can extend over part of the oxide region on both sides without affecting the device geometry. Similarly, the n^+ mask for the emitter and the collector contact does not have the usual close registration tolerance. Each of these implant or diffusion areas can extend over the adjacent oxide region without affecting the geometry of the device. Since space is not wasted adjacent to the isolation and mask registration is less critical, the dimensions of the device can be made smaller. This process, introduced by Fairchild and called *Isoplanar,* has important advantages in bipolar integrated circuits. The area of an Isoplanar transistor can be about one-half that of an equivalent junction-isolated device. This reduction in area, along with the reduced sidewall capacitance provided by oxide isolation, produces a much faster device with high packing density on the chip.

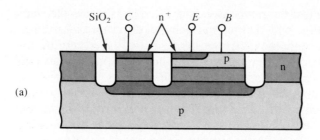

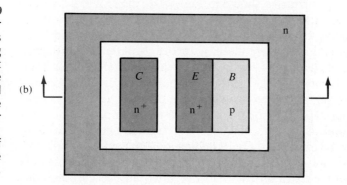

Figure 9-9
An Isoplanar transistor: (a) cross section showing local oxidations at the device perimeter and between the base and the collector contact region; (b) a top view of the device before metallization.

9.3.2 Merged Transistors

Some device structures useful in integrated circuits have no real analogues in discrete form. We will discuss one such structure, the charge transfer device, in Section 9.4. Another example is an integrated structure that is used in a type of digital logic called *merged transistor logic (MTL)*, or *integrated injection logic (I^2L)*. This bipolar device structure has been used in applications where low power consumption is important.

Let us begin by considering the simple transistor logic circuit shown in Fig. 9-10a. The ouput terminal is assumed to be connected to the input of another such circuit, and similarly the input comes from the output of a previous stage in a logic array. The p-n-p device provides a source of base current to the n-p-n, and the input terminal can be used to short-circuit this current to

Figure 9-10
A simple merged transistor cell: (a) circuit diagram; (b) cross section of the merged device, having a lateral p-n-p and an "inverted" n-p-n.

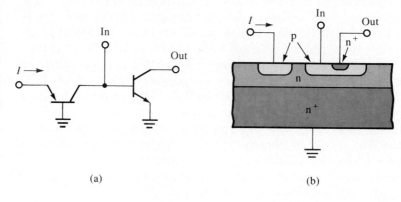

ground. With the input open, the n-p-n can be driven into saturation by the current source, and the output voltage is the small (≤ 100 mV) saturation voltage across the n-p-n. With the input terminal shorted to ground, on the other hand, the n-p-n is in the off state. This circuit can be used in positive logic, in which a logic 0 is the collector-to-emitter saturation voltage and a logic 1 is the collector voltage of the n-p-n in cutoff. In terms of voltages, the 0 is the very small $V_{CE} \sim 100$ mV for a saturated device and the 1 is the forward drop across the emitter junction ($V_{BE} \sim 700$ mV) of the next n-p-n. Therefore, the logic swing between "0" and "1" is only about 600 mV.

Figure 9-10b illustrates how one might build such a structure in a particularly simple way. The p-n-p current source is a lateral device, and the n-p-n is built "upside-down." That is, the n-p-n emitter is below its base and is common with the base of the p-n-p. Furthermore, the collector of the p-n-p is the same as the base of the n-p-n. Thus, the two transistors are *merged* into an integrated unit.

Two such merged transistor units can be connected as in Fig. 9-11 to form a NOR gate, which requires the output to be in the "0" state whenever *either* input is in the "1" state. To simplify the diagram, we represent the p-n-p current source schematically. If both *A* and *B* are "0," both current sources lead directly to ground and both transistors are off. But if either *A* or *B* is in the "1" state, the corresponding transistor is saturated and the output is "0."

Clearly, large logic arrays can be built up from this type of element. In fact, the possibilities are enhanced if the n-p-n is a multicollector device as in Fig. 9-12. Furthermore, the emitter p region of the current source can be used to drive a number of such n-p-n elements. This sharing of the current source p region (now called an *injector rail*) not only enhances the integration but also helps to spread the source current evenly among the various gates. The injector rail is ultimately led to a biasing circuit to control the source current. As a result of the use of current source transistors, the individual gates do not require biasing resistors. This is a major space-saving feature, since resistors use up considerable space on a chip (Section 9.3.4).

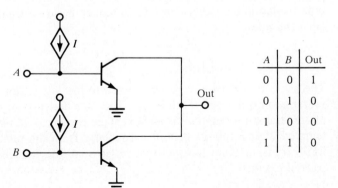

A	B	Out
0	0	1
0	1	0
1	0	0
1	1	0

(a)   (b)

Figure 9-11
A NOR circuit that is easily implemented using I^2L: (a) circuit diagram, with the current source p-n-p shown as a current generator; (b) truth table for the NOR gate.

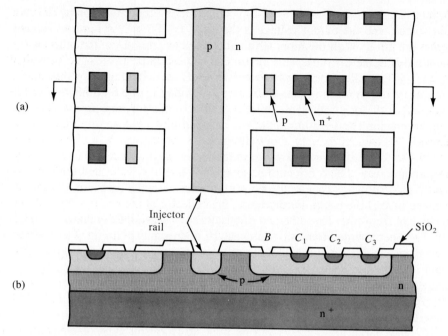

Figure 9-12
I^2L cell layout:
(a) several I^2L cells
fed by a common
p-type injector rail;
(b) cross section of
the device.
Metallizations are
not shown, since
these vary
depending upon
how the cells are to
be interconnected.

Since the I^2L cell does not require resistors, the entire cell is about the size of a single integrated transistor. The packing density of I^2L gates on the chip is therefore very large. Furthermore, the power consumption is quite low. In a typical switching operation only a fraction of a picojoule is used. Another advantage of I^2L is that other types of bipolar devices can be placed on the same chip. Thus buffer stages and other peripheral circuitry can be included easily.

The speed of the device can be increased by using more complex structures than that shown in Fig. 9-12, including Schottky clamps between the collectors and base of the n-p-n, shaped doping profiles obtained by ion implantation, and other refinements. It is also possible to employ Isoplanar or other isolation technology in I^2L to reduce parasitic transistor action between adjacent gates.

9.3.3 MOS Transistors

The MOS transistor is very well suited to the integration of large circuits and systems onto a single chip of Si. The fact that digital circuits require only on–off transistor response is an advantage for the MOS device. Large-scale integrated digital circuits benefit particularly from the relative simplicity of the MOS transistor and the fact that its size can be reduced for use in densely packed circuits.

Channel Stops. The importance of avoiding the formation of parasitic inversion in the field between transistors in an integrated circuit was discussed in Section 8.3.6. This problem in MOS circuits is analogous to the isolation problem for bipolar circuits. We mentioned in Section 8.3.6 that using a field oxide about ten times as thick as the gate oxide increases V_T in these areas. We cannot make the field oxide arbitrarily thick, however, because of practical problems of deposition and the fact that metallization patterns must be deposited over the "hills and valleys" of the resulting surface. A further increase in the field threshold voltage can be obtained by doping the region between transistors with the impurity type of the substrate (Fig. 9-13). For example, a masking and implant or diffusion step can be used to dope the field region prior to forming the source and drain regions. Such doping increases V_T in the field by increasing Q_d (see Fig. 8-17). Combined with a small value of C_i due to the thick field oxide, the Q_d/C_i term can lead to very high threshold voltage between devices. The disadvantage of channel stop doping is that the breakdown voltage of the source and drain junctions is reduced by the more heavily doped channel stop. Therefore, it is necessary to control the field doping carefully to increase V_T appropriately without significant lowering of the junction breakdown voltage. This close control of doping makes ion implantation particularly attractive for this application.

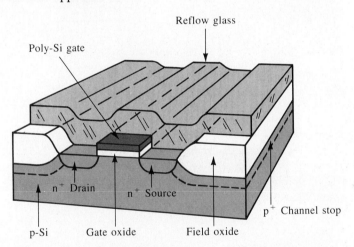

The use of thick field oxide and channel stops to prevent inadvertent channel formation outside the active region of the MOS transistor. Shown also is a chemical-vapor deposited SiO_2 glass layer that will be patterned prior to metallization. This glass overlayer is often doped with phosphorus and boron and heated to the softening point to flatten out the surface before patterning for metallization (a process called *reflow* glass).

Figure 9-13

LDD and Sidewall Spacers.   The use of the lightly doped drain (LDD) structure was described in Section 8.3.9 as a way of reducing the high field in the drain junction of small-geometry devices. Figure 9-14 illustrates the formation of an LDD transistor for application in integrated circuits. An important aspect of LDD fabrication is the use of *sidewall spacers* on each side of the gate. After formation of the thin gate oxide and polysilicon gate, an n-type implant forms the shallow, lightly doped source and drain regions (a). Then a thick oxide layer is deposited by a low-temperature chemical-vapor-deposition pro-

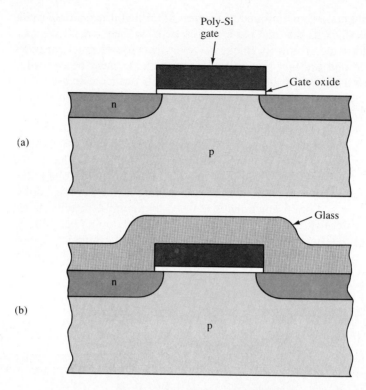

(a)

(b)

Figure 9-14
Fabrication of the lightly doped drain structure, using sidewall spacers. The polysilicon gate covers the thin gate oxide and masks the first low-dose implant (a). A thick oxide layer is deposited by low temperature CVD (b) and is anisotropically etched away to leave only the sidewall spacers (c). These spacers serve as a mask for the second, high-dose implant. After a drive-in diffusion, the LDD structure results (d).

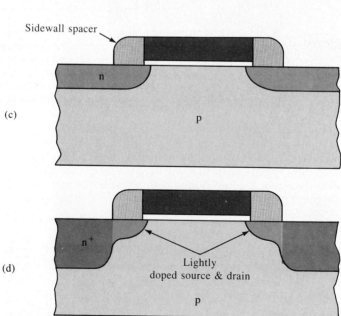

(c)

(d)

cess (b). The oxide is then removed by reactive ion etching, leaving only the sidewall spacers (c), which serve as a mask for the second implant. This high-dose n^+ implant is then driven in by a diffusion step. The result is heavily doped source and drain regions, separated from the channel region by small lightly doped extensions.

Self-aligned Silicide.   Since it is very important to reduce the series resistance of the gate and the source and drain regions for small-geometry devices, several techniques have been developed to improve the contact resistance. This is important not only because the areas of these regions decrease with smaller geometries, but also because shallower source and drain junctions are necessary as the device is scaled down, resulting in higher resistance regions. One approach to reducing substantially the resistance of the source and drain regions, and also the polysilicon gate region, is to use a refractory metal silicide to contact these regions. In Figure 9-15 the source–drain implant and sidewall spacer steps are followed by formation of a silicide simultaneously in the source, drain, and polysilicon gate regions. In this process, a thin layer of refractory metal is deposited and heated to form silicide wherever it touches ex-

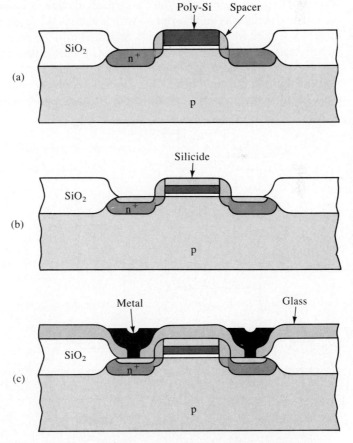

Figure 9-15
The formation of silicided source–drain and gate regions for low-resistance contacts: (a) the source-drain implant is followed by formation of the sidewall spacers as in Fig. 9-14; (b) a layer of refractory metal is deposited and reacted in regions of exposed Si to form a conducting silicide layer; (c) the unreacted metal is removed and a CVD glass is deposited and patterned for contact metallization.

posed silicon. Various silicides, including PtSi, MoSi$_2$, CoSi$_2$, and TiSi$_2$ have been used for this process. An advantage of the process shown in Fig. 9-15 is the fact that the source, drain, and gate silicide regions are formed simultaneously, with the sidewall spacers serving to align the gate edges. This self-aligned silicide process is sometimes called *salicide*.

Complementary MOS Devices.   A particularly useful device for digital applications is a combination of n-channel and p-channel MOS transistors on adjacent regions of the chip. This *complementary MOS* (commonly called *CMOS*) combination is illustrated in the basic inverter circuit of Fig. 9-16a. In this circuit the drains of the two transistors are connected together and form the output, while the input terminal is the common connection to the transistor gates. The p-channel device has a negative threshold voltage, and the n-channel transistor has a positive threshold voltage. Therefore, a zero voltage input ($V_{in} = 0$) gives zero gate voltage for the n-channel device, but the voltage between the gate and source of the p-channel device is $-V$. Thus the p-channel device is on, the n-channel device is off, and the full voltage V is measured at V_{out} (i.e., V appears across the nonconducting n-channel transistor). Alternatively, a positive value of V_{in} turns the n-channel transistor on the p-channel off. The output voltage measured across the "on" n-channel device is essentially zero. Thus, the circuit operates as an inverter — with a binary "1" at the input, the output is in the "0" state, whereas a "0" input produces a "1" output. The beauty of this circuit is that one of the devices is turned off for either condition. Since the devices are connected in series, no drain current flows, except for a small charging current during the switching process from one state to the other. Since the CMOS inverter uses very little power, it is particularly useful in appli-

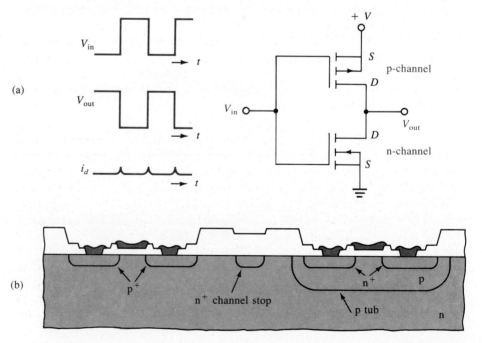

Figure 9-16
Complementary MOS structure: (a) CMOS inverter; (b) formation of p-channel and n-channel devices together.

cations such as electronic watch circuits which depend on very low power consumption. CMOS is also advantageous in very large-scale integrated circuits (Section 9.5), since even small power dissipation in each transistor becomes a problem when thousands or millions of them are integrated on a chip.

The device technology for achieving CMOS circuits consists mainly in arranging for both n- and p-channel devices with similar threshold voltages on the same chip. To achieve this goal, a diffusion or implantation must be performed in certain areas to obtain n and p regions for the fabrication of each type of device. Figure 9-16b illustrates the formation of n-channel devices in an otherwise p-channel surface. The diffused p region that is to contain the n-channel device is often called a "tub." The critical parameter of the p tub is its net acceptor concentration, which must be closely controlled to achieve a predictable threshold voltage for the n-channel transistor. Since it is difficult to form lightly doped diffused regions with precise doping concentrations by standard diffusion methods, ion implantation is generally used in forming the tubs. In Fig. 9-16b the p tub would be formed by a controlled implantation of boron into the surface, followed by a redistribution of the boron by diffusion. With the p tub in place, source and drain implants and diffusions (and perhaps a channel stop) are performed to make the n-channel and p-channel transistors. Matching of the two transistors is achieved by control of the surface doping in the p tub and by additional threshold adjustment of both transistors by ion implantation.

Attention must be paid in CMOS design to the fact that combining n-channel and p-channel devices in close proximity can lead to inadvertent (*parasitic*) bipolar structures. In fact, a p-n-p-n structure can be found in Fig. 9-16, which can serve as an inefficient but troublesome *thyristor* (see Ch. 11). Under certain biasing conditions the p-n-p part of the structure can supply base current to the n-p-n structure, causing a large current to flow. This process, called *latchup,* can be a serious problem in CMOS circuits. Several methods have been used to eliminate the latchup problem, including using both n-type and p-type tubs, separated by trench isolation (Fig. 9-7). The use of two separate tubs (wells) also allows independent control of threshold voltages in both types of transistor.

Silicon on Insulator (SOI).   An interesting and useful extension of the Si MOS process can be achieved by growing very thin films of single crystal Si on insulating substrates. Two such substrates which have the appropriate thermal expansion match to Si are sapphire and spinel ($MgO-Al_2O_3$). Epitaxial Si films can be grown on these substrates by chemical deposition (e.g., the pyrolysis of silane), with typical film thickness of about 1 μm. The film can be etched by standard photolithographic techniques into islands for each transistor (Fig. 9-17). Diffusion of p^+ and n^+ areas into these islands for source and drain regions result in the MOS devices illustrated in Fig. 9-17. Since the film is so thin, the source and drain regions can be made to extend entirely through the film to the sapphire substrate. As a result, the junction capacitance is reduced to the very small capacitance associated with the sidewalls between the diffu-

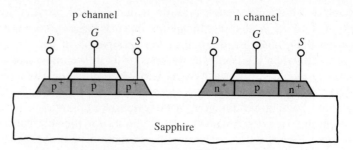

Figure 9-17 Silicon on sapphire. Both n-channel and p-channel enhancement transistors are made in islands of Si film on the insulating sapphire substrate. These devices can be interconnected for CMOS applications.

sions and the channel region. In addition, since interconnections between devices pass over the insulating substrate, the usual interconnection-Si substrate capacitance is eliminated (along with the possibility of parasitic induced channels in the field between devices). These capacitance reductions improve considerably the high-frequency operation of circuits using such devices.

Other insulators can be used for SOI devices, including SiO_2. Since oxide can easily by grown on Si substrates, it serves as an attractive insulator for subsequent growth of thin-film Si. Since polycrystalline Si can be deposited directly over SiO_2, devices can be made in thin poly-Si films. However, to avoid grain boundaries and other defects typical of polycrystalline material, a variety of techniques have been developed to grow single crystal Si on oxide. For example, the oxide layer can be formed beneath the surface of a Si wafer by high-dose oxygen implantation. The thin Si layer remaining on the surface above the implanted oxide can be used as a seed for growth of a Si crystalline layer. In another approach, transient annealing techniques are used to grow a crystalline film of Si over an oxide layer by first depositing poly-Si and then laterally seeding the crystal as it regrows across the surface in crystalline form.

Examining the two devices of Fig. 9-17 more closely, one of them is unlike the transistors we have considered thus far. The thin Si film is lightly doped p-type, and therefore the device labeled "p-channel" appears *junctionless*. Such a device is able to operate in the enhancement mode (normally off) because of the equilibrium effects of the work function difference and the interface charge. With the usual Φ_{ms} and Q_i a depletion region is formed in the central p material of each device with zero gate voltage. In fact, for a Si film of about 1 μm or less, this depletion region can extend all the way through the Si to the insulator. Such a condition is called *deep depletion*,[†] and no drain-to-source current flows. In the n-channel device a positive gate voltage greater than V_T induces an inverted region at the surface, as usual for an n-channel enhancement device. For the p-channel case a small negative voltage V_G removes the deple-

---

[†] This term is also used in a different context to mean $W > W_m$ in a transient gate bias condition. We will deal with this use of "deep depletion" in Section 9.4.1.

tion and causes hole accumulation beneath the gate. The result is the formation of a conducting channel by a small negative gate voltage, as is the case for a conventional p-channel enhancement device. Although the deep depletion type of p-channel device operates by a somewhat different mechanism, its current–voltage characteristics are similar to the conventional device.

Since both p-channel and n-channel transistors can be made on the same insulating surface, the silicon-on-insulator technique is quite compatible with CMOS circuit fabrication.

As an alternative to the deep depletion device approach of Fig. 9-17, conventional enhancement devices of both channel types can be placed on the same substrate by multiple deposition of Si films. Two epitaxial growth steps can be used with appropriate masking to obtain lightly doped p and n islands for the fabrication of each type of device.

The SOI approach also benefits from modern MOS methods such as ion implantation and Si gate technology. For circuits requiring high speeds, low standby power (due to the elimination of junction leakage to the substrate), and radiation tolerance (due to the elimination of the Si substrate), the extra expense of preparing sapphire substrates or growing crystalline Si films on oxide is compensated by increased performance.

9.3.4   Integration of Other Circuit Elements

One of the most revolutionary developments of integrated circuit technology is the fact that integrated transistors are cheaper to make than are more mundane elements such as resistors and capacitors. There are, however, numerous applications calling for diodes, resistors, and capacitors in integrated form. In this section we discuss briefly how these circuit elements can be implemented on the chip. We will also discuss a very important circuit element — the interconnection pattern which ties all of the integrated devices together in a working system.

Diodes.   It is simple to build p-n junction diodes in a monolithic circuit. It is also common practice to use transistors to perform diode functions. Since many transistors are included in a monolithic circuit, no special diffusion step is required to fabricate the diode elements. There are a number of ways in which a transistor can be connected as a diode (Prob. 7.9). Perhaps the most common method is to use the emitter junction as the diode, with the collector and base shorted. This configuration is essentially the narrow base diode structure, which has high switching speed with little charge storage. Since all the transistors can be made simultaneously, the proper connections can be included in the metallization pattern to convert some of the transistors into diodes.

Resistors.   Diffused or implanted resistors can be obtained in monolithic circuits by using the shallow junctions used in forming the transistor regions (Fig. 9-18a). For example, during the base diffusion, a resistor can be diffused which is made up of a thin p-type layer within one of the n-type islands. Alternatively, a p region can be made during the base diffusion, and an n-type resis-

Figure 9-18
Monolithic
resistors: (a) cross
section showing
use of base and
emitter diffusions
for resistors;
(b) top view of two
resistor patterns.

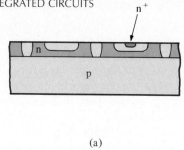

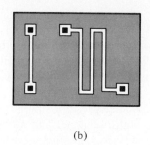

(a)   (b)

tor channel can be included within the resulting p region during the emitter diffusion step. In either case, the resistance channel can be isolated from the rest of the circuit by proper biasing of the surrounding material. For example, if the resistor is a p-type channel obtained during the base diffusion, the surrounding n material can be connected to the most positive potential in the circuit to provide reverse-bias isolation. The resistance of the channel depends on its length, width, depth of the diffusion, and resistivity of the diffused material. Since the depth and resistivity are determined by the requirements of the base or emitter diffusion, the variable parameters are the length and width. Two typical resistor geometries are shown in Fig. 9-18b. In each case the resistor is long compared with its width, and a provision is made on each end for making contact to the metallization pattern.

Design of diffused resistors begins with a quantity called the *sheet resistance* of the diffused layer. If the average resistivity of a diffused region is ρ, the resistance of a given length L is $R = \rho L/w\text{t}$, where w is the width and t is the thickness of the layer. Now if we consider one square of the material, such that $L = w$, we have the sheet resistance $R_s \equiv \rho/\text{t}$ in units of ohms per square. We notice that R_s measured for a given layer is numerically the same for any size square. This quantity is simple to measure for a thin diffused layer by a four-point probe technique.[†] Therefore, for a given diffusion, the sheet resistance is generally known with good accuracy. The resistance then can be calculated from the known value of R_s and the ratio L/w (the *aspect ratio*) for the resistor. We can make the width w as small as possible within the requirements of heat dissipation and photolithographic limitations and then calculate the required length from w and R_s. Design criteria for diffused resistors include geometrical factors, such as the presence of high current density at the inside corner of a sharp turn. In some cases it is necessary to round corners slightly in a folded or zigzag resistor (Fig. 9-18b) to reduce this problem.

To reduce the amount of space used for resistors or to obtain larger resistance values, it is often necessary to obtain surface layers having larger sheet resistance than is available during the standard base or emitter diffusions. A useful alternative is to use ion implantation to form shallow regions having very high sheet resistance ($\sim 10^5$ Ω/square). This procedure can provide a considerable

---

[†]This is a very useful method, in which current is introduced into a wafer at one probe, collected at another probe, and the voltage is measured by two probes in between. Special formulas are required to calculate resistivity or sheet resistance from these measurements.

saving of space on the chip. In integrated FET circuits it is common to replace load resistors with depletion-mode transistors, as mentioned in Section 8.3.6.

Capacitors.   One of the most important elements of an integrated circuit is the capacitor. This is particularly true in the case of memory circuits, where charge is stored in a capacitor for each bit of information. In *random access memory* (RAM) the memory cells are organized in matrix form and are accessed along rows and columns to store or retrieve data. Particularly dense memories can be made using a one-transistor, one-capacitor memory cell, in which the transistor serves as a switch to deliver and access charge in the capacitor. Using MOS capacitors, the charge tends to leak away in a few milliseconds, so it is necessary to periodically refresh the information. This type of memory is called a *dynamic* RAM, or DRAM. Although the refeshing of the DRAM requires extra circuitry, this is a popular memory because of its very small cell area (allowing several million cells on a single chip) and low power consumption.

Figure 9-19a illustrates a one-transistor DRAM cell, in which the n-channel MOS transistor provides access to the adjacent MOS capacitor. The top plate of the capacitor is polysilicon, and the bottom plate is an inversion charge contacted by an n^+ region of the transistor. The terms *bit line* and *word line* refer to the row and column organization of the memory.

The simple MOS capacitor shown in Fig. 9-19a suffers from the fact that the area of the capacitor depends on the available surface area in the cell design. Obviously, the drive to smaller memory cells for high packing density on the chip has motivated creative approaches to increasing the capacitance (and therefore storage capacity) while decreasing the surface area. A particularly useful approach employs the trench technology described in Fig. 9-7, making use

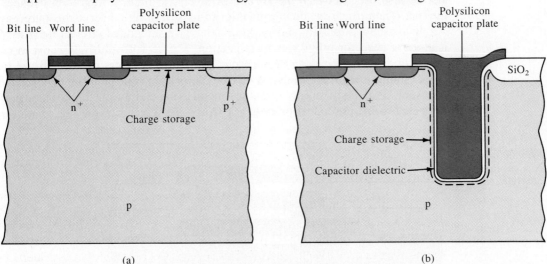

(a)   (b)

Integrated capacitors for DRAM cells: (a) one-transistor memory cell in which the transistor stores and accesses charge in an adjacent planar MOS capacitor; (b) replacement of the planar capacitor with a trench capacitor, in which charge is stored in the inversion region surrounding the trench.

Figure 9-19

of the thin oxide in the trench as a capacitor dielectric. Figure 9-19b illustrates the use of a trench capacitor to substitute for the planar capacitor in (a). In this case the "top plate" of the capacitor is the polysilicon in the trench, and the charge is stored in the inversion region surrounding the trench. Since a much larger area is obtained around the trench, this capacitor requires only a very small area at the surface of the chip.

The trench capacitor is only one of a number of inventive capacitor designs, many of which use multilayer metallization and dielectrics with larger permittivity than SiO_2. In fact, as much effort goes into developing integrated capacitor designs as in designing better transistors for high-density ICs.

Contacts and Interconnections. During the metallization step, the various regions of each circuit element are contacted and proper interconnection of the circuit elements is made. Aluminum is commonly used for the top metallization, since it adheres well to Si and to SiO_2 if the temperature is raised briefly to about 550°C after deposition. Gold is used on some monolithic devices, but the adherence properties of Au to Si and SiO_2 are poor. Therefore, intermediate metallizations must be made to bond the Au to the surface.

As mentioned throughout this chapter, silicide contacts and doped polysilicon conductors are commonly used in integrated circuits. By opening windows through the oxide layers to these conductors, Al metallization can be used to contact them and connect them to other parts of the circuit. In cases where Al is used to contact the Si surface, it is usually necessary to use Al containing about 1 percent Si to prevent the metal from incorporating Si from the layer being contacted, thereby causing "spikes" in the surface. Thin diffusion barriers are also used between the Al and Si layers, to prevent migration between the two. The refractory silicides mentioned in Section 9.3.3 serve this purpose.

Increased complexity and packing density in integrated circuits inevitably leads to a need for multilayer metallization. Figure 9-20 illustrates how three levels of metallization can be incorporated with interspersing dielectrics. In this example, the three metals may all be Al, or they may be different conductors such as polysilicon or refractory metals (depending on the heat each is subjected to in subsequent processing). Also, the dielectrics may be grown or

Figure 9-20
An example of multilayer metallization.

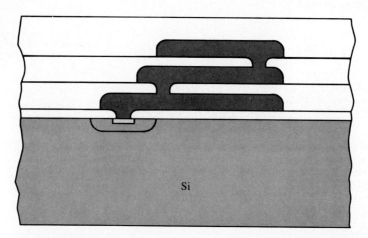

deposited oxides, boro-phospho-silicate glass for reflow planarization, nitrides, etc. The planarization of the surface is extremely important to prevent breaks in the metallization, which can occur in traversing a step on the surface. Various approaches using reflow glass, polyimide, and other materials to achieve planarization have been used.

In designing the layout of elements for a monolithic circuit, topological problems must be solved to provide efficient interconnection without *crossovers* — points at which one conductor crosses another conductor. If crossovers must be made on the Si surface, they can be accomplished easily at a resistor. Since the implanted or diffused resistor is covered by SiO_2, a conductor can be deposited crossing the insulated resistor. In cases requiring crossovers where no resistor is available, a low-value diffused resistor can be inserted in one of the conductor paths. For example, a short n^+ region can be diffused during the emitter step and contacted at each end by one of the conductors. The other conductor can then cross over the oxide layer above the n^+ region. Usually, this can be accomplished without appreciable increase in resistance, since the n^+ region is heavily doped and its length can be made small.

During the metallization step, appropriate points in the circuit are connected to relatively large *pads* to provide for external contacts. These metal pads are visible in photographs of monolithic circuits as rectangular areas spaced around the periphery of the device. In the mounting and packaging process, these pads are contacted by small Au or Al wires or by special techniques such as those discussed in Section 9.6.

One of the most interesting and broadly useful integrated devices is the *charge-coupled device* (*CCD*). The CCD is part of a broader class of structures known generally as *charge transfer devices*. These are dynamic devices that move charge along a predetermined path under the control of clock pulses. These devices find applications in memories, various logic functions, signal processing, and imaging. In this section we lay the groundwork for understanding these devices, but their present forms and variety of applications must be found in the current literature.

**9.4
CHARGE TRANSFER
DEVICES**

9.4.1 Dynamic Effects in MOS Capacitors

The basis of the CCD is the dynamic storage and withdrawal of charge in a series of MOS capacitors. Thus we must begin by extending the MOS discussion of Chapter 8 to include the basics of dynamic effects. Figure 9-21 shows an MOS capacitor on a p-type substrate with a large positive gate pulse applied. A depletion region exists under the gate, and the surface potential increases considerably under the gate electrode. In effect, the surface potential forms a *potential well,* which can be exploited for the storage of charge.

If the positive gate bias has been applied for a sufficiently long time, electrons accumulate at the surface and the steady state inversion condition is established. The source of these carriers is the thermal generation of electrons at or near the suface. In effect, the inversion charge tells us the capacity of the well

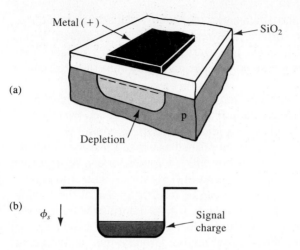

Figure 9-21
An MOS capacitor with a positive gate pulse: (a) depletion region and surface charge; (b) potential well at the interface, partially filled with electrons corresponding to the surface charge shown in (a).

for storing charge. The time required to fill the well thermally is called the *thermal relaxation time,* and it depends on the quality of the semiconductor material and its interface with the insulator. For good materials the thermal relaxation time can be much longer than the charge storage times involved in CCD operation.

If instead of a steady state bias we apply a large positive pulse to the MOS gate electrode, a deep potential well is first created. Before inversion has occurred by thermal generation, the depletion width is greater than it would be at equilibrium ($W > W_m$). This transient condition is sometimes called *deep depletion,* a different use of the term from that of Section 9.3.3. If we can inject electrons into this potential well electrically or optically, they will be stored there.[†] The storage is temporary, however, because we must move the electrons out to another storage location before thermal generation becomes appreciable.

What is needed is a simple method for allowing charge to flow from one potential well to an adjacent one quickly and without losing much charge in the process. If this is accomplished, we can inject, move, and collect packets of charge dynamically to do a variety of electronic functions.

9.4.2 The Basic CCD

The original CCD structure proposed in 1969 by Boyle and Smith of Bell Laboratories consisted of a series of metal electrodes forming an array of MOS capacitors as shown in Fig. 9-22. Voltage pulses are supplied in three lines (L_1, L_2, L_3), each connected to every third electrode in the row (G_1, G_2, G_3). These voltages are clocked to provide potential wells, which vary with time as

---

[†]The potential well should not be confused with the depletion region, which extends into the bulk of the semiconductor. The "depth" of the well is measured in electrostatic potential, not distance. Electrons stored in the potential well are in fact located very near the semiconductor surface.

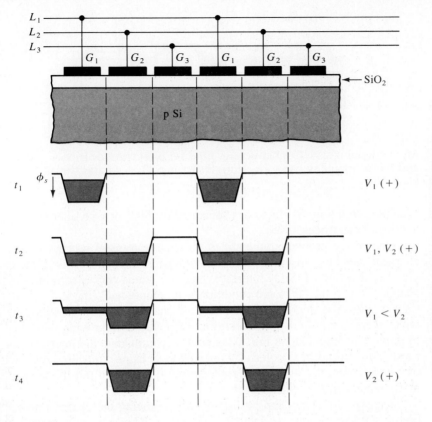

Figure 9-22
The basic CCD, composed of a linear array of MOS capacitors. At time t_1 the G_1 electrodes are positive, and the charge packet is stored in the G_1 potential well. At t_2 both G_1 and G_2 are positive, and the charge is distributed between the two wells. At t_3 the potential on G_1 is reduced, and the charge flows to the second well. At t_4 the transfer of charge to the G_2 well is completed.

in Fig. 9-22. At time t_1 a potential well exists under each G_1 electrode, and we assume this well contains a packet of electrons from a previous operation. At time t_2 a potential is applied also to the adjacent electrode G_2, and the charge equalizes across the common G_1–G_2 well. It is easy to visualize this process by thinking of the mobile charge in analogy with a fluid, which flows to equalize its level in the expanding container. This fluid model continues at t_3 when V_1 is reduced, thus decreasing the potential well under G_1. Now the charge flows into the G_2 well, and this process is completed at t_4 when V_1 is zero. By this process the packet of charge has been moved from under G_1 to G_2. As the procedure is continued, the charge is next passed to the G_3 position, and continues down the line as time proceeds. In this way charge can be injected using an input diode, transported down the line, and detected at the other end.

9.4.3 Improvements on the Basic Structure

Several problems arise in the implementation of the CCD structure of Fig. 9-22. For example, the separation between electrodes must be very small to allow coupling between the wells. An improvement can be made by using an overlapping gate structure such as that shown in Fig. 9-23. This can be done, for

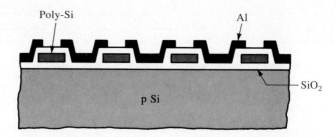

Figure 9-23 An overlapping gate CCD structure. One set of electrodes is polycrystalline Si, and the overlapping gates are Al in this case. SiO$_2$ separates the adjacent electrodes.

example, with poly-Si electrodes separated by SiO$_2$ or with alternating poly-Si and metal electrodes.

One of the problems inherent to the charge transfer process is that some charge is inevitably lost during the many transfers along the CCD. If the charges are stored at the Si–SiO$_2$ interface, surface states trap a certain amount of charge. Thus if the "0" logic condition is an empty well, the leading edge of a train of pulses is degraded by the loss of charge required in filling the traps which were empty in the "0" condition. One way of improving this situation is to provide enough bias in the "0" state to accommodate the interface and bulk traps. This procedure is colorfully referred to as using a *fat zero*. Even with the use of fat zeros, the signal is degraded after a number of transfers, by inherent inefficiencies in the transfer process.

Transfer efficiency can be improved by moving the charge transfer layer below the semiconductor–insulator interface. This can be accomplished by using ion implantation or epitaxial growth to create a layer of opposite type than the substrate. This shifts the maximum potential under each electrode into the semiconductor bulk, thus avoiding the semiconductor–insulator interface. This type of device is referred to as a *buried channel* CCD.

The three-phase CCD shown in Fig. 9-22 is only one example of a variety of CCD structures. Figure 9-24 illustrates one method for achieving a two-phase system, in which voltages are sequentially applied to alternating gate electrodes from two lines. A two-level poly-Si gate structure is used, in which the gate electrodes overlap, and a donor implant near the Si surface creates a built-in well under half of each electrode. When both gates are turned off (b), potential wells exist only under the implanted regions, and charge can be stored in any of these wells. With electrode G_2 pulsed positively, the charge packet shown in (b) is transferred to the deepest well under G_2, which is its implanted region as (c) indicates. Then with both gates off, the wells appear as in (b) again, except that the charge is now under the G_2 electrode. The next step in the transfer process is obviously to pulse G_1 positively, so that the charge moves to the implanted region under the G_1 electrode to the right. This type of CCD is implemented in a linear shift register in Fig. 9-25. Shown in this figure is the output section of the device, showing contacting metallization and pads

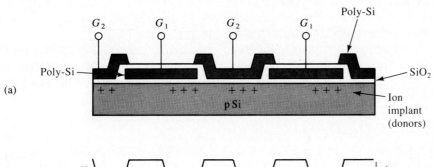

(a)

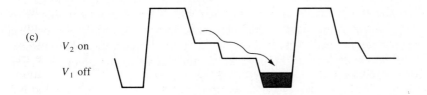

(b) V_1, V_2 off

(c)

V_2 on

V_1 off

Figure 9-24
A two-phase CCD
with an extra
potential well built
in under the right
half of each
electrode by donor
implantation.

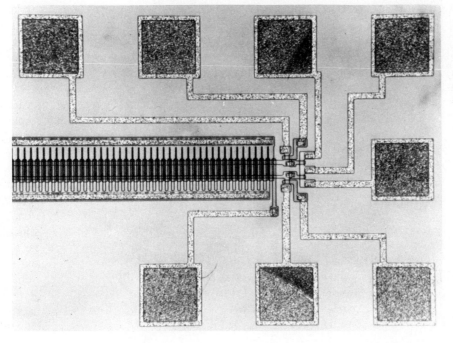

Figure 9-25
Photograph of the
output section of a
shift register made
using the CCD
structure of
Fig. 9-24.
(Photograph
courtesy of Texas
Instruments, Inc.)

for lead bonding. This device is fabricated using electron-beam lithography to achieve electrode lengths smaller than 4 μm. The packing density for this device corresponds to 10–20 million bits per square inch.

Other improvements to the basic structure are important in various applications. These include channel stops or other methods for achieving lateral confinement for the stored charge. Regeneration points must be included in the array to refresh the signal after it has been degraded.

9.4.4 Applications of CCDs

CCDs are used in a number of ways, including signal processing functions such as delay, filtering, and multiplexing several signals. Other applications include digital memories and logic arrays. Another interesting application of CCDs is in imaging, in which an array of photosensors is used to form charge packets proportional to light intensity, and these packets are shifted to a detector point for readout. There are numerous ways of accomplishing this in CCDs, including the linear array line scanner, in which the second dimension is obtained by moving the scanner relative to the image. Alternatively, an area image sensor can be made which scans the image electronically in both dimensions. The latter device can be used as an alternative to the electron beam-addressed television imaging tube (Fig. 9-26).

Figure 9-26
A charge-coupled device image sensor with an area of 5.5 × 5.5 cm and 2048 × 2048 image-sensing elements (pixels). (Photograph courtesy of Tektronix, Inc.)

In the early development of integrated circuits it was felt that the inevitable defects that occur in processing would prevent the fabrication of devices containing more than a few dozen logic gates. One approach to integration on a larger scale tried in the late 1960s involved fabricating many identical logic gates on a wafer, testing them, and interconnecting the good ones (a process called *discretionary wiring*). While this approach was being developed, however, radical improvements were made in device processing which increased the yield of good chips on a wafer dramatically. By the early 1970s it was possible to build circuits with many hundreds of components per chip, with reasonable yield. These improvements made discretionary wiring obsolete almost as soon as it was developed. By reducing the number of processing defects, improving the packing density of components, and increasing the wafer size, it is now possible to place millions of device elements on a single chip of silicon and to obtain many perfect chips per wafer. The cost savings are impressive, as is the changing image of the electronic system itself. For example, Fig. 9-27 illustrates three stages in the process of achieving a 16 k-bit (actually $2^{14} = 16,384$) dynamic random access memory (DRAM). The savings in cost per bit are considerable if one 16 k-bit chip can be used instead of sixteen 1 k-bit chips. There are many sources of these savings, including reduced interconnection and packaging costs. However, the obvious advantage for device fabrication illustrated by Fig. 9-27 is that many more 16 k-bit chips (c) can be achieved on a Si wafer than would be the case for the collection of 1 k-bit chips shown in (a). Because of this, much of the improvement in integrated circuit economics comes from using element dimensions near the limit of the photolithography and using large Si wafers (6–8 inches in diameter). As a result, size reductions such as that in Fig. 9-27 have led to memory chips sized at 64K, 256K, 1M, 4M, etc.

The layout of a 4M-bit memory can be seen in Fig. 9-28, which illustrates $2^{22} = 4,194,304$ bits of dynamic random access memory. This device employs

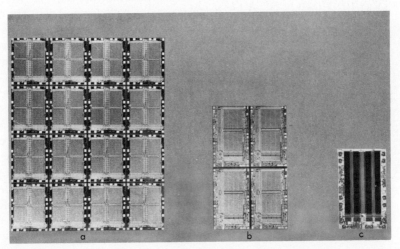

Figure 9-27
Three ways of obtaining 16 kilobits of memory: (a) sixteen 1 k-bit chips; (b) four 4 k-bit chips; (c) one 16 k-bit chip. (Photograph courtesy of Intel Corp.)

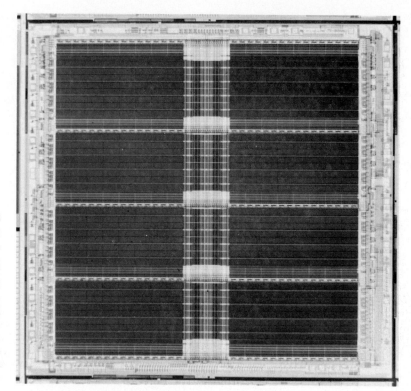

Figure 9-28
A 4M-bit DRAM chip. This chip, 1 cm on a side, contains 2^{22} memory cells and employs a trench transistor and capacitor design shown in Fig. 9-29. (Photograph courtesy of Texas Instruments, Inc.)

trench capacitor CMOS technology, with 1-μm design rules (minimum feature sizes) and double-level metallization. To accommodate the 8.4 million components (transistors and capacitors) in a chip 1 cm square, the surface area of each memory cell is about 9 μm^2. This reduced cell area is accomplished by putting part of the transistor inside the top of the trench (Fig. 9-29). This device is an example of *very large-scale integration* (*VLSI*), a term that was invented when *large*-scale integration (LSI) began to seem inadequate. There are no precise definitions of LSI and VLSI, since devices continue to increase in complexity, and yesterday's very large scale device is modest compared with today's. Some people use *ultra* large-scale integration (ULSI) to refer to chips as large as the one shown in Fig. 9-28.

The use of VLSI in memories is widespread, and this ready availability of memory capability has created a revolution in modern electronics. An important feature of many memory devices is that they are *programmable* on the chip. There are several ways of achieving this programming feature. For example, an MOS transistor can be made with two gate electrodes, the lower of which is insulated and "floating" electrically. The threshold voltage for the transistor depends on whatever charge may be stored on this floating gate. A large voltage ($\sim$25 V) between the top gate electrode and the substrate allows some electrons to cross the insulator and charge the floating gate. This capability allows the memory to be programmed by selectively altering the threshold of

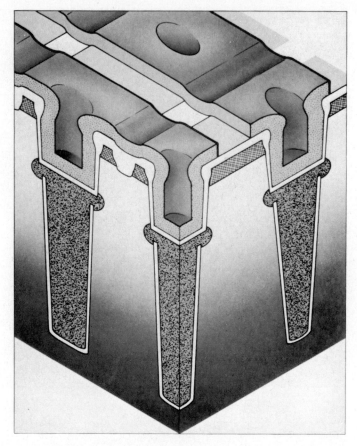

Figure 9-29
An isometric cross section of the trench cell used in the 4M-bit DRAM shown in Fig. 9-28. The charge is stored on the inside of the trench, and the p^+ substrate serves as the capacitor plate in this design. The transistor is located primarily in the top of the trench, so that the total cell surface is only about $9\mu m^2$. (Illustration courtesy of Texas Instruments, Inc.)

individual transistors. Exposure of the memory to ultraviolet light increases the conductivity of the insulator and allows the charge to leak away from the floating gate. Thus the memory is *erasable*. Another form of erasable memory takes advantage of the fact that charges can be stored at the interface between SiO_2 and Si_3N_4 layers forming the gate insulator. These charges can be *electrically altered* without erasing the entire memory. Programmable features are particularly useful in read-only memories (ROM), and programmable ROM is referred to in general as PROM. An erasable PROM is called EPROM, and an electrically alterable ROM is called EAROM.

Although the achievement of many powers of two in memory is impressive and important, other LSI chips are important for the integration of many different system functions. A *microprocessor* includes functions for a computer central processing unit (CPU). A *microcomputer* includes memory along with the processing, control, timing, and interface circuits required to perform very complex computing functions. Figure 9-30 illustrates a microprocessor chip with various areas outlined by function.

Before leaving this section it might be useful to provide some calibration regarding the dimensions we have been discussing. It is difficult to visualize

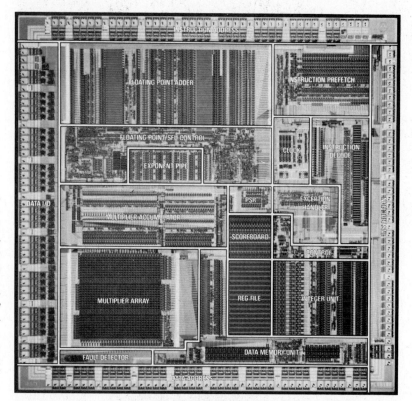

Figure 9-30
A microprocessor chip with various functional areas delineated. This microprocessor uses the *reduced instruction set computing (RISC)* architecture. (Photograph courtesy of Motorola, Inc.)

Figure 9-31
Size comparison of FET integrated circuit elements with an amoeba. The two photomicrographs are superimposed to the same scale. (Photograph courtesy of IBM.)

the size of transistor–capacitor cells or CCD elements appropriate for VLSI. Perhaps the micrograph of Fig. 9-31 can provide this calibration. It superimposes a portion of an 8 k-bit LSI memory with the image of an amoeba to the same scale. It is interesting that the amoeba, considered rather small on a biological scale, is really quite large on the scale of microelectronics. About 40 memory cells fit within the area taken up by the amoeba. If a more densely packed memory such as that shown in Fig. 9-28 were used, the amoeba would completely dwarf the memory cells.

After the preceding discussions of rather dramatic fabrication steps in monolithic circuit technology, the processes of attaching leads and packaging the devices could seem rather mundane. Such an impression would be far from accurate, however, since the techniques discussed in this section are crucial to the overall fabrication process. In fact, the handling and packaging of individual circuits can be the most critical steps of all from the viewpoints of cost and reliability. The individual IC chip must be connected properly to outside leads and packaged in a way that is convenient for use in a larger circuit or system. Since the devices are handled individually once they are separated from the wafer, bonding and packaging are expensive processes. Considerable work has been done to reduce the steps required in bonding. We shall discuss the most straightforward technique first, which involves bonding individual leads from the contact pads on the circuit to terminals in the package. Then we shall consider two important methods for making all bonds simultaneously. Finally, we shall discuss a few typical packaging methods for ICs.

**9.6
TESTING,
BONDING,
AND PACKAGING**

9.6.1 Testing

After the wafer of monolithic circuits has been processed and the final metallization pattern defined, it is placed in a holder under a microscope and is aligned for testing by a multiple-point probe (Fig. 9-32). The probe contacts the various pads on an individual circuit, and a series of tests are made of the electrical properties of the device. The various tests are programmed to be made automatically in a very short time. These tests may take only milliseconds for a simple circuit, or up to 30 s or more for a complex LSI chip. The information from these tests is fed into a computer, which compares the results with information stored in its memory, and a decision is made regarding the acceptability of the circuit. If there is some defect so that the circuit falls below specifications, the computer instructs the test probe to mark the circuit with a dot of ink. The probe automatically steps the prescribed distance to the next circuit on the wafer and repeats the process. After all of the circuits have been tested and the substandard ones marked, the wafer is removed from the testing machine, scribed between the circuits, and broken apart. Then the individual chips are mounted on their substrates for packaging, after the inked circuits have been discarded. In the testing process, information from tests on each circuit can be printed out to facilitate analysis of the rejected circuits or to evaluate the fabrication process for possible changes.

Figure 9-32
A wafer of monolithic circuits under test by a multiple-point probe. In this example 40 test probes touch the contact pads of a microprocessor chip about $\frac{1}{2}$ cm on a side. The probes are rigidly fixed in position, so that the wafer can be stepped to the next circuit with little realignment. (Photograph courtesy of AT&T Bell Laboratories.)

9.6.2 Wire Bonding

The earliest method used for making contacts from the monolithic chip to the package was the bonding of fine Au wires. Later techniques expanded wire bonding to include Al wires and several types of bonding processes. Here we shall outline only a few of the most important aspects of wire bonding.

If the chip is to be wire bonded, it is first mounted solidly on a metal lead frame (Fig. 9-33) or on a metallized region of an insulating substrate. In this process a thin layer of Au (perhaps combined with Ge or other elements to improve the metallurgy of the bond) is placed between the bottom of the chip and the substrate; heat and a slight scrubbing motion are applied, forming an alloyed bond which holds the chip firmly to the substrate. This process is called *die bonding* (die, the singular form of dice, is used interchangeably with chip to refer to the individual monolithic device). Once the chip is mounted, the interconnecting wires are attached from the various contact pads to posts on the lead frame (Fig. 9-34).

In Au wire bonding, a spool of fine Au wire (about 0.0007–0.002-in. diameter) is mounted in a *lead bonder* apparatus, and the wire is fed through a glass or tungsten carbide *capillary* (Fig. 9-35a). A hydrogen gas flame jet is swept past the wire to form a ball on the end. In *thermocompression bonding* the chip (or in some cases the capillary) is heated to about 360°C, and the capillary is brought down over the contact pad. When pressure is exerted by the capillary on the ball, a bond is formed between the Au ball and the Al pad (Fig. 9-35b). Then the capillary is raised and moved to a post on the lead frame.

Figure 9-33
Mounting of chips in metal lead frames, in preparation for die bonding and contacting steps. (Photograph courtesy of Texas Instruments, Inc.)

Figure 9-34
Attachment of leads from the Al pads on the periphery of the chip to posts on the package. (Photograph courtesy of Motorola, Inc.)

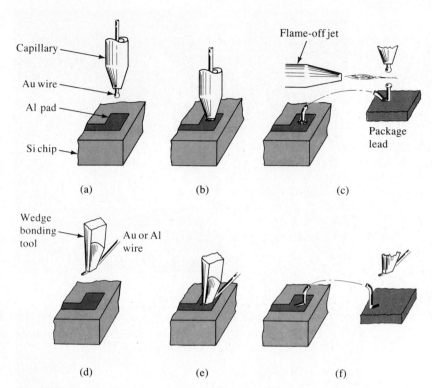

Figure 9-35
Wire bonding techniques: (a) capillary positioned over one of the contact pads for a ball (nail-head) bond; (b) pressure exerted to bond the wire to the pad; (c) post bond and flame-off; (d) wedge bonding tool; (e) pressure and ultrasonic energy applied; (f) post bond completed and wire broken or cut for next bond.

The capillary is brought down again, and the combination of force and temperature bonds the wire to the post. After raising the capillary again, the hydrogen flame is swept past, forming a new ball (Fig. 9-35c); then the process is repeated for the other pads on the chip.

There are many variations in this basic method. For example, the substrate heating can be eliminated by *ultrasonic bonding*. In this method a tungsten carbide capillary is held by a tool connected to an ultrasonic transducer. When it is in contact with a pad or a post, the wire is vibrated under pressure to form a bond. Other variations include techniques for automatically removing the "tail," which is left on the post in Fig. 9-35c. When the bond to the chip is made by exerting pressure on a ball at the end of the Au wire, it is called a *ball bond* or a *nail-head bond*, because of the shape of the deformed ball after the bond is made.

Aluminum wire can be used in ultrasonic bonding; it has several advantages over Au, including the absence of possible metallurgical problems in bonds between Au and Al pads. When Al wire is used, the flame-off step is replaced by cutting or breaking the wire at appropriate points in the process. In forming a bond, the wire is bent under the edge of a wedge-shaped bonding tool (Fig. 9-35d). The tool then applies pressure and ultrasonic vibration, forming the bond (Fig. 9-35e and f). The resulting flat bond, formed by the bent wire wedged between the tool and the bonding surface, is called a *wedge bond*. A closeup view of ball and wedge bonds is given in Fig. 9-36.

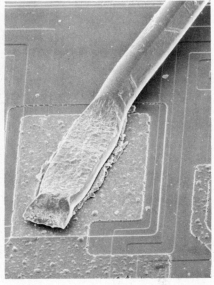

(a)   (b)

Figure 9-36
Scanning electron
micrographs of a
ball bond (a) and a
wedge bond (b).
(Photographs
courtesy of the
National Institute of
Standards and
Technology.)

9.6.3   Flip-chip and Beam-lead Techniques

The time consumed in bonding wires individually to each pad on the chip can be overcome by several methods of simultaneous bonding. The *flip-chip* and the *beam-lead* approaches are typical of these methods. In each case, relatively thick metal is deposited on the contact pads before the devices are separated from the wafer. After separation, the deposited metal is used to contact a matching metallized pattern on the package substrate.

In the flip-chip method, "bumps" of solder or special metal alloys are deposited on each contact pad. These metal bumps rise about 50 μm above the surface of the monolithic chip. After separation from the wafer, each chip is turned upside down, and the bumps are properly aligned with the metallization pattern on the substrate. At this point, ultrasonic bonding or solder alloying attaches each bump to its corresponding connector on the substrate (Fig. 9-37). An obvious advantage of this method is that all connections are made simultaneously. Disadvantages include the fact that the bonds are made under the chip and therefore cannot be inspected visually. Furthermore, it is necessary to heat and/or exert pressure on the chip.

In beam-lead technology, bonds to the substrate pattern are made external to the Si chip. The process is more complicated than flip-chip methods, but there are important compensations. Basically, the beam-lead technique calls for thick (about 10 μm) metal tabs on the wafer, leading away from the circuit at each contact pad. These relatively large ("beam") leads are commonly made

Figure 9-37 Flip-chip bonding. The integrated circuit chip is mounted directly on the thick-film substrate by simultaneously soldering a number of metallized "bumps" on the chip to the interconnection patterns on the substrate. (Photograph courtesy of Delco Electronics Corp.)

by electroplating Au onto the wafer in regions defined by a photoresist pattern. The beam leads extend from the metallized interconnection pattern to outside the active area of each circuit. Then the wafer is mounted face down in wax on a flat disk and is lapped from the back side to a total wafer thickness of about 50 μm. A photoresist pattern is used on the back surface to delineate the circuits, and the Si is then etched away from between the individual chips. After the Si has been removed in the desired pattern, the beam leads are left protruding from each circuit. Then the circuit can be mounted face down on the substrate, and the protruding beam leads can be bonded simultaneously to the substrate metallization pattern. Alternatively, the chip can be placed face up into a cavity in the substrate, with the beam leads over the cavity edge. In either case, all bonds to the connector pattern can be made simultaneously without heating or exerting pressure on the chip itself.

9.6.4 Packaging

The final step in IC fabrication is packaging the device in a suitable medium that can protect it from the environment of its intended application. In most cases this means the surface of the device must be isolated from moisture and contaminants and the bonds and other elements must be protected from corrosion and mechanical shock. The problems of surface protection are greatly minimized by modern passivation techniques, but it is still necessary to provide some protection in the packaging. In every case, the choice of package type must be made within the requirements of the application and cost considerations. There are many techniques for encapsulating devices, and the various methods are constantly refined and changed. Here we shall consider just a few general methods for the purpose of illustration.

In the early days of IC technology, all devices were packaged in metal headers. In this method the device is alloyed to the surface of the header, wire bonds are made to the header posts, and a metal lid is welded over the device and wiring. Although this method has several drawbacks, it does provide complete sealing of the unit from the outside environment. This is often called a *hermetically sealed* device. After the chip is mounted on the header and bonds are made to the posts, the header cap can be welded shut in a controlled environment (e.g., an inert gas), which maintains the device in a prescribed atmosphere.

Integrated circuits are now mounted in packages with many output leads. For example, Fig. 9-38 shows a dual in-line package (DIP), in which connectors are brought out along two sides. After mounting the chip on a stamped metal lead frame such as that shown in Fig. 9-33 and forming the contacts, the package is formed by applying a ceramic or plastic case and trimming away the unwanted parts of the lead frame. In the particular example of Fig. 9-38, a window is provided over the chip to allow erasure of the EPROM by ultraviolet excitation.

Other package designs bring interconnections through multiple patterned ceramic layers to an array of pins. This package type (Fig. 9-39) is called a *pin grid array* (*PGA*). Such packages are extremely complex and require manufacturing methods approaching those of the semiconductor chip itself.

Since a sizable fraction of the cost of an IC is due to bonding and packaging, there have been a number of innovations for automating the process. These include the use of film reels that contain the metal contact pattern onto which the chips can be bonded. The film can then be fed into packaging equipment, where the position registration capabilities of a film reel can be used for automated handling. This process, called *tape-automated bonding* (*TAB*), is particularly useful in mounting several chips on a large ceramic substrate having multilevel interconnection patterns (called a *multichip module*).

Figure 9-38
Packaging of a microcomputer chip in a DIP. A transparent lid is included over the chip for erasure of the EPROM by ultraviolet light. (Photograph courtesy of Intel Corp.)

PROBLEMS

9.1. Assume that boron is diffused into a uniform n-type Si sample, resulting in a net doping profile $N_a(x) - N_d$. Set up an expression relating the sheet resistance of the diffused layer to the acceptor profile $N_a(x)$ and the junction depth x_j. Assume that $N_a(x)$ is much greater than the background doping N_d over most of the diffused layer.

9.2. A typical sheet resistance of a base diffusion layer is 100 Ω/square.
   (a) What should be the aspect ratio of a 5-kΩ resistor, using this diffusion?
   (b) Draw a pattern for this resistor (see Fig. 9-18b) which uses little area for a width $w = 10$ μm.

9.3. A 5-μm n-type epitaxial layer ($N_d = 10^{16}$ cm^{-3}) is grown on a p-type Si substrate. Areas of the n layer are to be junction isolated by a boron diffusion at 1200°C ($D = 2.5 \times 10^{-12}$ cm^2/s). The surface boron concentration is held constant at 10^{20} cm^{-3} (see Prob. 5.22).
   (a) What time is required for this isolation diffusion?
   (b) How far does an Sb-doped buried layer ($D = 2 \times 10^{-13}$ cm^2/s) diffuse into the epitaxial layer during this time, assuming the concentration at the substrate-epitaxial boundary is constant at 10^{20} cm^{-3}?

9.4. In Section 9.2.1 ten steps are listed for the fabrication of a double-diffused transistor. Make a list of this type for the double-diffused epitaxial transistor illustrated in Fig. 9-8d, including metallization.

9.5. We wish to find the overall values of α and β for the composite p-n-p transistor of Fig. P9-5. Assume that the terminal currents of the lateral p-n-p device are related by α_1 and β_1 (neglect saturation currents); similarly, let α_2 and β_2 describe the n-p-n.
   (a) Find α and β for the composite p-n-p.
   (b) Show that when $\alpha_2 = 1$, the overall $\alpha = 1$ and $\beta = \beta_1\beta_2$.

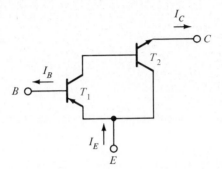

Figure P9-5

9.6. Figure P9-6 shows a top view and a cross section of a portion of a monolithic circuit consisting of a transistor and a resistor. Each element is junction isolated.
   (a) Identify the various regions in the two drawings.
   (b) Sketch the processing steps in fabricating this circuit, using Fig. 9-8 a–d as an example.

Figure P9-6

 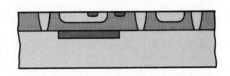

9.7. The two-input TTL gate shown in Fig. P9-7 employs two emitters for transistor T_1. The transistors are double-diffused epitaxial devices (with buried layers), and are junction isolated. The resistors are obtained during the base diffusion. Design an appropriate layout, with contact pads $A-E$ on the periphery of the chip, and draw the six masks required for fabrication of this circuit.

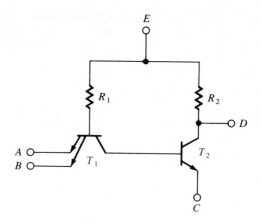

Figure P9-7

READING LIST

Armstrong, J. A. "The Science of VLSI." *Physics Today* 39, no. 10 (October 1986): 24–83.

Blogett, A. J. "Microelectronic Packaging." *Scientific American* 249, no. 1 (July 1983): 86–96.

Ghandi, S. K. *VLSI Fabrication Principles.* New York: John Wiley, 1983.

Gupta, D. C., and P. H. Langer. *Emerging Semiconductor Technology.* Philadelphia: ASTM Publications, 1987.

Jaeger, R. C. *The Modular Series on Solid State Devices. Vol. V: Introduction to Microelectronic Fabrication.* Reading, Mass.: Addison-Wesley, 1988.

Lu, N. C. C. "Advanced Cell Structure for Dynamic RAMs." *IEEE Circuits and Devices Magazine* (January, 1989): 27–36.

McGurie, G. E. ed. *Semiconductor Materials and Process Technology Handbook.* Park Ridge, N.J.: Noyes Publishers, 1988.

Special Issue on Micron and Submicron Circuit Engineering, *Proceedings of the IEEE* 71, no. 5 (May 1983): 547–681.

Muller, R. S. and T. I. Kamins. *Device Electronics for Integrated Circuits.* 2d ed. New York: John Wiley, 1986.

Seraphim, D. P., R. Lasky, and C-Y Li. *Principles of Electronic Packaging.* New York: McGraw-Hill, 1989.

Special Issue on Submicron Lithography. *Solid State Technology.* (September 1984 and September 1987).

Sze, S. M. ed. 2d ed. *VLSI Technology.* New York: McGraw-Hill, 1988.

Sze, S. M. *Semiconductor Devices.* New York: John Wiley, 1985.

Tarui, Y. ed. *VLSI Technology.* New York: Springer-Verlag, 1986.

Tasch, A. F. and L. H. Parker. "Memory Cell and Technology Issues for 64 Mb and 256 Mb One-Transistor Cell MOS DRAMs." *Proceedings of the IEEE* 77, no. 3 (March 1989): 374–384.

Special Issue on Thin Films. *Physics Today* 33, no. 5 (May 1980): 26–55.

Troutman, R. R. *Latchup in CMOS Technology.* Boston: Kluwer Academic Publ., 1986.

Wolf, S. and R. N. Tauber. *Silicon Processing for the VLSI Era.* Sunset Beach, Calif.: Lattice Press, 1986.

Yang, E. S. *Microelectronic Devices.* New York: McGraw-Hill, 1988.

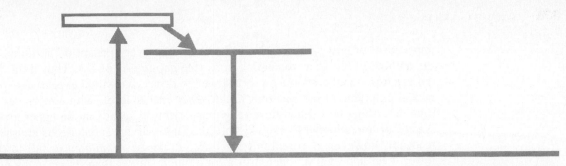

chapter 10

LASERS

The word LASER is an acronym for *light amplification by stimulated emission of radiation*, which sums up the operation of an important optical and electronic device. The laser is a source of highly directional, monochromatic, coherent light, and as such it has revolutionized some longstanding optical problems and has created some new fields of basic and applied optics. The light from a laser, depending on the type, can be a continuous beam of low or medium power, or it can be a short burst of intense light delivering millions of watts. Light has always been a primary communications link between man and his environment, but until the invention of the laser, the light sources available for transmitting information and performing experiments were generally neither monochromatic nor coherent, and were of relatively low intensity. Thus the laser is of great interest in optics; but it is equally important in optoelectronics, particularly in fiber–optic communications. The effort of creating a useful laser system involves the techniques of solid state (or gaseous) electronics, and the transmission and detection of laser signals makes use of a broad range of electronics.

The last three letters in the word *laser* are intended to imply how the device operates: by the *stimulated emission* of *radiation*. In Chapter 2 we discussed the emission of radiation when excited electrons fall to lower energy states; but generally, these processes occur randomly and can therefore be classed as *spontaneous emission*. This means that the rate at which electrons fall from an

**10.1
STIMULATED
EMISSION**

upper level of energy E_2 to a lower level E_1 is at every instant proportional to the number of electrons remaining in E_2 (the *population* of E_2). Thus if an initial electron population in E_2 were allowed to decay, we would expect an exponential emptying of the electrons to the lower energy level, with a mean decay time describing how much time an average electron spends in the upper level. An electron in a higher or excited state need not wait for spontaneous emission to occur, however; if conditions are right, it can be *stimulated* to fall to the lower level and emit its photon in a time much shorter than its mean spontaneous decay time. The stimulus is provided by the presence of photons of the proper wavelength. Let us visualize an electron in state E_2 waiting to drop spontaneously to E_1 with the emission of a photon of energy $h\nu_{12} = E_2 - E_1$ (Fig. 10-1). Now we assume that this electron in the upper state is immersed in an intense field of photons, each having energy $h\nu_{12} = E_2 - E_1$, and in phase with the other photons. The electron is induced to drop in energy from E_2 to E_1, contributing a photon whose wave is *in phase* with the radiation field. If this process continues and other electrons are stimulated to emit photons in the same fashion, a large radiation field can build up. This radiation will be *monochromatic* since each photon will have an energy of precisely $h\nu_{12} = E_2 - E_1$ and will be *coherent*, because all the photons released will be in phase and reinforcing. This process of stimulated emission can be described quantum mechanically to relate the probability of emission to the intensity of the radiation field. Without quantum mechanics we can make a few observations here about the relative rates at which the absorption and emission processes occur. Let us assume the instantaneous populations of E_1 and E_2 to be n_1 and n_2, respectively. We know from earlier discussions of distributions and the Boltzmann factor that at *thermal equilibrium* the relative population will be

$$\frac{n_2}{n_1} = e^{-(E_2 - E_1)/kT} = e^{-h\nu_{12}/kT} \tag{10-1}$$

if the two levels contain an equal number of available states.

The negative exponent in this equation indicates that $n_2 \ll n_1$ at equilibrium; that is, most electrons are in the lower energy level as expected. If the atoms exist in a radiation field of photons with energy $h\nu_{12}$, such that the energy density of the field is $\rho(\nu_{12})$,[†] then stimulated emission can occur along with

Figure 10-1
Stimulated transition of an electron from an upper state to a lower state, with accompanying photon emission.

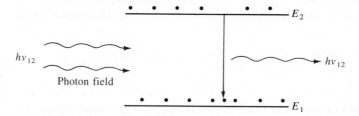

hv_{12}

Photon field

hv_{12}

E_2

E_1

[†]The energy density $\rho(\nu_{12})$ indicates the total energy in the radiation field per unit volume and per unit frequency, due to photons with $h\nu_{12} = E_2 - E_1$.

absorption and spontaneous emission. The rate of stimulated emission is proportional to the instantaneous number of electrons in the upper level n_2 and to the energy density of the stimulating field $\rho(\nu_{12})$. Thus we can write the stimulated emission rate as $B_{21}n_2\rho(\nu_{12})$, where B_{21} is a proportionality factor. The rate at which the electrons in E_1 absorb photons should also be proportional to $\rho(\nu_{12})$, and to the electron population in E_1. Therefore, the absorption rate is $B_{12}n_1\rho(\nu_{12})$, where B_{12} is a proportionality factor for absorption. Finally, the rate of spontaneous emission is proportional only to the population of the upper level. Introducing still another coefficient, we can write the rate of spontaneous emission as $A_{21}n_2$. For steady state the two emission rates must balance the rate of absorption to maintain constant populations n_1 and n_2 (Fig. 10-2).

$$
\begin{aligned}
B_{12}n_1\rho(\nu_{12}) &= A_{21}n_2 + B_{21}n_2\rho(\nu_{12}) \\
\text{Absorption} &= \text{spontaneous} + \text{stimulated} \\
&\quad\;\; \text{emission} \qquad\;\; \text{emission}
\end{aligned}
\tag{10-2}
$$

This relation was described by Einstein, and the coefficients B_{12}, A_{21}, B_{21} are called the *Einstein coefficients*. We notice from Eq. (10-2) that no energy density ρ is required to cause a transition from an upper to a lower state; spontaneous emission occurs without an energy density to drive it. The reverse is not true, however; exciting an electron to a higher state (absorption) requires the application of energy, as we would expect thermodynamically.

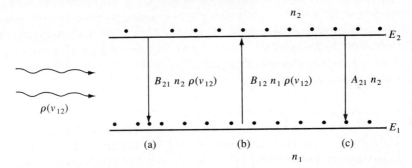

Figure 10-2
Balance of absorption and emission in steady state: (a) stimulated emission; (b) absorption; (c) spontaneous emission.

At thermal equilibrium, the ratio of the stimulated to spontaneous emission rates is generally very small, and the contribution of stimulated emission is negligible. With a photon field present,

$$
\frac{\text{Stimulated emission rate}}{\text{Spontaneous emission rate}} = \frac{B_{21}n_2\rho(\nu_{12})}{A_{21}n_2} = \frac{B_{21}}{A_{21}}\rho(\nu_{12})
\tag{10-3}
$$

As Eq. (10-3) indicates, the way to enhance the stimulated emission over spontaneous emission is to have a very large photon field energy density $\rho(\nu_{12})$. In the laser, this is encouraged by providing an *optical resonant cavity* in which the photon density can build up to a large value through multiple internal reflections at certain frequencies (ν).

Similarly, to obtain more stimulated emission than absorption we must have $n_2 > n_1$:

$$\frac{\text{Stimulated emission rate}}{\text{Absorption rate}} = \frac{B_{21}n_2\rho(\nu_{12})}{B_{12}n_1\rho(\nu_{12})} = \frac{B_{21}}{B_{12}}\frac{n_2}{n_1} \qquad (10\text{-}4)$$

Thus if stimulated emission is to dominate over absorption of photons from the radiation field, we must have a way of maintaining more electrons in the upper level than in the lower level. This condition is quite unnatural, since Eq. (10-1) indicates that n_2/n_1 is less than unity for any equilibrium case. Because of its unusual nature, the condition $n_2 > n_1$ is called *population inversion*. It is also referred to as a condition of *negative temperature*. This rather startling terminology emphasizes the nonequilibrium nature of population inversion, and refers to the fact that the ratio n_2/n_1 in Eq. (10-1) could be larger than unity only if the temperature were negative. Of course, this manner of speaking does not imply anything about temperature in the usual sense of that word. The fact is that Eq. (10-1) is a thermal equilibrium equation and cannot be applied to the situation of population inversion without invoking the concept of negative temperature.

In summary, Eqs. (10-3) and (10-4) indicate that if the photon density is to build up through a predominance of stimulated emission over both spontaneous emission and absorption, two requirements must be met. We must provide (1) an optical resonant cavity to encourage the photon field to build up and (2) a means of obtaining population inversion. A large part of our disussion in the following sections will be devoted to examining the second requirement for various energy level structures. There are ways of obtaining population inversion in the atomic levels of many solids, liquids, and gases, and in the energy bands of semiconductors. Thus the possibilities for laser systems with various materials are quite extensive.

10.2
THE RUBY LASER

The first working laser was built in 1960 by Maiman,[†] using a ruby crystal. Ruby belongs to the family of gems consisting of Al_2O_3 with various types of impurities. For example, pink ruby contains about 0.05 percent Cr atoms. Similarly, Al_2O_3 doped with Ti, Fe, or Mn results in variously colored sapphire. Most of these materials can be grown by modern techniques as single crystals.

10.2.1 The Resonant Cavity

Ruby crystals are available in rods several inches long, convenient for forming an optical cavity (Fig. 10-3). The crystal is cut and polished so that the ends are flat and parallel, with the end planes perpendicular to the axis of the rod. These ends are coated with a highly reflective material, such as Al or Ag, producing a resonant cavity in which light intensity can build up by multiple re-

---

[†]T. H. Maiman, *Nature*, vol. 187, pp. 493–494 (August 6, 1960).

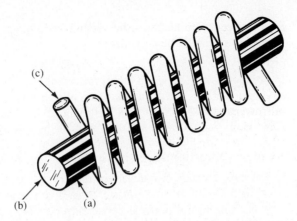

Figure 10-3
Schematic diagram
of a ruby laser:
(a) ruby rod;
(b) end mirror;
(c) flash tube.

flections. One of the end mirrors is constructed to be partially transmitting so that a fraction of the light will "leak out" of the resonant system. This transmitted light is the output of the laser. Of course, in designing such a laser one must choose the amount of transmission to be a small perturbation on the resonant system. The gain in photons per pass between the end plates must be larger than the transmission at the ends, scattering from impurities, absorption, and other losses. The arrangement of parallel plates providing multiple internal reflections is similar to that used in the Fabry–Perot interferometer;[†] thus the silvered ends of the laser cavity are often referred to as Fabry–Perot faces. As Fig. 10-4 indicates, light of a particular frequency can be reflected back and forth within the resonant cavity in a reinforcing (coherent) manner if an integral number of half-wavelengths fit between the end mirrors. Thus the length of the cavity for stimulated emission must be

$$L = \frac{m\lambda}{2} \tag{10-5}$$

where **m** is an integer. In this equation λ is the photon wavelength within the laser material. If we wish to use the wavelength λ_0 of the output light in the

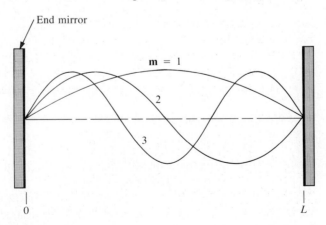

End mirror

m = 1

2

3

0

L

Figure 10-4
Resonant modes
within a laser
cavity.

[†]Interferometers are discussed in many sophomore physics texts.

atmosphere (often taken as the vacuum value), the index of refraction **n** of the laser material must be considered

$$\lambda_0 = \lambda\mathbf{n} \tag{10-6}$$

In practice, the distance in a half-wavelength of visible or near infrared light is so small that Eq. (10-5) is automatically satisfied over some portion of a real mirror spacing. Mechanical adjustment of L to some $\mathbf{m}\lambda/2$ is not necessary, since many values of $\mathbf{m}$ and $\lambda/2$ will fit the resonant condition.

10.2.2 Population Inversion in Ruby

In the case of ruby, Cr atoms in the crystal have energy levels as shown in Fig. 10-5. Without attempting to indicate fine points of the energy level scheme, this figure portrays those ruby levels important for stimulated emission. This is basically a three-level system. Absorption occurs in a rather broad range in the green part of the spectrum, raising electrons from the ground state E_1 to the band of levels designated E_3 in the figure. These excited levels are highly unstable, and the electrons decay rapidly to the level E_2. This transition occurs with the energy difference $E_3 - E_2$ given up as heat. The level E_2 is very important for the stimulated emission process since electrons in this level have a mean lifetime of about 5 ms before they fall to the ground state. Because this lifetime is relatively long, E_2 is called a *metastable* state. If electrons are excited from E_1 to E_3 at a rate faster than the decay rate from E_2 back to E_1, the population of the metastable state E_2 becomes larger than that of the ground state E_1 (we assume that electrons fall from E_3 to E_2 in a negligibly short time). This buildup of electrons in the metastable level can be compared to the filling of a sink with an open drain. When water is poured into the sink faster than it drains out, the water level must increase. Similarly, population inversion can be established between E_2 and E_1 if electrons arrive at E_2 faster than they leave.

Figure 10-5
Energy levels for
chromium ions in
ruby.

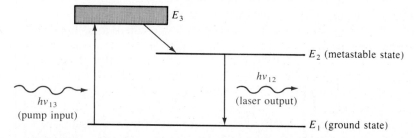

Population inversion is obtained by *optical pumping* of the ruby rod with a flash lamp such as the one shown in Fig. 10-3c. A common type of flash lamp is a glass tube wrapped around the ruby rod and filled with xenon gas. A capacitor can be discharged through the xenon-filled tube, creating a pulse of very intense light over a broad spectral range.

If the light pulse from the flash tube is several milliseconds in duration, we might expect an output from the ruby laser over a large fraction of that time. However, the laser does not operate continuously during the light pulse but instead emits a series of very short spikes (Fig. 10-6). When the flash lamp in-

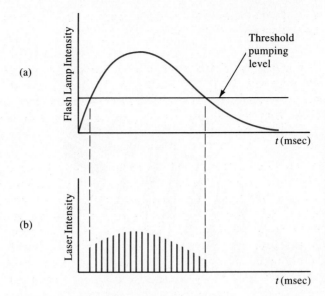

Figure 10-6
Laser spikes in the output of a ruby laser: (a) typical variation of flash lamp intensity with time; (b) laser spikes occurring while the flash lamp intensity is above the threshold pumping level.

tensity becomes large enough to create population inversion (the *threshold pumping level*), stimulated emission from the metastable level to the ground level occurs, with a resulting laser output. Once the stimulated emission begins, however, the metastable level is depopulated very quickly. Thus the laser output consists of an intense spike lasting from a few nanoseconds to microseconds. After the stimulated emission spike, population inversion builds up again and a second spike results. This process continues as long as the flash lamp intensity is above the threshold pumping level.

10.2.3   Giant Pulse Lasers

We conclude from Fig. 10-6 that the metastable level never receives a highly inverted population of electrons. Whenever the population of E_2 reaches the minimum required for stimulated emission, these electrons are depleted quickly in one of the spikes. To prevent this, we must somehow keep the coherent photon field in the ruby rod from building up (and thus prevent stimulated emission) until after a larger population inversion is obtained. This can be accomplished if we temporarily interrupt the resonant character of the optical cavity. This process is called *Q-spoiling* or *Q-switching,* where Q is the quality factor of the resonant structure. A straightforward method for doing this is illustrated in Fig. 10-7. The front face of the ruby rod is silvered to be partially reflecting, but the back face is left unsilvered. The back reflector of the optical cavity is provided by an external mirror, which can be rotated at high speeds. When the mirror plane is aligned exactly perpendicular to the laser axis, a resonant structure exists; but as the mirror rotates away from this position, there is no buildup of photons by multiple reflection, and no laser action can occur. Thus during a flash from the xenon lamp, a very large inverted population builds up as the mirror rotates off-axis. When the mirror finally returns to the

Figure 10-7
Q-switched ruby laser in which one face of the resonant cavity is an external rotating mirror.

position at which light reflects back into the rod, stimulated emission can occur, and the large population of the metastable level is given up in one intense laser pulse. This structure is called a *giant pulse laser* or a *Q-switched laser*. By saving the electron population for a single pulse, a large amount of energy is given up in a very short time. For example, if the total energy in the pulse is 1 J and the pulse width is 100 nsec (10^{-7} s), the pulse power is 10^7 J/s = 10 MW.

A number of alternative methods of *Q*-spoiling are available. For example, in the *Kerr cell* a transverse electric field is used to control the transmission of light longitudinally through a material in the laser path. By controlling the polarization of light transmitted through the cell, the *Q* of the laser cavity can be reduced until the cell is pulsed to the "on" state by the applied electric field. Another example is the use of bleachable dyes placed between the rod and one end mirror. The dye solution is opaque until the population inversion reaches a point at which the spontaneous emission is intense enough to optically bleach the dye. Then the path to the mirror is clear and the resonant cavity is established. The reading list at the end of this chapter includes discussions of other giant pulse techniques.

10.3
OTHER LASER
SYSTEMS

One important disadvantage of a three-level laser system, such as the ruby laser, is that no significant stimulated emission occurs until at least half the ground state electrons have been excited to the metastable state. That is, population inversion exists only when the metastable state is more densely populated than the ground state. This makes for a high threshold for optical pumping and a considerable waste of energy. Better performance may be expected from

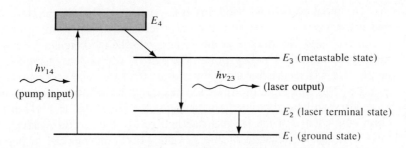

Figure 10-8
Typical four-level
laser system.

a system with an extra level above the ground state (Fig. 10-8). In such a four-level laser system, population inversion begins between the metastable state (E_3) and the laser terminal state (E_2) with very little pumping. Of course, the emission rate from E_2 to the ground state must be faster than the rate from E_3 to E_2; otherwise, the electron population of E_2 could build up and exceed that of E_3, destroying the population inversion. In addition, E_2 must be substantially empty at equilibrium; that is, it must be several kT above E_1 to prevent its being populated by thermal excitation of ground state electrons.

10.3.1 Rare Earth Systems

The most common four-level solid state lasers utilize host materials doped with atoms from the *rare earth* section of the periodic table. For example, neo-dymium (Nd) is used as the active atom in a number of host materials. The literature abounds with descriptions of four-level laser systems, since the active spectroscopic levels of each rare earth atom are somewhat different when placed in various host lattices. For example, the laser wavelengths of Nd in the yttrium-aluminum garnet (YAG) lattice are different from those obtained for the same dopant in various glasses. The number of useful doping atoms is continually being expanded, and laser host materials now include not only exotic crystals and glasses but also plastics and even liquid solutions.

10.3.2 Gas Lasers

Even the four-level solid state laser is quite inefficient, in that a large amount of optical energy is required to create population inversion. Most of the light from the flash lamp is of the wrong wavelength for the desired pumping and is therefore wasted. Generally, it is necessary to operate these lasers on a pulsed basis, allowing the system to cool between pulses. Many laser applications, however, require a steady, long-term output of light (often called *continuous wave,* or *cw* operation). A laser system that can be excited by means other than optical pumping is needed. A good alternative is provided by the gas laser systems, in which electric discharges can be created to excite atoms. In a discharge, energy is transferred to the atoms of the gas by electron impact and collisions between atoms. The resulting mixture of electrons and gas ions (*plasma*) is a good source of spectral emission and, in many cases, can be maintained continuously as long as power is supplied.

The most commonly used gas laser system employs a mixture of helium and neon gases. This mixture uses a process of resonant energy transfer between colliding He and Ne atoms. Energy exchange between atoms becomes highly probable when they possess a pair of similar energy levels. For example, He has metastable levels at almost the same energies as the 2s and 3s levels of Ne. Since the lifetimes for electrons in these He levels are relatively long, they will be populated for many of the He atoms at a given time. The close coincidence of the He metastable levels to the 2s and 3s levels of Ne allows a resonant exchange of energy and a resultant increase in population of the corresponding Ne levels. The laser emissions of interest are due to electronic transitions in the Ne atoms; thus the primary purpose of He in the mixture is to enhance the excitation process.

The Ne levels important for laser action are shown in Fig. 10-9. The strongest emission line occurs between a 2s level and a 2p level, with a wavelength of 1.1 μm (in the infrared). Another important transition occurs between a 3s level and a 2p level, with photons given off at 0.6328 μm (red). This is the most popular operating mode of the He–Ne laser, since photons are created in the red part of the visible spectrum. Thus the He–Ne laser in the 6328-Å mode is a natural choice in many applications which require visible light and cw operation.

Figure 10-9
Schematic of laser transitions in Ne.

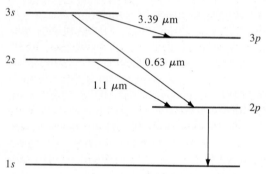

A typical He–Ne laser consists of a glass tube to contain the gas mixture, a means of creating a discharge in the gas, and end mirrors for forming the resonant cavity (Fig. 10-10). It is common to place the mirrors outside the glass envelope, or *plasma tube,* to make alignment easier. However, when this is done, it is necessary to arrange for the light to be transmitted through the end of the tube with a minimum of reflection. This is accomplished by terminating the plasma tube with an optically flat glass plate, oriented at an angle to the tube axis. There is a particular angle, called the *Brewster angle,* for which internal reflection within the plasma tube is zero for light with a given polarization. Excitation of the system is commonly accomplished by a d-c discharge between two electrodes within the plasma tube; alternatively, rf power can be applied to generate the plasma. Although it is capable of cw operation, the He–Ne laser is rather inefficient (typically less than 1 percent). Thus for a He–Ne gas laser with a cw light output of 10 mW, we should expect to supply at least several watts of power to the gas discharge. Other gas systems are better suited for

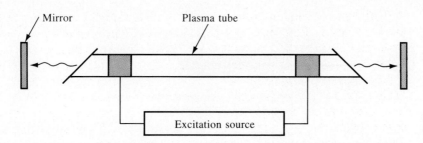

Figure 10-10
Gas laser structure.

generating large amounts of optical power output; for example, CO_2 lasers can generate thousands of watts of continuous power at rather high conversion efficiencies.

The laser became an important part of semiconductor device technology in 1962 when the first p-n junction lasers were built in GaAs (infrared)[†] and GaAsP (visible).[‡] We have already discussed the incoherent light emission from p-n junctions (LEDs), generated by the spontaneous recombination of electrons and holes injected across the junction. In this section we shall concentrate on the requirements for population inversion due to these injected carriers and the nature of the coherent light from p-n junction lasers. These devices differ from the solid, gas, and liquid lasers discussed previously in several important respects. Junction lasers are remarkably small (typically on the order of $0.1 \times 0.1 \times 0.3$ mm), they exhibit high efficiency, and the laser output is easily modulated by controlling the junction current. Semiconductor lasers operate at low power compared, for example, with ruby or CO_2 lasers; on the other hand, these junction lasers compete with He–Ne lasers in power output. Thus the function of the semiconductor laser is to provide a portable and easily controlled source of low-power coherent radiation. They are particularly suitable for fiber optic communication systems (Section 6.4.2).

**10.4
SEMICONDUCTOR
LASERS**

10.4.1   Population Inversion at a Junction

If a p-n junction is formed between degenerate materials, the bands under forward bias appear as shown in Fig. 10-11. If the bias (and thus the current) is large enough, electrons and holes are injected into and across the transition region in considerable concentrations. As a result, the region about the junction is far from being depleted of carriers. This region contains a large concentration of electrons within the conduction band and a large concentration of holes within the valence band. If these population densities are high enough, a con-

[†]R. N. Hall et al., *Physical Review Letters* 9, pp. 366–368 (November 1, 1962); M. I. Nathan et al., *Applied Physics Letters* 1, pp. 62–64 (November 1, 1962); T. M. Quist et al., *Applied Physics Letters* 1, pp. 91–92 (December 1, 1962).
[‡]N. Holonyak, Jr., and S. F. Bevacqua, *Applied Physics Letters* 1, pp. 82–83 (December 1, 1962).

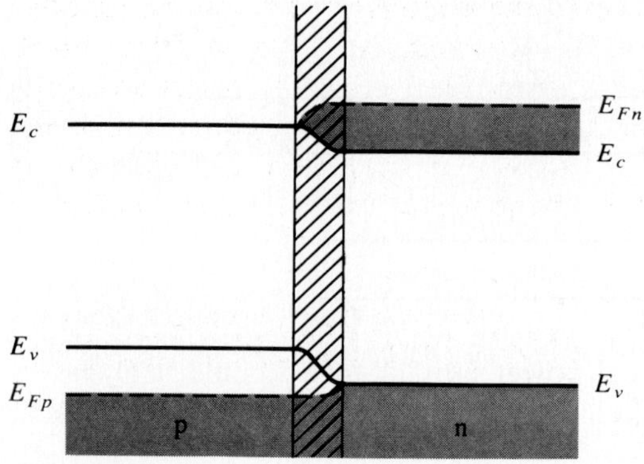

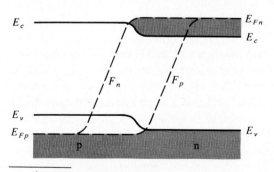

Figure 10-11
Band diagram of a
p-n junction laser
under forward bias.
The cross-hatched
region indicates the
inversion region at
the junction.

dition of population inversion results, and the region about the junction over which it occurs is called an *inversion region.*[†]

Population inversion at a junction is best described by the use of the concept of *quasi-Fermi levels* (Section 4.3.3). Since the forward-biased condition of Fig. 10-11 is a distinctly nonequilibrium state, the equilibrium equations defining the Fermi level are not applicable. In particular, the concentration of electrons in the inversion region (and for several diffusion lengths into the p material) is larger than equilibrium statistics would imply; the same is also true for the injected holes in the n material. We can use Eqs. (4-15) to describe the carrier concentrations in terms of the quasi-Fermi levels for electrons and holes in steady state. Thus

$$n = N_c e^{-(E_c - F_n)/kT} = n_i e^{(F_n - E_i)/kT} \tag{10-7a}$$

$$p = N_v e^{-(F_p - E_v)/kT} = n_i e^{(E_i - F_p)/kT} \tag{10-7b}$$

Using Eqs. (10-7a) and (10-7b), we can draw F_n and F_p on any band diagram for which we know the electron and hole distributions. For example, in Fig. 10-12, F_n in the neutral n region is essentially the same as the equilibrium Fermi level E_{Fn}. This is true to the extent that the electron concentration on the n side is equal to its equilibrium value. However, since large numbers of electrons are injected across the junction, the electron concentration begins at a

Figure 10-12
Quasi-Fermi levels
in a laser junction
under forward bias.

[†]This is a different meaning of the term from that used in reference to MOS transistors.

high value near the junction and decays exponentially to its equilibrium value n_p deep in the p material. Therefore, F_n drops from E_{Fn} as shown in Fig. 10-12. We notice that, deep in the neutral regions, the quasi-Fermi levels are essentially equal. The separation of F_n and F_p at any point is a measure of the departure from equilibrium at that point. Obviously, this departure is considerable in the inversion region, since F_n and F_p are separated by an energy greater than the band gap (Fig. 10-13).

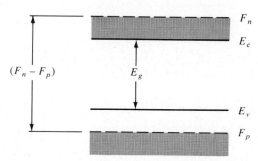

Figure 10-13 Expanded view of the inversion region.

Unlike the case of the two-level system discussed in Section 10.1, the condition for population inversion in semiconductors must take into account the distribution of energies available for transitions between the bands. The basic definition of population inversion holds — for dominance of stimulated emission between two energy levels separated by energy $h\nu$, the electron population of the upper level must be greater than that of the lower level. The unusual aspect of a semiconductor is that bands of levels are available for such transitions. Population inversion obviously exists for transitions between the bottom of the conduction band E_c and the top of the valence band E_v in Fig. 10-13. In fact, transitions between levels in the conduction band up to F_n and levels in the valence band down to F_p take place under conditions of population inversion. For any given transition energy $h\nu$ in a semiconductor, population inversion exists when

$$(F_n - F_p) > h\nu \qquad (10\text{-}8a)$$

For band-to-band transitions, the minimum requirement for population inversion occurs for photons with $h\nu = E_c - E_v = E_g$

$$(F_n - F_p) > E_g \qquad (10\text{-}8b)$$

When F_n and F_p lie within their respective bands (as in Fig. 10-13), stimulated emission can dominate over a range of transitions, from $h\nu = (F_n - F_p)$ to $h\nu = E_g$. As we shall see below, the dominant transitions for laser action are determined largely by the resonant cavity and the strong recombination radiation occurring near $h\nu = E_g$.

In choosing a material for junction laser fabrication, it is necessary that electron-hole recombination occur directly, rather than through trapping processes such as are dominant in Si or Ge. Gallium arsenide is an example of such a "direct" semiconductor. Furthermore, we must be able to dope the material n-type or p-type to form a junction. If an appropriate resonant cavity can be constructed in the junction region, a laser results in which population inversion is accomplished by the bias current applied to the junction (Fig. 10-14).

Figure 10-14
Variation of
inversion region
width with forward
bias: $V(a) < V(b)$.

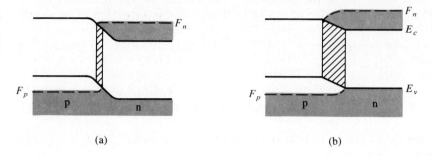

(a)   (b)

10.4.2 Emission Spectra for p-n Junction Lasers

Under forward bias, an inversion layer can be obtained along the plane of the junction, where a large population of electrons exists at the same location as a large hole population. A second look at Fig. 10-13 indicates that spontaneous emission of photons can occur due to direct recombination of electrons and holes, releasing energies ranging from approximately $F_n - F_p$ to E_g. That is, an electron can recombine over an energy from F_n to F_p, yielding a photon of energy $h\nu = F_n - F_p$, or an electron can recombine from the bottom of the conduction band to the top of the valence band, releasing a photon with $h\nu = E_c - E_v = E_g$. These two energies serve as the approximate outside limits of the laser spectra.

The photon wavelengths which participate in stimulated emission are determined by the length of the resonant cavity as in Eq. (10-5). Figure 10-15 illustrates a typical plot of emission intensity vs. photon energy for a semiconductor laser. At low current levels (Fig. 10-15a), a spontaneous emission spectrum containing energies in the range $E_g < h\nu < (F_n - F_p)$ is obtained. As the current is increased to the point that significant population inversion exists, stimulated emission occurs at frequencies corresponding to the cavity modes as shown in Fig. 10-15b. These modes correspond to successive numbers of integral half-wavelengths fitted within the cavity, as described by

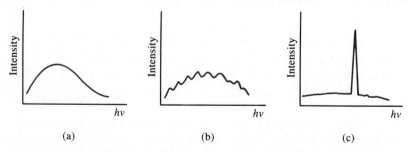

(a)   (b)   (c)

Figure 10-15   Light intensity vs. photon energy $h\nu$ for a junction laser: (a) incoherent emission below threshold; (b) laser modes at threshold; (c) dominant laser mode above threshold. The intensity scales are greatly compressed from (a) to (b) to (c).

Eq. (10-5). Finally, at a still higher current level, a most preferred mode or set of modes will dominate the spectral output (Fig. 10-15c). This very intense mode represents the main laser output of the device; the output light will be composed of almost monochromatic radiation superimposed on a relatively weak radiation background, due primarily to spontaneous emission.

The separation of the modes in Fig. 10-15b is complicated by the fact that the index of refraction **n** for GaAs depends on wavelength λ. From Eq. (10-5) we have

$$\mathbf{m} = \frac{2L\mathbf{n}}{\lambda_0} \tag{10-9}$$

If **m** (the number of half-wavelengths in L) is large, we can use the derivative to find its rate of change with λ_0:

$$\frac{d\mathbf{m}}{d\lambda_0} = -\frac{2L\mathbf{n}}{\lambda_0^2} + \frac{2L}{\lambda_0}\frac{d\mathbf{n}}{d\lambda_0} \tag{10-10}$$

Now reverting to discrete changes in **m** and λ_0, we can write

$$-\Delta\lambda_0 = \frac{\lambda_0^2}{2L\mathbf{n}}\left(1 - \frac{\lambda_0}{\mathbf{n}}\frac{d\mathbf{n}}{d\lambda_0}\right)^{-1}\Delta\mathbf{m} \tag{10-11}$$

If we let $\Delta\mathbf{m} = -1$, we can calculate the change in wavelength $\Delta\lambda_0$ between adjacent modes (i.e., between modes **m** and **m** − 1).

10.4.3 The Basic Semiconductor Laser

To build a p-n junction laser, we need to form a junction in a highly doped, direct semiconductor (GaAs, for example), construct a resonant cavity in the proper geometrical relationship to the junction, and make contact to the junction in a mounting which allows for efficient heat transfer. A simple fabrication technique is outlined in Fig. 10-16. Beginning with a degenerate n-type sample, a p region is formed on one side, for example by diffusing Zn into the

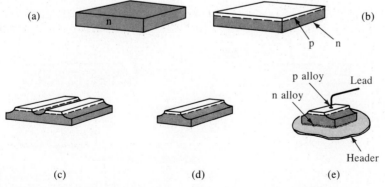

Fabrication of a simple junction laser: (a) degenerate n-type sample; (b) diffused p layer; (c) isolation of junctions by cutting or etching; (d) individual junction to be cut or cleaved into devices; (e) mounted laser structure.

Figure 10-16

n-type GaAs. Since Zn is in column II of the periodic table and is introduced substitutionally on Ga sites, it serves as an acceptor in GaAs; therefore, the heavily doped Zn diffused layer forms a p$^+$ region (Fig. 10-16b). At this point we have a large-area planar p-n junction. Next, grooves are cut or etched along the length of the sample as in Fig. 10-16c, leaving a series of long p regions isolated from each other. These p-n junctions can be cut or broken apart (Fig. 10-16d) and then cleaved into devices of the desired length.

At this point in the fabrication process, the very important requirement of a resonant cavity must be considered. It is necessary that the front and back faces (Fig. 10-16e) be flat and parallel. This can be accomplished by cleaving. If the sample has been oriented so that the long junctions of Fig. 10-16d are perpendicular to a crystal plane of the material, it is possible to cleave the sample along this plane into laser devices, letting the crystal structure itself provide the parallel faces. The device is then mounted on a suitable header, and contact is made to the p region. Various techniques are used to provide adequate heat sinking of the device for large forward current levels.

10.4.4 Heterojunction Lasers

The device described above was the first type used in the early development of semiconductor lasers. Since the device contains only one junction in a single type of material, it is referred to as a *homojunction* laser. To obtain more efficient lasers, and particularly to build lasers that operate at room temperature, it is necessary to use multiple layers in the laser structure. Such devices, called *heterojunction lasers,* can be made to operate continuously at room temperature to satisfy the requirements of optical communications. An example of a heterojunction laser is shown in Fig. 10-17. In this structure the injected carriers are confined to a narrow region so that population inversion can be built up

Figure 10-17
Use of a single heterojunction for carrier confinement in laser diodes: (a) AlGaAs heterojunction grown on the thin p-type GaAs layer; (b) band diagrams for the structure of (a), showing confinement of electrons to the thin p region under bias.

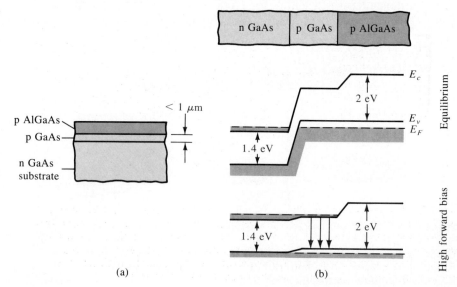

at lower current levels. The result is a lowering of the *threshold current* at which laser action begins. Carrier confinement is obtained in this single-heterojunction laser by the layer of AlGaAs grown epitaxially on the GaAs.

In GaAs the laser action occurs primarily on the p side of the junction due to a higher efficiency for electron injection than for hole injection. In a normal p-n junction the injected electrons diffuse into the p material such that population inversion occurs for only part of the electron distribution near the junction. However, if the p material is narrow and terminated in a barrier, the injected electrons can be confined near the junction. In Fig. 10-17a, an epitaxial layer of p-type AlGaAs ($E_g \simeq 2$ eV) is grown on top of the thin p-type GaAs region. The wider band gap of AlGaAs effectively terminates the p-type GaAs layer, since injected electrons do not surmount the barrier at the GaAs-AlGaAs heterojunction (Fig. 10-17b). As a result of the confinement of injected electrons, laser action begins at a substantially lower current than for simple p-n junctions. In addition to the effects of carrier confinement, the change of refractive index at the heterojunction provides a waveguide effect for optical confinement of the photons.

A further improvement can be obtained by sandwiching the active GaAs layer between two AlGaAs layers (Fig. 10-18). This *double-heterojunction* structure further confines injected carriers to the active region, and the change in refractive index at the GaAs–AlGaAs boundaries helps to confine the generated light waves. In the double-heterojunction laser shown in Fig. 10-18b the injected current is restricted to a narrow stripe along the lasing direction, to re-

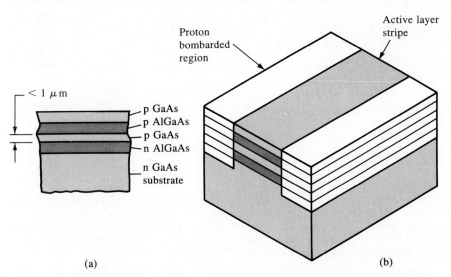

(a)   (b)

A double-heterojunction laser structure: (a) multiple layers used to confine injected carriers and provide waveguiding for the light; (b) a stripe geometry designed to restrict the current injection to a narrow stripe along the lasing direction. One of many methods for obtaining the stripe geometry, this example is obtained by proton bombardment of the shaded regions in (b), which converts the GaAs and AlGaAs to semi-insulating form.

Figure 10-18

duce the total current required to drive the device. This type of laser was a major step forward in the development of lasers for fiber–optic communications.

Separate Confinement and Graded Index Channels. One of the disadvantages of the double-heterostructure laser shown in Fig. 10-18 is the fact that the carrier confinement and the optical waveguiding both depend on the same heterojunctions. It is much better to optimize these two functions by using a narrow confinement region for keeping the carriers in a region of high recombination, and a somewhat wider optical waveguide region. In Fig. 10-19a we show a *separate confinement* laser in which the width of the optical waveguiding region (*w*) is optimized by using the refractive index step at a separate heterojunction from that used to confine the carriers. For example, in the GaAs/$Al_xGa_{1-x}As$ system the optical confinement (waveguiding) occurs at a boundary with much larger composition *x* (and therefore smaller refractive index) than is the case for the carrier confinement barrier. By grading the composition of the AlGaAs it is possible to obtain even better waveguiding. For example, in Fig. 10-19b a parabolic grading of the refractive index leads to a waveguide within the laser analogous to that shown in Fig. 6-20 for a fiber. This *graded index separate confinement heterostructure* (*GRINSCH*) laser also provides built-in fields for better electron confinement.

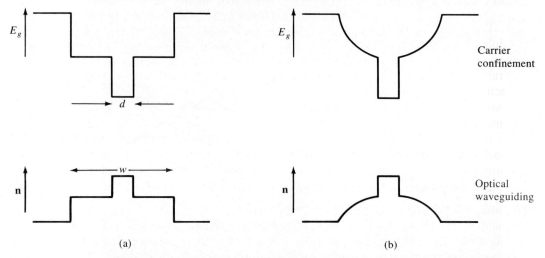

Figure 10-19 Separate confinement of carriers and waveguiding: (a) use of separate changes in AlGaAs alloy composition to confine carriers in the region (*d*) of smallest band gap, and to obtain waveguiding (*w*) at the larger step in refractive index; (b) grading the alloy composition, and therefore the refractive index, for better waveguiding and carrier confinement.

Distributed Feedback and Buried Heterostructure. In continual efforts to improve semiconductor laser performance, very complex structures requiring imaginative use of epitaxial growth have been employed. As an example of this development, Fig. 10-20 shows a cross section of a multilayer GaAs–AlGaAs device, obtained by scanning electron microscopy on a cleaved sec-

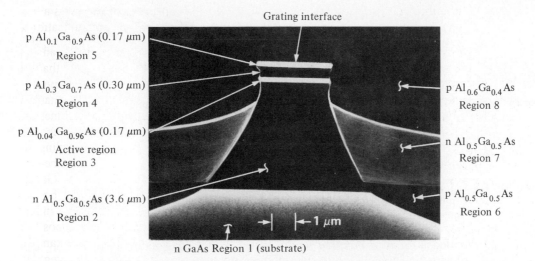

Grating interface

p $Al_{0.1}Ga_{0.9}As$ (0.17 μm)
Region 5

p $Al_{0.3}Ga_{0.7}As$ (0.30 μm)
Region 4

p $Al_{0.04}Ga_{0.96}As$ (0.17 μm)
Active region
Region 3

n $Al_{0.5}Ga_{0.5}As$ (3.6 μm)
Region 2

p $Al_{0.6}Ga_{0.4}As$
Region 8

n $Al_{0.5}Ga_{0.5}As$
Region 7

p $Al_{0.5}Ga_{0.5}As$
Region 6

1 μm

n GaAs Region 1 (substrate)

Cross section (a scanning electron micrograph) of a buried heterostructure laser that includes lateral as well as vertical confinement of the injected current by the grown layers and distributed feedback by a grating. The structure was grown by liquid phase epitaxy. (Photograph courtesy of Xerox Corp.)   **Figure 10-20**

tion. This device, containing eight separate crystal regions, was made by multiple liquid-phase epitaxial growths on a GaAs substrate. The active AlGaAs region is 0.17 μm thick and is surrounded by wider band gap layers, which (laterally and vertically) confine injected carriers, restrict injected current to the active portion of the device, and provide waveguiding. This type of *buried heterostructure* device is much more efficient than types with only vertical confinement. Another special feature of this device is called *distributed feedback,* in which a corrugation (or *grating*) is used along the length of the laser to provide Bragg reflection instead of reflection from end mirrors. This lack of end mirrors allows the integration of lasers on a chip with optical waveguides and other components. This type of laser also enhances wavelength stability.

Quantum Well Lasers and Other Advanced Structures. As the carrier confinement region of Fig. 10-19 is decreased to a few hundred angstroms, the resulting well takes on the quantum well properties described in Sections 2.4.3 and 3.2.5. Since transitions occur via discrete states higher in the well (Fig. 3-13), it is possible to obtain laser photon energies larger than the GaAs band gap. Other advanced structures are designed to produce single-wavelength lasing and more power output. For example, very short cavities can lead to lasing on a single longitudinal mode. Similarly, a laser can be cleaved into two separate parts through which the light propagates, with the two parts independently controlled electrically (Fig. 10-21a). The first diode is operated above threshold while the second one can be above or below threshold for tuning purposes. A change in injection current in the second diode causes a change in carrier density, and therefore a slight change in refractive index. By this means, a single wavelength common to both cavities can be selected. Such a structure is

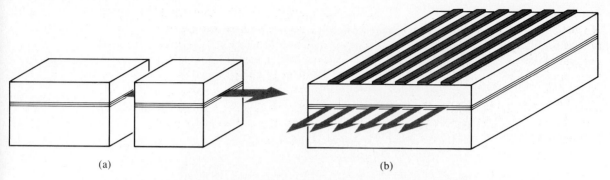

Figure 10-21 Examples of advanced laser structures: (a) cleaved-coupled cavity laser for single mode operation; (b) phase-locked laser array for low-divergence, high-power output.

called a *cleaved-coupled cavity* laser. By forming an array of buried heterostructure lasers closely spaced along the heterojunction plane, it is possible to obtain phase-locked emission over a large number of stripes (Fig. 10-21b) and considerable output power.

10.4.5 Materials for Semiconductor Lasers

We have discussed the properties of the junction laser largely in terms of GaAs and AlGaAs. However, as discussed in Section 6.4.2, the InGaAsP/InP system is particularly well suited for the type of lasers used in fiber optic communication systems. Lattice matching (Section 1.4.1) is important in creating heterostructures by epitaxial growth. The fact that the AlGaAs band gap can be varied by choice of composition on the column III sublattice allows the formation of barriers and confining layers such as those shown in Section 10.4.4. The quaternary alloy InGaAsP is particularly versatile in the fabrication of laser diodes, allowing considerable choice of wavelength and flexibility in lattice matching. By choice of composition, lasers can be made in the infrared range 1.3–1.55 μm required for fiber optics. Since four components can be varied in choosing an alloy composition, InGaAsP allows simultaneous choice of energy gap (and therefore emission wavelength) and lattice constant (for lattice matched growth on convenient substrates). In many applications, however, other wavelength ranges are required for laser output. For example, the use of lasers in pollution diagnostics requires wavelengths farther in the infrared than are available from InGaAsP and AlGaAs. In this application the ternary alloy PbSnTe provides laser output wavelengths from about 7 μm to more than 30 μm at low temperatures, depending on the material composition. For intermediate wavelengths, the InGaSb system can be used.

Materials chosen for the fabrication of semiconductor lasers must be efficient light emitters and also be amenable to the formation of p-n junctions and in most cases the formation of heterojunction barriers. These requirements eliminate some materials from practical use in laser diodes. For example, semiconductors with indirect band gaps are not sufficiently efficient light emitters for practical laser fabrication. The II–VI compounds, on the other hand, are generally very efficient at emitting light but are not well suited for the formation of junctions. As a result, these materials can be made to lase by photopumping, but not by injection with a p-n junction. As the requirements for laser junctions in new applications expand, other materials will undoubtedly emerge. Most likely these new laser materials will be ternary or quaternary alloys of III–V compounds. Should a method be devised for efficient carrier injection in II–VI compounds, however, a host of new laser materials would become available.

PROBLEMS

10.1. Assume that the system described by Eq. (10-2) is in thermal equilibrium at an extremely high temperature such that the energy density $\rho(\nu_{12})$ is essentially infinite. Show that $B_{12} = B_{21}$.

10.2. The system described by Eq. (10-2) interacts with a blackbody radiation field whose energy density per unit frequency at ν_{12} is

$$\rho(\nu_{12}) = \frac{8\pi h\nu_{12}^3}{c^3}[e^{h\nu_{12}/kT} - 1]^{-1}$$

from Planck's radiation law. Given the result of Prob. 10.1, find the value of the ratio A_{21}/B_{12}.

10.3. Assuming equal electron and hole concentrations and band-to-band transitions, calculate the minimum carrier concentration $n = p$ for population inversion in GaAs at 300 K. The intrinsic carrier concentration in GaAs is about 10^6 cm^{-3}.

10.4. The degenerate occupation of bands shown in Fig. 10-13 helps maintain the laser requirement that emission must overcome absorption. Explain how the degeneracy prevents band-to-band absorption at the emission wavelength.

READING LIST

Botez, D. "Laser Diodes Are Power-Packed," *IEEE Spectrum* (June, 1985): 43–54.

Burns, G. and M. I. Nathan. "P-N Junction Lasers." *Proceedings of the IEEE* 52 (July 1964): 770–94.

Casey, Jr., H. C. and M. B. Panish. *Heterostructure Lasers Part A: Fundamental Principles.* New York: Academic Press, Inc., 1978.

Chemal, D. D. "Quantum Wells for Photonics." *Physics Today* 38, no. 5 (May 1985): 56–64.

Miller, S. E. and I. P. Kaminow, eds. *Optical Fiber Telecommunications II.* San Diego: Academic Press, 1988.

Special Issue on Lasers, *Physics Today* 41, no. 10 (October 1988): 24–63.

Special Issue on Lasers, *IEEE Journal of Quantum Electronics* QE-23, no. 6 (June, 1987): 650–695.

Special Issue on Lasers in Research, *Physics Today* 30, no. 5 (May 1977).

Special Issue on Optoelectronics, *Physics Today* 38, no. 5 (May 1985): 23–64.

Rowell, J. M. "Photonic Materials." *Scientific American* 255, no. 4 (October 1986): 147–57.

Shimony, A. "The Reality of the Quantum World," *Scientific American* (January, 1988): 46–57.

Tien, P. K. "Integrated Optics and New Wave Phenomena in Optical Waveguides." *Reviews of Modern Physics* 49, no. 2 (April 1977): 361–420.

Tsang, W. T. "Semiconductor Lasers and Photodetectors by Molecular Beam Epitaxy." in L. L. Chang and K. Ploog, eds., *Molecular Beam Epitaxy and Heterostructures*. Boston: Martinus Nijhoff Publishers, 1985.

Verdeyen, J.T. *Laser Electronics*. Englewood Cliffs, N.J.: Prentice Hall 1981.

Wilson, J. and J. F. B. Hawkes. *Optoelectronics,* 2nd ed. Cambridge, Prentice Hall International, 1989.

chapter 11

p-n-p-n SWITCHING DEVICES

One of the most common applications of electronic devices is in switching, which requires the device to change from an "off" or *blocking* state to an "on" or *conducting* state. We have discussed the use of transistors in this application, in which base current drives the device from cutoff to saturation. Similarly, diodes and other devices can be used to serve as certain types of switches. There are a number of important switching applications that require a device remain in the blocking state under forward bias until switched to the conducting state by an external signal. Several devices which fulfill this requirement have been developed, and we shall discuss a family of switches in this chapter, the *semiconductor controlled rectifier (SCR)*[†] and related devices. These devices are typified by a high impedance ("off" condition) under forward bias until a switching signal is applied; after switching they exhibit low impedance ("on" condition). The signal required for switching can be varied externally; therefore, these devices can be used to block or pass currents at predetermined levels. In this chapter we shall discuss the physical operation of these devices, as well as certain other switching devices related to the SCR.

The SCR is a four-layer (p-n-p-n) structure that effectively blocks current through two terminals until it is turned on by a small signal at a third terminal. There are many varieties of the basic p-n-p-n structure, and we shall not attempt to cover all of them; however, we can discuss the basic operation and physical mechanisms involved in these devices. We shall begin by investigat-

---

[†]Since Si is the material commonly used for this device, it is often called a *silicon controlled rectifier*.

ing the current flow in a two-terminal p-n-p-n device and then extend the discussion to include triggering by a third terminal. We shall see that the p-n-p-n structure can be considered for many purposes as a combination of p-n-p and n-p-n transistors, and the analysis in Chapter 7 can be used as an aid in understanding its behavior.

11.1 SWITCHING MECHANISMS

Before discussing the control of an SCR using a third terminal, it is important to understand the basic transistor action at work in a p-n-p-n structure. Therefore, in this section we analyze the four-layer structure with only two terminals.

11.1.1 The p-n-p-n Diode

First we consider a four-layer diode structure with an *anode* terminal A at the outside p region and a *cathode* terminal K at the outside n region (Fig. 11.1a). We shall refer to the junction nearest the anode as j_1, the center junction as j_2, and the junction nearest the cathode as j_3. When the anode is biased positively with respect to the cathode (v positive), the device is forward biased. However, as the I–V characteristic of Fig. 11-1b indicates, the forward-biased condition of this diode can be considered in two separate states, the high-impedance or *forward-blocking* state and the low-impedance or *forward-conducting* state. In the device illustrated here the forward I–V characteristic switches from the blocking to the conducting states at a critical peak forward voltage V_P.

We can anticipate the discussion of conduction mechanisms to follow by noting that an initial positive voltage v places j_1 and j_3 under forward bias and the center junction j_2 under reverse bias. As v is increased, most of the forward voltage in the blocking state must appear across the reverse-biased junction j_2. After switching to the conducting state, the voltage from A to K is very small (less than 1 V), and we conclude that in this condition all three junctions must be forward biased. The mechanism by which j_2 switches from reverse bias to forward bias is the subject of much of the discussion to follow.

In the *reverse-blocking* state (v negative), j_1 and j_3 are reverse biased and j_2 is forward biased. Since the supply of electrons and holes to j_2 is restricted by

Figure 11-1
A two-terminal p-n-p-n device: (a) basic structure and common circuit symbol; (b) *I*–*V* characteristic.

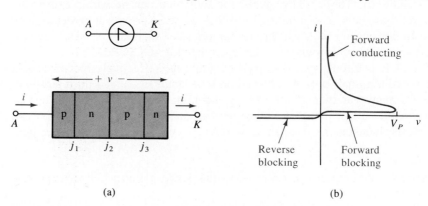

(a)

(b)

the reverse-biased junctions on either side, the device current is limited to a small saturation current arising from thermal generation of EHPs near j_1 and j_3. The current remains small in the reverse-blocking condition until avalanche breakdown occurs at a large reverse bias. In a properly designed device, with guards against surface breakdown (see Fig. 6-2), the reverse breakdown voltage can be several thousand volts.

We shall now consider the mechanism by which this device, often called a *Shockley diode,* switches from the forward-blocking state to the forward-conducting state.

11.1.2   The Two-Transistor Analogy

The four-layer configuration of Fig. 11-1a suggests that the p-n-p-n diode can be considered as two coupled transistors: j_1 and j_2 form the emitter and collector junctions, respectively, of a p-n-p transistor; similarly, j_2 and j_3 form the collector and emitter junctions of an n-p-n (note the emitter of the n-p-n is on the right, which is the reverse of what we usually draw). In this analogy, the collector region of the n-p-n is in common with the base of the p-n-p, and the base of the n-p-n serves as the collector region of the p-n-p. The center junction j_2 serves as the collector junction for both transistors.

This two-transistor analogy is illustrated in Fig. 11-2. The collector current i_{C1} of the p-n-p transistor drives the base of the n-p-n, and the base current i_{B1} of the p-n-p is dictated by the collector current i_{C2} of the n-p-n. If we associate an emitter-to-collector current transfer ratio α with each transistor, we can use the analysis in Chapter 7 to solve for the current i. Using Eq. (7-37b) with $\alpha_1 = \alpha_N$ for the p-n-p, $\alpha_2 = \alpha_N$ for the n-p-n, and with I_{CO1} and I_{CO2} for the respective collector saturation currents, we have

$$i_{C1} = \alpha_1 i + I_{CO1} = i_{B2} \tag{11-1a}$$

$$i_{C2} = \alpha_2 i + I_{CO2} = i_{B1} \tag{11-1b}$$

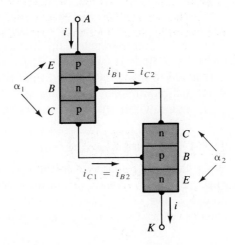

Figure 11-2
Two-transistor analogy of the p-n-p-n diode.

But the sum of i_{C1} and i_{C2} is the total current through the device:

$$i_{C1} + i_{C2} = i \tag{11-2}$$

Taking this sum in Eq. (11-1) we have

$$i(\alpha_1 + \alpha_2) + I_{CO1} + I_{CO2} = i$$

$$i = \frac{I_{CO1} + I_{CO2}}{1 - (\alpha_1 + \alpha_2)} \tag{11-3}$$

As. Eq. (11-3) indicates, the current i through the devices is small (approximately the combined collector saturation currents of the two equivalent transistors) as long as the sum $\alpha_1 + \alpha_2$ is small compared with unity. As the sum of the alphas approaches unity, the current i increases rapidly. The current does not increase without limit as Eq. (11-3) implies, however, because the derivation is no longer valid as $\alpha_1 + \alpha_2$ approaches unity. Since j_2 becomes forward biased in the forward-conducting state, both transistors become saturated after switching. The two transistors remain in saturation while the device is in the forward-conducting state, being held in saturation by the device current.

11.1.3   Variation of α with Injection

Since the two-transistor analogy implies that switching involves an increase in the alphas to the point that $\alpha_1 + \alpha_2$ approaches unity, it may be helpful to review how alpha varies with injection for a transistor. The emitter-to-collector current transfer ratio α is given in Section 7.2.2 as the product of the emitter injection efficiency γ and the base transport factor B. An increase in α with injection can be caused by increases in either of these factors, or both. At very low currents (such as in the forward-blocking state of the p-n-p-n diodes), γ is usually dominated by recombination in the transition region of the emitter junction (Section 7.7.4). As the current is increased, injection across the junction begins to dominate over recombination within the transition region (Section 5.6.2) and γ increases. There are several mechanisms by which the base transport factor B increases with injection, including the saturation of recombination centers as the excess carrier concentration becomes large. Whichever mechanism dominates, the increase in $\alpha_1 + \alpha_2$ required for switching of the p-n-p-n diode is automatically accomplished. In general, no special design is required to maintain $\alpha_1 + \alpha_2$ smaller than unity during the forward-blocking state; this requirement is usually met at low currents by the dominance of recombination within the transition regions of j_1 and j_3.

11.1.4   Forward-Blocking State

When the device is biased in the forward-blocking state (Fig. 11-3a), the applied voltage v appears primarily across the reverse-biased junction j_2. Although j_1 and j_3 are forward biased, the current is small. The reason for this becomes clear if we consider the supply of electrons available to n_1 and holes to

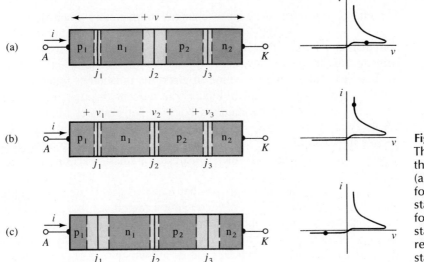

Figure 11-3
Three bias states of the p-n-p-n diode: (a) the forward-blocking state; (b) the forward-conducting state; (c) the reverse-blocking state.

p_2. Focusing attention first upon j_1, let us assume a hole is injected from p_1 into n_1. If the hole recombines with an electron in n_1 (or in the j_1 transition region), that electron must be resupplied to the n_1 region to maintain space charge neutrality. The supply of electrons in this case is severely restricted, however, by the fact that n_1 is terminated in j_2, a reverse-biased junction. In a normal p-n diode the n region is terminated in an ohmic contact, so that the supply of electrons required to match recombination (and injection into p) is unlimited. In this case, however, the electron supply is restricted essentially to those electrons generated thermally within a diffusion length of j_2. As a result, the current passing through the j_1 junction is approximately the same as the reverse saturation current of j_2. A similar argument holds for the current through j_3; holes required for injection into n_2 and to feed recombination in p_2 must originate in the saturation current of the center junction j_2. The applied voltage v divides appropriately among the three junctions to accommodate this small current throughout the device.

In this discussion we have tacitly assumed that the current crossing j_2 is strictly the thermally generated saturation current. This implies that electrons injected by the forward-biased junction j_3 do not diffuse across p_2 in any substantial numbers, to be swept across the reverse-biased junction into n_1 by transistor action. This is another way of saying that α_2 (for the "n-p-n transistor") is small. Similarly, the supply of holes to p_2 is primarily thermally generated, since few holes injected at j_1 reach j_2 without recombination (i.e., α_1 is small for the "p-n-p"). Now we can see physically why Eq. (11-3) implies a small current while $\alpha_1 + \alpha_2$ is small: Without the transport of charge provided by transistor action, the thermal generation of carriers is the only significant source of electrons to n_1 and holes to p_2.

11.1.5   Conducting State

The charge transport mechanism changes dramatically when transistor action begins. As $\alpha_1 + \alpha_2$ approaches unity by one of the mechanisms described above, many holes injected at j_1 survive to be swept across j_2 into p_2. This helps to feed the recombination in p_2 and to support the injection of holes into n_2. Similarly, the transistor action of electrons injected at j_3 and collected at j_2 supplies electrons for n_1. Obviously, the current through the device can be much larger once this mechanism begins. The transfer of injected carriers across j_2 is regenerative, in that a greater supply of electrons to n_1 allows greater injection of holes at j_1 while maintaining space charge neutrality; this greater injection of holes further feeds p_2 by transistor action, and the process continues to repeat itself.

If $\alpha_1 + \alpha_2$ is large enough, so that many electrons are collected in n_1 and many holes are collected in p_2, the depletion region at j_2 begins to decrease. Finally the reverse bias disappears across j_2 and is replaced by a forward bias, in analogy with a transistor biased deep in saturation. When this occurs, the three small forward-bias voltages appear as shown in Fig. 11-3b. Two of these voltages essentially cancel in the overall v, so that the forward voltage drop of the device from anode to cathode in the conducting state is not much greater than that of a single p-n junction. For Si this forward drop is less than 1 V, until ohmic losses become important at high current levels.

We have discussed the current transport mechanisms in the forward-blocking and forward-conducting states, but we have not indicated how switching is initiated from one state to the other. Basically, the requirement is that the carrier injection at j_1 and j_3 must somehow be increased so that significant transport of injected carriers across j_2 occurs. Once this transport begins, the regenerative nature of the process takes over and switching is completed.

11.1.6   Triggering Mechanisms

There are several methods by which a p-n-p-n diode can be switched (or *triggered*) from the forward-blocking state to the forward-conducting state. For example, an increase in the device temperature can cause triggering, by sufficiently increasing the carrier generation rate and the carrier lifetimes. These effects cause a corresponding increase in device current and in the alphas discussed above. Similarly, optical excitation can be used to trigger a device by increasing the current through EHP generation.[†] The most common method of triggering a two-terminal p-n-p-n, however, is simply to raise the bias voltage to the peak value V_P. This type of *voltage triggering* results in a breakdown (or significant leakage) of the reverse-biased junction j_2; the accompanying increase in current provides the injection at j_1 and j_3 and transport required for switching to the conducting state. The breakdown mechanism commonly occurs by a combination of *base-width narrowing* and *avalanche multiplication*.

[†]Four-layer devices that can be triggered by a pulse of light are useful in many optoelectronic systems. This type of device is often called a *light-activated SCR*, or *LASCR*.

When carrier multiplication occurs in j_2, many electrons are swept into n_1 and holes into p_2. This process provides the majority carriers to these regions needed for increased injection by the emitter junctions. Because of transistor action, the full breakdown voltage of j_2 need not be reached. As we showed in Eq. (7-58), breakdown occurs in the collector junction of a transistor with $i_B = 0$ when $M\alpha = 1$. In the coupled transistor case of the p-n-p-n diode, breakdown occurs at j_2 when

$$M_p\alpha_1 + M_n\alpha_2 = 1 \qquad (11-4)$$

where M_p is the hole multiplication factor and M_n is the multiplication factor for electrons.

As the bias v increases in the forward-blocking state, the depletion region about j_2 spreads to accommodate the increased reverse bias on the center junction. This spreading means that the neutral base regions on either side (n_1 and p_2) become thinner. Since α_1 and α_2 increase as these base widths decrease, triggering can occur by the effect of base-width narrowing. A true punch-through of the base regions is seldom required, since moderate narrowing of these regions can increase the alphas enough to cause switching. Furthermore, switching may be the result of a combination of avalanche multiplication and base-width narrowing, along with possible leakage current through j_2 at high voltage. From Eq. (11-4) it is clear that with avalanche multiplication present, the sum $\alpha_1 + \alpha_2$ need not approach unity to initiate breakdown of j_2. Once breakdown begins, the increase of carriers in n_1 and p_2 drives the device to the forward-conducting state by the regenerative process of coupled transistor action. As switching proceeds, the reverse bias is lost across j_2 and the junction breakdown mechanisms are no longer active. Therefore, base narrowing and avalanche multiplication serve only to start the switching process.

If a forward-bias voltage is applied rapidly to the device, switching can occur by a mechanism commonly called *dv/dt triggering*. Basically, this type of triggering occurs as the depletion region of j_2 adjusts to accommodate the increasing voltage. As the depletion width of j_2 increases, electrons are removed from the n_1 side and holes are removed from the p_2 side of the junction. For a slow increase in voltage, the resulting flow of electrons toward j_1 and holes toward j_3 does not constitute a significant current. If dv/dt is large, however, the rate of charge removal from each side of j_2 can cause the current to increase significantly. In terms of the junction capacitance (C_{j2}) of the reverse-biased junction, the transient current is given by

$$i(t) = \frac{dC_{j2}v_{j2}}{dt} = C_{j2}\frac{dv_{j2}}{dt} + v_{j2}\frac{dC_{j2}}{dt} \qquad (11-5)$$

where v_{j2} is the instantaneous voltage across j_2. This type of current flow is often called *displacement current*. The rate of change of C_{j2} must be included in calculating current, since the capacitance varies with time as the depletion width changes.

The increase in current due to a rapid rise in voltage can cause switching well below the steady state triggering voltage V_P. Therefore, a *dv/dt* rating is usually specified along with V_P for p-n-p-n diodes. Obviously, dv/dt triggering can be a disadvantage in circuits subjected to unpredictable voltage transients.

The various triggering mechanisms discussed in this section apply to the two-terminal p-n-p-n diode. As we shall see in the following section, the semiconductor controlled rectifier is triggered by an external signal applied to a third terminal.

11.2
THE
SEMICONDUCTOR
CONTROLLED
RECTIFIER

The semiconductor controlled rectifier (SCR) is useful in many applications, such as in power switching and in various control circuits. This device can handle currents from a few milliamperes to hundreds of amperes. Since it can be turned on externally, the SCR can be used to regulate the amount of power delivered to a load simply by passing current only during selected portions of the line cycle. A common example of this application is the light-dimmer switch used in many homes. At a given setting of this switch, an SCR is turned on and off repetitively, such that all or only part of each power cycle is delivered to the lights. As a result, the light intensity can be varied continuously from full intensity to dark. The same control principle can be applied to motors, heaters, and many other systems. We shall discuss this type of application in this section, after first establishing the fundamentals of device operation.

11.2.1 Gate Control

The most important four-layer device in power circuit applications is the three-terminal SCR[†] (Fig. 11-4). This device is similar to the p-n-p-n diode, except

Figure 11-4
A semiconductor
controlled rectifier:
(a) four-layer
geometry and
common circuit
symbols; (b) I–V
characteristics.

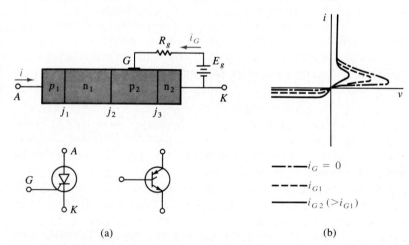

(a)   (b)

---

that a third lead (*gate*) is attached to one of the base regions. When the SCR is biased in the forward-blocking state, a small current supplied to the gate can initiate switching to the conducting state. As a result, the anode switching voltage V_P decreases as the current i_G applied to the gate is increased (Fig. 11-4b). This type of turn-on control makes the SCR a useful and versatile device in switching and control circuits.

To visualize the *gate triggering* mechanism, let us assume the device is in the forward-blocking state, with a small saturation current flowing from anode to cathode. A positive gate current causes holes to flow from the gate into p_2, the base of the n-p-n transistor. This added supply of holes and the accompanying injection of electrons from n_2 into p_2 initiates transistor action in the n-p-n. After a transit time τ_{t2}, the electrons injected by j_3 arrive at the center junction and are swept into n_1, the base of the p-n-p. This causes an increase of hole injection by j_1, and these holes diffuse across the base n_1 in a transit time τ_{t1}. Thus, after a delay time of approximately $\tau_{t1} + \tau_{t2}$, transistor action is established across the entire p-n-p-n and the device is driven into the forward-conducting state. In most SCRs the delay time is less than a few microseconds, and the required gate current for turn-on is only a few milliamperes. Therefore, the SCR can be turned on by a very small amount of power in the gate circuit. On the other hand, the device current i can be many amperes, and the power controlled by the device may be very large.

It is not necessary to maintain the gate current once the SCR switches to the conducting state; in fact, the gate essentially loses control of the device after regenerative transistor action is initiated. For most devices a gate current pulse lasting a few microseconds is sufficient to ensure switching. Ratings of minimum gate pulse height and duration are generally provided for particular SCR devices.

11.2.2 Turning off the SCR

Turning off the SCR, changing it from the conducting state to the blocking state, can be accomplished by reducing the current i below a critical value (called the *holding current*) required to maintain the $\alpha_1 + \alpha_2 = 1$ condition. In some SCR devices, gate turn-off can be used to reduce the alpha sum below unity. For example, if the gate voltage is reversed in Fig. 11-4, holes are extracted from the p_2 base region. If the rate of hole extraction by the gate is sufficient to remove the n-p-n transistor from saturation, the device turns off. However, there are often problems involving the lateral flow of current in p_2 to the gate; nonuniform biasing of j_3 can result from the fact that the bias on this emitter junction varies with position when a lateral current flows. Therefore, SCR devices must be specifically designed for turn-off control; at best, this turn-off capability can be utilized only over a limited range for a given device.

Some four-layer devices have two gate leads, one attached to n_1 and the other to p_2 (Fig. 11-5). This type of device is often called a *semiconductor controlled switch* (*SCS*). The availability of the second gate electrode provides additional flexibility in circuit design. The SCS biased in the forward-blocking

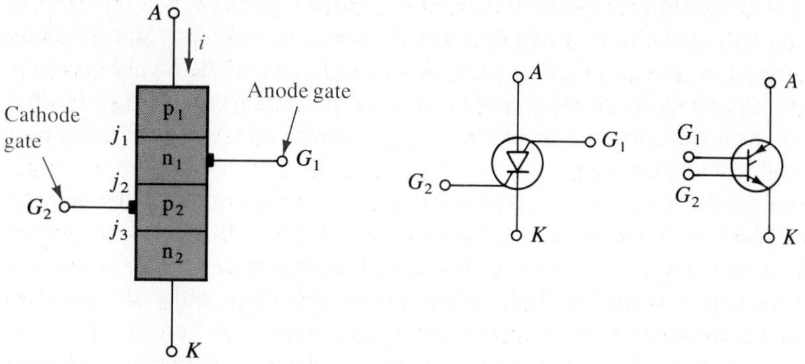

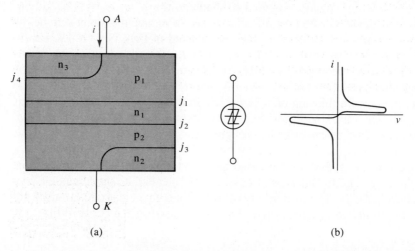

Figure 11-5
Semiconductor
controlled switch:
schematic
configuration and
common circuit
symbols.

state can be switched to the conducting state by a positive current pulse applied to the *cathode gate* (at p_2) or by a negative pulse at the *anode gate* (n_1). If the device is designed for turn-off capability, separate circuits can be employed for turn-on at one gate and turn-off at the other. Other advantages of the SCS configuration include the possibility of minimizing unwanted dv/dt switching; for example, the gate not used for triggering can be capacitively coupled to the nearest current terminal (G_1 to A or G_2 to K) to allow a charging path for j_2 during a voltage transient without causing inadvertent switching.

11.2.3 Bilateral Devices

In many applications it is useful to employ devices which switch symmetrically with forward and reverse bias. This type of *bilateral* device is particularly useful in a-c circuits in which sinusoidal signals are switched on and off during positive and negative portions of the cycle. A typical bilateral p-n-p-n diode configuration is shown in Fig. 11-6a. This device differs from the p-n-p-n diode of Fig. 11-1 in that the n_2 region extends over only half the width of the cathode, and a new region n_3 is diffused into half of the anode region. In effect, this *bilateral diode switch* consists of two separate p-n-p-n diodes: the p_1-n_1-p_2-n_2 section and the p_2-n_1-p_1-n_3 section. We notice that the device shown in

Figure 11-6
A bilateral
diode switch:
(a) schematic
of the device
configuration and a
common circuit
symbol; (b) typical
I–V characteristic.

(a)

(b)

Fig. 11-6a is symmetrical. With the anode A biased positively with respect to the cathode K, junction j_1 is forward biased, while j_2 is reverse biased. Junction j_3 is shorted at one end (as is j_4) by the metal contact. When j_2 is biased to breakdown, however, a lateral current flows in p_2 biasing the left edge of j_3 into injection, and the device switches. This *shorted-emitter* design is commonly used in SCRs to enhance triggering control. During this operation junction j_4 remains dormant. Because the device is symmetrical, j_4 serves the shorted emitter function when the polarity is reversed (K positive with respect to A), and j_1 is the junction which is biased to breakdown in initiating the switching operation. If the bilateral diode is constructed properly, the forward and reverse characteristics are symmetrical as shown in Fig. 11-6b.

Q   *Bilateral triode switches* (sometimes called *triacs*) can be constructed with SCR characteristics that can be triggered in either the forward- or reverse-bias mode. A good discussion of these devices is presented in the book by Gentry et al. in the reading list for this chapter.

11.2.4   Fabrication and Applications

Many variations of diffusion, implantation, alloying, and epitaxial growth are used in the fabrication of p-n-p-n devices. The type of fabrication process depends largely on the power rating and intended use of the device. We can gain some insight into SCR fabrication by considering the example of Fig. 11.7, which employs a combination of alloying and diffusion. In this example, a p-n-p structure is formed by diffusing p regions into both sides of an n-type Si wafer. The cathode is then formed by an n-type alloyed junction (using an alloy such as Au-Sb), and contact is made to the gate region by alloying an Al wire to the top p region. Contact is made to the anode terminal by alloying the anode p region to a metal substrate with an Al preform (Fig. 11-8a). For high-current devices the anode is attached to a heavy copper stud, and the cathode is contacted by a large cable. In high-current operation, heat is carried away from the junction by the massive metal substrate. In the device shown in Fig. 11-8, tungsten disks are soldered to the anode and cathode regions to take advantage of the similarity of the coefficient of thermal expansion between tungsten and Si. The entire device is hermetically sealed in a housing, which provides protection from the atmosphere and from thermal and mechanical shock (Fig. 11-8b). Devices with this type of mounting can be rated at several hundred amperes in the conducting state. Of course, SCR devices intended for small-signal applications can be made in simpler and smaller packages.

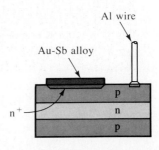

Figure 11-7
An alloy-diffused power SCR.

Al wire

Au-Sb alloy

n^+

p

n

p

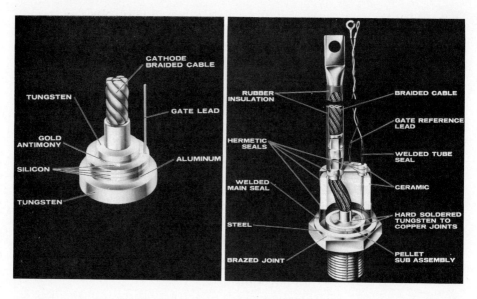

Figure 11-8
A high-current SCR, fabricated by the alloy-diffused method of Fig. 11-7: (a) mounting of the anode and cathode regions to tungsten disks; (b) cutaway view of the encapsulated device. (Illustrations courtesy of General Electric Company.)

Applications of SCRs and other four-layer devices are quite varied and extend into many fields of electronics, switching, and control. As a simple example, let us consider the problem of delivering variable power to a load from a constant line source (Fig. 11-9). The load may be the heater windings of a furnace, a light bulb, or another circuit. The amount of power delivered to the

Figure 11-9
Example of the use of an SCR to control the power delivered to a load: (a) schematic diagram of the circuit; (b) waveforms of the delivered signal and the phase-variable trigger pulse.

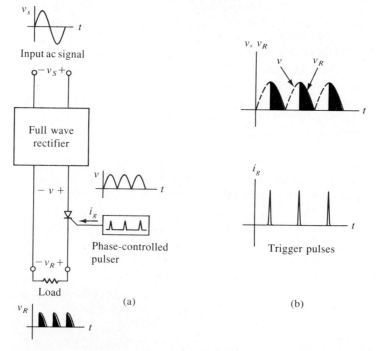

load during each half-cycle depends on the switching of the SCR. If pulses are delivered to the gate near the beginning of each half-cycle, essentially the full power of the input is delivered to the load. On the other hand, if the trigger pulses are delayed, the SCR does not turn on until later in the half-cycle. As a result, the amount of power delivered to the load can be varied from almost full power to no power.

Many examples of SCR and other four-layer device applications can be found in the reading list of this chapter and in the current literature.

PROBLEMS

11.1. Explain why two separate transistors cannot be connected as in Fig. 11-2 to achieve the p-n-p-n switching action of Fig. 11-1.

11.2. In the p-n-p-n diode (Fig. 11-3a), the junction j_3 is forward biased during the forward-blocking state. Why, then, does the forward bias provided by the gate-to-cathode voltage in Fig. 11-4 cause switching?

11.3. (a) Sketch the energy band diagrams for the p-n-p-n diode in equilibrium; in the forward-blocking state; and in the forward-conducting state.
(b) Sketch the excess minority carrier distributions in regions n_1 and p_2 when the p-n-p-n diode is in the forward-conducting state.

11.4. Use schematic techniques such as those illustrated in Fig. 7-5 to describe the hole flow and electron flow in a p-n-p-n diode for the forward-blocking state and for the forward-conducting state. Explain the diagrams and be careful to define any new symbols (e.g., those representing EHP generation and recombination).

11.5. Using the coupled transistor model, rewrite Eqs. (11-1) to include avalanche multiplication in j_2, and show that Eq. (11-4) is valid for the p-n-p-n diode.

READING LIST

Blicher, A. *Thyristor Physics.* New York: Springer-Verlag, 1976.

Gentry, F. E., F. W. Gutzwiller, N. Holonyak, Jr., and E. E. Von Zastrow. *Semiconductor Controlled Rectifiers: Principles and Application of p-n-p-n Devices.* Englewood Cliffs, N.J.: Prentice Hall, 1964.

Ghandi, S. K. *Semiconductor Power Devices,* New York: John Wiley, 1977.

Special Issue on High-Power Semiconductor Devices, *IEEE Transactions on Electron Devices* ED-23, no. 8 (August 1976).

Laster, C. *Thyristor Theory and Application.* Blue Ridge Summit, Pa.: Tab Books, 1986.

Sze, S. M. *Physics of Semiconductor Devices.* New York: John Wiley, 1981.

Taylor, P. D. *Thyristor Design and Realization.* Chichester: John Wiley, 1987.

chapter 12

NEGATIVE CONDUCTANCE MICROWAVE DEVICES

We have discussed a number of devices that are useful in microwave circuits, such as the varactor and tunnel diode. We also studied specially designed high-frequency transistors, which can provide amplification and other functions at the lower microwave frequencies. However, transit time and other effects limit the application of transistors beyond the 10^9-Hz range. Therefore, other devices are required to perform electronic functions such as amplification and d-c to microwave power conversion at higher frequencies.

Several important devices for high-frequency applications use the instabilities that occur in semiconductors. An important type of instability involves *negative conductance*. Here we shall concentrate on two of the most commonly used negative conductance devices: transit time diodes, which depend on a combination of carrier injection and transit time effects, and *Gunn* diodes, which depend on the transfer of electrons from a high-mobility state to a low-mobility state. Each is a two-terminal device that can be operated in a negative conductance mode to provide amplification or oscillation at microwave frequencies in a proper circuit.

We have discussed negative resistance in our study of the tunnel diode (Section 6.2). In the case of the tunnel diode, the *I–V* characteristic exhibits a clear negative resistance region. However, the functions of amplification and oscillation can occur when the local a-c current density is properly out of phase with the local a-c electric field within the device. In this chapter we shall investigate how negative conductance arises and how it can be used in microwave applications.

In this section we describe a type of microwave negative conductance device that operates by a combination of carrier injection and transit time effects. Diodes with simple p-n junction structure, or with variations on that structure, are biased to achieve tunneling or avalanche breakdown, with an a-c voltage superimposed on the d-c bias. The carriers generated by the injection process are swept through a drift region to the terminals of the device. We shall see that the a-c component of the resulting current can be approximately 180° out of phase with the applied voltage under proper conditions of bias and device configuration, giving rise to negative conductance and oscillation in a resonant circuit. Transit time devices can convert d-c to microwave a-c signals with high efficiency and are very useful in the generation of microwave power for many applications.

12.1
TRANSIT TIME
DEVICES

12.1.1 The IMPATT Diode

The original suggestion for a microwave device employing transit time effects was made by W. T. Read and involved an n^+-p-i-p^+ structure such as that shown in Fig. 12-1. This device operates by injecting carriers into the drift region and is called an *impact avalanche transit time* (*IMPATT*) diode. Although IMPATT operation can be obtained in simpler structures, the Read diode is best suited for illustration of the basic principles. The device consists essentially of two regions: (1) the n^+-p region at which avalanche multiplication occurs and (2) the i (essentially intrinsic) region through which generated holes must drift in moving to the p^+ contact. Similar devices can be built in the p^+-n-i-n^+ configuration, in which electrons resulting from avalanche multiplication drift through the i region, taking advantage of the higher mobility of electrons compared to holes.

Although detailed calculations of IMPATT operation are complicated and generally require computer solutions, the basic physical mechanism is simple. Essentially, the device operates in a negative conductance mode when the a-c component of current is negative over a portion of the cycle during which the a-c voltage is positive, and vice versa. The negative conductance occurs because of two processes, causing the current to lag behind the voltage in time:

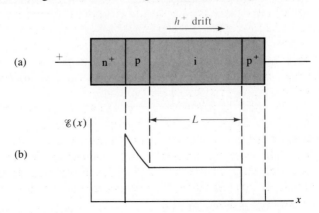

Figure 12-1
The Read diode:
(a) basic device
configuration;
(b) electric field
distribution in the
device under
reverse bias.

(1) a delay due to the avalanche process and (2) a further delay due to the transit time of the carriers across the drift region. If the sum of these delay times is approximately one-half cycle of the operating frequency, negative conductance occurs and the device can be used for oscillation and amplification.

From another point of view, the a-c conductance is negative if the a-c component of carrier flow drifts opposite to the influence of the a-c electric field. For example, with a d-c reverse bias on the device of Fig. 12-1, holes drift from left to right (in the direction of the field) as expected. Now, if we superimpose an a-c voltage such that $\mathcal{E}$ decreases during the negative half-cycle, we would normally expect the drift of holes to decrease also. However, in IMPATT operation the drift of holes through the i region actually increases while the a-c field is decreasing. To see how this happens, let us consider the effects of avalanche and drift for various points in the cycle of applied voltage (Fig. 12-2).

To simplify the discussion, we shall assume that the p region is very narrow and that all the avalanche multiplication takes place in a thin region near the n^{+}-p junction. We shall approximate the field in the narrow p region by a

Figure 12-2
Time dependence of the growth and drift of holes during a cycle of applied voltage for the Read diode:
(a) $\omega t = 0$;
(b) $\omega t = \pi/2$;
(c) $\omega t = \pi$;
(d) $\omega t = 3\pi/2$. The hole pulse is sketched as a dotted line on the field diagram.

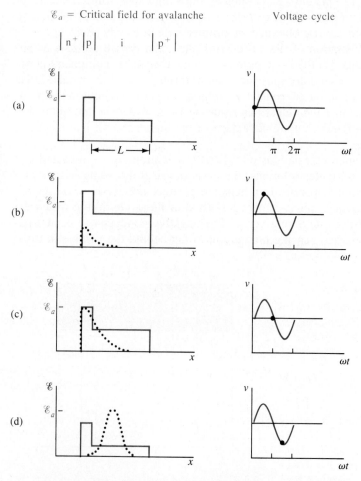

uniform value. If the d-c bias is such that the critical field for avalanche $\mathcal{E}_a$ is just met in the n$^+$-p space charge region (Fig. 12-2a), avalanche multiplication begins at $t = 0$. Electrons generated in the avalanche move to the n$^+$ region, and holes enter the i drift region. We assume the device is mounted in a resonant microwave circuit so that an a-c signal can be maintained at a given frequency. As the applied a-c voltage goes positive, more and more holes are generated in the avalanche region. In fact, the pulse of holes (dotted line) generated by the multiplication process continues to grow as long as the electric field is above $\mathcal{E}_a$ (Fig. 12-2b). It can be shown that the particle current due to avalanche increases exponentially with time while the field is above the critical value. The important result of this growth is that the hole pulse reaches its peak value not at $\pi/2$ when the voltage is maximum, but at π (Fig. 12-2c). Therefore, there is a phase delay of $\pi/2$ inherent in the avalanche process itself. A further delay is provided by the drift region. Once the avalanche multiplication stops ($\omega t > \pi$), the pulse of holes simply drifts toward the p$^+$ contact (Fig. 12-2d). But during this period the a-c terminal voltage is negative. Therefore, the dynamic conductance is negative, and energy is supplied to the a-c field.

If the length of the drift region is chosen properly, the pulse of holes is collected at the p$^+$ contact just as the voltage cycle is completed, and the cycle then repeats itself. The pulse will drift through the length L of the i region during the negative half-cycle if we choose the transit time to be one-half the oscillation period

$$\frac{L}{v_d} = \frac{1}{2}\frac{1}{f}, \qquad f = \frac{v_d}{2L} \qquad (12\text{-}1)$$

where f is the operating frequency and v_d is the drift velocity for holes.[†] Therefore, for a Read diode the optimum frequency is one-half the inverse transit time of holes across the drift region v_d/L. In choosing an appropriate resonant circuit for this device, the parameter L is critical. For example, taking $v_d = 10^7$ cm/s for Si, the optimum operating frequency for a device with an i region length of 5 μm is $f = 10^7/2(5 \times 10^{-4}) = 10^{10}$ Hz. Negative resistance is exhibited by an IMPATT diode for frequencies somewhat above and below this optimum frequency for exact 180° phase delay. A careful analysis of the small-signal impedance shows that the minimum frequency for negative conductance varies as the square root of the d-c bias current for frequencies in the neighborhood of that described by Eq. (12-1).

Although the Read diode of Fig. 12-1 displays most directly the operation of IMPATT devices, simpler structures can be used, and in some cases they may be more efficient. Negative conductance can be obtained in simple p-n junctions or in p-i-n devices. In the case of the p-i-n, most of the applied volt-

---

[†]In general, v_d is a function of the local electric field. However, these devices are normally operated with fields in the i region sufficiently large that holes drift at their scattering limited velocity (Fig. 3-24). For this case the drift velocity does not vary appreciably with the a-c variations in the field.

age occurs across the i region, which serves as a uniform avalanche region and also as a drift region. Therefore, the two processes of delay due to avalanche and drift, which were separate in the case of the Read diode, are distributed within the i region of the p-i-n. This means that both electrons and holes participate in the avalanche and drift processes.

12.1.2 The QWITT Diode

A major drawback of the IMPATT diode for very high frequency operation is the fact that the avalanche process, which depends on random impact ionization events, is inherently noisy. A variety of approaches have been investigated to find alternative methods for injecting carriers into the drift region without relying on the avalanche mechanism. A particularly interesting device (invented and developed at The University of Texas at Austin[†]) employs resonant tunneling through a quantum well to inject electrons into the drift region. This device, called the *quantum well injection transit time (QWITT)* diode, is shown schematically in Fig. 12-3. The device structure consists of a single GaAs quantum well between two $Al_xGa_{1-x}As$ barriers, in series with a drift region of undoped GaAs. This structure is then placed between two n^+-GaAs regions to form contacts. The conduction band diagram of the device at equilibrium is shown in Fig. 12-3b. For an isolated quantum well, when the voltage across the well equals $2E_1/q$, a large peak in the current through the device occurs (Fig. 12-4a). In the QWITT diode, we can achieve maximum resonant tunneling of electrons through the well if the dc bias is adjusted so that the following expression is satisfied [Fig. 12-4b]:

$$2E_1/q = V_{RF} \sin \omega t + V_i \qquad (12-2)$$

Figure 12-3
(a) Physical structure of the QWITT diode; (b) energy band diagram of the device when no bias is applied. The length of the transit-time region *W* is much greater than the quantum-well thickness.

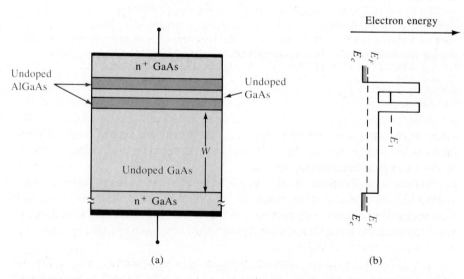

(a)

(b)

[†]V. P. Kesan, et al., "A New Transit-Time Device Using Quantum-Well Injection," *IEEE Electron Device Letters*, EDL-8, no. 4, p. 129 (April 1987).

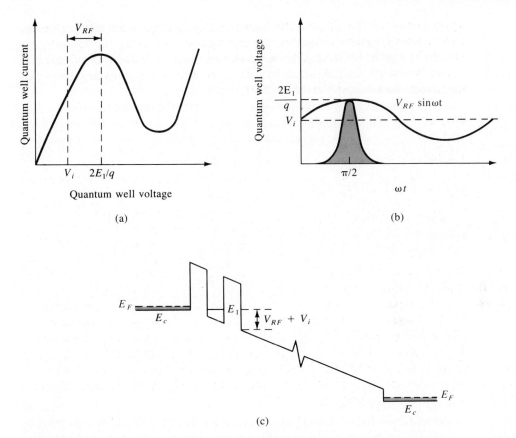

Operation of the QWITT with RF voltage applied: (a) the bias point for the QWITT diode is adjusted so that the bias voltage across the quantum-well injection region V_i is less than the resonant bias by an amount equal to the RF voltage across the well V_{RF}; (b) under this bias condition current injection through the well will peak at $\omega t = \pi/2$ when the total instantaneous voltage across the well is equal to $2E_1/q$; (c) schematic energy band diagram when the voltage across the quantum well is in resonance ($\omega t = \pi/2$).

Figure 12-4

where V_{RF} is the amplitude of the ac voltage across the quantum well, V_i is the effective dc voltage across the well, and E_1 is the energy level in the well.

If an appropriate bias is applied so that the Fermi level in the n^+-GaAs is below the energy level E_1 by an amount V_{RF} and an ac signal $V_{RF} \sin \omega t$ is superimposed on this bias, then resonant tunneling of electrons through the well will peak at an injection angle of $\pi/2$ in the ac cycle [Fig. 12-4]. The electrons will then traverse through the drift region at their saturation velocity, resulting in a dynamic negative resistance.

The QWITT diode can also be biased to move the Fermi level in the n^+-GaAs above the energy state in the GaAs well so that current injection peaks around $3\pi/2$ in the negative half of the ac voltage cycle. Compared to injection near $\pi/2$, this mode should have higher efficiency, since electrons injected at $3\pi/2$ drift while the RF voltage is negative. No other transit-time device exhibits an injection angle of greater than π.

It is possible to operate the resonant tunneling quantum well diode as an oscillator without the benefit of the drift region by biasing in the negative resistance region of Fig. 12-4a. Without the transit time effect, however, such a *resonant tunneling diode (RTD)* is less efficient. Compared to other transit-time devices and quantum-well oscillators, the QWITT diode has several advantages. Quantum mechanical tunneling in the QWITT diode is a low-noise injection mechanism with superior high-frequency characteristics. Depending on the bias level, the device may permit injection of carriers into the drift region at more favorable phase angles than other transit-time devices, so higher efficiencies can be expected. The characteristics of the injection region, and therefore the injected current pulse, can be adjusted by changing the physical dimensions of the quantum-well structure. Thus both the length of the transit-time region and the shape of the current pulse can be optimized to obtain the best power-frequency performance from the device. The QWITT diode should extend the normal frequency limit associated with transit-time devices, while providing higher output power than simple quantum-well RTD oscillators.

12.1.3   The TRAPATT Diode

An important and highly efficient mode of operation for a transit time device is illustrated in Fig. 12-5. If a large current is suddenly applied to the sample, the displacement current $\epsilon \, d\mathscr{E}/dt$ can be so large that the point $\mathscr{E} > \mathscr{E}_a$ propagates through the device faster than the maximum carrier velocity. Therefore, an *avalanche zone* moves through the i region, filling it with an electrostatically neutral electron–hole population, or *plasma*. The plasma is created very rap-

Figure 12-5
TRAPATT cycle:
(a) electron–hole
plasma created by a
rapidly propagating
field; (b) depletion
of the plasma.

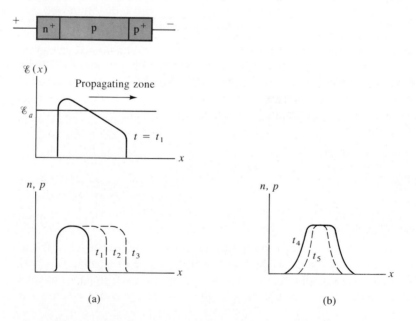

(a)   (b)

idly, before the carriers have a chance to drift appreciably. The presence of these carriers in large numbers greatly reduces the terminal voltage across the diode. The electron–hole plasma collapses as holes drift to the right and electrons drift to the left. As the EHP plasma is depleted during the recovery transient, the terminal voltage builds up to a high value and the current goes toward a low value. In a proper resonant circuit, the cyclic buildup and discharge of the plasma gives rise to efficient microwave power generation. Since the electron–hole plasma is initiated during the transit of an avalanche zone, this type of operation is generally called a TRAPATT cycle, for *TRA*pped *Plasma Avalanche Triggered Transit*.

Microwave devices that operate by the *transferred electron* mechanism are often called *Gunn diodes* after J. B. Gunn, who first demonstrated one of the forms of oscillation. As we shall see, there are many modes of operation for these devices. In the transferred electron mechanism, the conduction electrons of some semiconductors are shifted from a state of high mobility to a state of low mobility by the influence of a strong electric field. Negative conductance operation can be achieved in a diode[†] for which this mechanism applies, and the results are varied and useful in microwave circuits.

First, we shall describe the process of electron transfer and the resulting change of mobility. Then we shall consider some of the modes of operation for diodes using this mechanism.

12.2.1   The Transferred Electron Mechanism

In Section 3.4.4 we discussed the nonlinearity of mobility at high electric fields. In most semiconductors the carriers reach a scattering limited velocity, and the velocity vs. field plot saturates at high fields (Fig. 3-24). In some materials, however, the energy of electrons can be raised by an applied field to the point that they transfer from one region of the conduction band to another, higher-energy region. For some band structures, negative conductivity can result from this electron transfer. To visualize this process, let us recall the discussion of energy bands in Section 3.1. The band diagrams we usually draw vs. distance in the sample are good approximations when the conduction electrons exist near the minimum energy of the conduction band. However, in the more complete band diagram, electron energy is plotted vs. the propagation vector **k**, as in Fig. 3-5. It was shown in Example 3-1 that the **k** vector is proportional to the electron momentum in the vector direction; therefore, energy bands such as those in Fig. 3-5 are said to be plotted in *momentum space*.

[†]These devices are called diodes, since they are two-terminal devices. No p-n junction is involved, however. Gunn effect and related devices utilize bulk instabilities, which do not require junctions.

A simplified band diagram for GaAs is shown in Fig. 12-6 for reference; some of the detail has been omitted in this diagram to isolate the essential features of electron transfer between bands. In n-type GaAs the valence band is filled, and the *central valley* (or *minimum*) of the conduction band at Γ ($\mathbf{k} = 0$) normally contains the conduction electrons. There is a set of *subsidiary minima* at L (sometimes called *satellite valleys*) at higher energy,[†] but these minima are many kT above the central valley and are normally unoccupied. Therefore, the direct band gap at Γ and the energy bands centered at $\mathbf{k} = 0$ are generally used to describe the conduction processes in GaAs. This was true of our discussion of GaAs lasers in Section 10.4, for example. The presence of the satellite valleys at L is crucial to the Gunn effect, however. If the material is subjected to an electric field above some critical value (about 3000 V/cm), the electrons in the central Γ valley of Fig. 12-6 gain more energy than the 0.30 eV separating the valleys; therefore, there is considerable scattering of electrons into the higher-energy satellite valley at L.

Once the electrons have gained enough energy from the field to be transferred into the higher-energy valley, they remain there as long as the field is greater than the critical value. The explanation for this involves the fact that the combined effective density of states for the upper valleys is much greater than for the central valley (by a factor of about 24). Although we shall not prove it here, it seems reasonable that the probability of electron scattering be-

Figure 12-6
Simplified band diagram for GaAs, illustrating the lower (Γ) and upper (L) valleys in the conduction band.

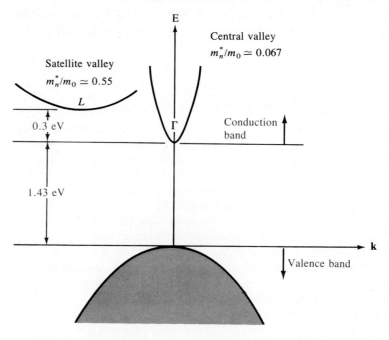

[†]We have shown only one satellite valley for convenience; there are other equivalent valleys for different directions in $\mathbf{k}$-space. The effective mass ratio of 0.55 refers to the combined satellite valleys.

tween valleys should depend on the density of states available in each case, and that scattering from a valley with many states into a valley with few states would be unlikely. As a result, once the field increases above the critical value, most conduction electrons in GaAs reside in the satellite valleys and exhibit properties typical of that region of the conduction band. In particular, the effective mass for electrons in the higher L valleys is almost eight times as great as in the central valley, and the electron mobility is much lower. This is an important result for the negative conductivity mechanism: As the electric field is increased, the electron velocity increases until a critical field is reached; then the electrons *slow down* with further increase in field. The electron transfer process allows electrons to gain energy at the expense of velocity over a range of values of the electric field. Taking current density as $q\mathrm{v}_d n$, it is clear that current also drops in this range of increasing field, giving rise to a negative differential conductivity $dJ/d\mathscr{E}$.

A possible dependence of electron velocity vs. electric field for a material capable of electron transfer is shown in Fig. 12-7. For low values of field, the electrons reside in the lower (Γ) valley of the conduction band, and the mobility ($\mu_\Gamma = \mathrm{v}_d/\mathscr{E}$) is high and constant with field. For high values of field, electrons transfer to the satellite valleys, where their velocity is smaller and their mobility lower. Between these two states is a region of negative slope on the v_d vs. $\mathscr{E}$ plot, indicating a negative differential mobility $d\mathrm{v}_d/d\mathscr{E} = -\mu^*$.

The actual dependence of electron drift velocity on electric field for GaAs and InP is shown in Fig. 12-8. The negative resistance due to electron transfer occurs at a higher field for InP, and the electrons achieve a higher peak velocity before transfer from Γ to L occurs.

The existence of a drop in mobility with increasing electric field and the resultant possibility of negative conductance were predicted by Ridley and Watkins and by Hilsum several years before Gunn demonstrated the effect in GaAs. The mechanism of electron transfer is therefore often called the Ridley–Watkins–Hilsum mechanism. This negative conductivity effect depends only on the bulk properties of the semiconductor and not on junction or surface effects. It is therefore called a *bulk negative differential conductivity* (*BNDC*) effect.

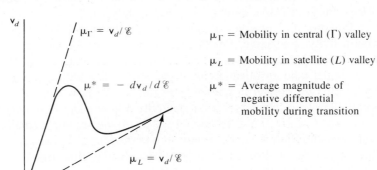

μ_Γ = Mobility in central (Γ) valley

μ_L = Mobility in satellite (L) valley

μ^* = Average magnitude of negative differential mobility during transition

Figure 12-7
A possible characteristic of electron drift velocity vs. field for a semiconductor exhibiting the transferred electron mechanism.

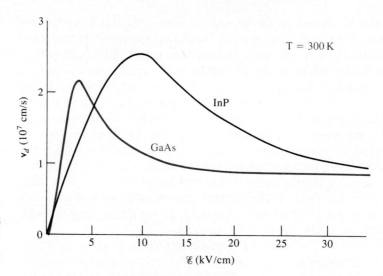

Figure 12-8
Electron drift velocity vs. field for GaAs and InP.

12.2.2 Formation and Drift of Space Charge Domains

If a sample of GaAs is biased such that the field falls in the negative conductivity region, space charge instabilities result, and the device cannot be maintained in a d-c stable condition. To understand the formation of these instabilities, let us consider first the dissipation of space charge in the usual semiconductor. It can be shown from treatment of the continuity equation that a localized space charge dies out exponentially with time in a homogeneous sample with positive resistance (Prob. 12.2). If the initial space charge is Q_0, the instantaneous charge is

$$Q(t) = Q_0 e^{-t/\tau_d} \tag{12-3}$$

where $\tau_d = \epsilon/\sigma$ is called the *dielectric relaxation time*. Because of this process, random fluctuations in carrier concentration are quickly neutralized, and space charge neutrality is a good approximation for most semiconductors in the usual range of conductivities. For example, the dielectric relaxation time for a 1.0 Ω-cm Si or GaAs sample is approximately 10^{-12} s.

Equation (12-3) gives a rather remarkable result for cases in which the conductivity is negative. For these cases τ_d is negative also, and *space charge fluctuations build up* exponentially in time rather than dying out. This means that normal random fluctuations in the carrier distribution can grow into large space charge regions in the sample. Let us see how this occurs in a GaAs sample biased in the negative conductivity regime. The velocity–field diagram for n-type GaAs is illustrated in Fig. 12-9a. If we assume a small shift of electron concentration in some region of the device, a dipole layer can form as shown in Fig. 12-9b. Under normal conditions this dipole would die out quickly. However, under conditions of negative conductivity the charge within the dipole, and therefore the local electric field, builds up as shown in Fig. 12-9c. Of course, this buildup takes place in a stream of electrons drifting from

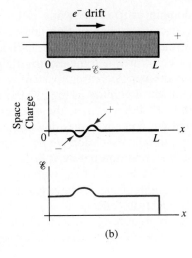

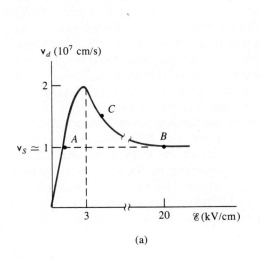

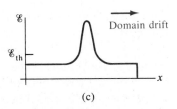

(a)

(b)

(c)

Figure 12-9
Buildup and drift of a space charge domain in GaAs: (a) velocity-field characteristic for n-type GaAs; (b) formation of a dipole; (c) growth and drift of a dipole for conditions of negative conductivity.

the cathode to the anode, and the dipole (now called a *domain*) drifts along with the stream as it grows. Eventually the drifting domain will reach the anode, where it gives up its energy as a pulse of current in the external circuit.

During the initial growth of the domain, an increasing fraction of the applied voltage appears across it, at the expense of electric field in the rest of the bar. As a result, it is unlikely that more than one domain will be present in the bar at a time; after the formation of one domain, the electric field in the rest of the bar quickly drops below the threshold value for negative conductivity. If the bias is d-c, the field outside the moving domain will stabilize at a positive conductivity point such as A in Fig. 12-9a, and the field in the domain will stabilize at the high-field value B.

Let us follow the motion of a single domain as illustrated by Fig. 12-9. A small dipole forms from a random noise fluctuation (or more likely at a permanent *nucleation site* such as a crystal defect, a doping inhomogeneity, or the cathode itself), and this dipole grows and drifts down the bar as a domain. During the early stages of domain development, we can assume a uniform electric field in the bar, except just at the small dipole layer. If the field is in the negative mobility region, such as point C of Fig. 12-9a, the slightly higher field within the dipole results in a lower value of electron drift velocity inside the dipole than outside. As a result, electrons on the right (downstream) of the domain drift away, while electrons pile up on the left (upstream) side. This causes the accumulation and depletion layers of the dipole to grow, thereby further increasing the electric field in the domain. This is obviously a runaway process, in which the electric field within the domain grows while that outside the domain decreases. A stable condition is realized when the domain field increases to point B in Fig. 12-9a, and the field outside drops to point A. When this condition is met, the electrons drift at a constant velocity v_s everywhere, and the domain moves down the bar without further growth.

In this discussion we have assumed that the domain has time to grow to its stable condition before it drifts out of the bar. This is not always the case; for example, in a short bar with a low concentration of electrons, a dipole can drift the length of the bar before it develops into a domain. We can specify limits on the electron concentration n_0 and sample length L for successful domain formation by requiring the transit time (L/v_s) to be greater than the dielectric relaxation time (absolute value) in the negative mobility region. This requirement gives

$$\frac{L}{v_s} > \frac{\epsilon}{q\mu^* n_0}$$

$$L n_0 > \frac{\epsilon v_s}{q\mu^*} \simeq 10^{12} \text{ cm}^{-2}$$

(12-4)

for n-type GaAs, where the average negative differential mobility[†] is taken to be -100 cm²/V-s. Therefore, for successful domain formation there is a critical product of electron concentration and sample length.

The type of domain motion we have described here was the first mode of operation observed by Gunn. In the observation of current vs. voltage for a GaAs sample, Gunn found a linear ohmic relation up to a critical bias, beyond which the current came in sharp pulses. The pulses were separated in time by an amount proportional to the sample length. This length dependence was due to the transit time L/v_s required for a domain nucleated at the cathode to drift the length of the bar. Gunn performed an interesting experiment in which he used a tiny capacitive probe to measure the electric field at various positions down the bar. By scanning the field distribution in the bar at various times in the cycle, he was able to plot out the growth and drift of the domains.

The formation of stable domains is not the only mode of operation for transferred electron devices. Nor is it the most desirable mode for most applications, since the resulting short pulses of current are inefficient sources of microwave power. We have discussed this mode first because it is simple to visualize; now we shall turn our attention to several more useful modes of operation.

[†]This is a rather crude approximation, since μ^* is not a constant but varies considerably with field; the negative dielectric relaxation time therefore changes with time as the domain grows.

12.2.3   Modes of Operation in Resonant Circuits

The simple drift of stable domains discussed above is attainable in resistive-loaded circuits with d-c bias, but this mode is not really typical of microwave applications. Negative conductivity devices are usually operated in resonant circuits, such as high-Q resonant microwave cavities. Since the applied voltage varies in time, we can expect the device behavior to depend on both the amplitude and the frequency of the voltage variations.

Although we cannot consider all of the possible modes of operation here, we can discuss several of the most important. Figure 12-10 summarizes several cases for which the bias voltage varies in amplitude and frequency. The plot of

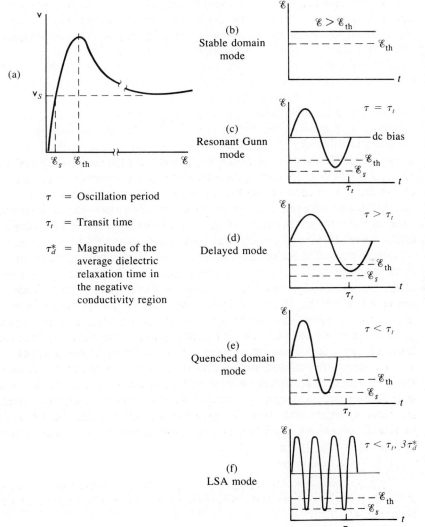

τ = Oscillation period

τ_t = Transit time

τ_d^* = Magnitude of the average dielectric relaxation time in the negative conductivity region

Figure 12-10
Modes of operation:
(a) velocity-field characteristic for n-type GaAs, with threshold and sustaining fields for domains indicated;
(b) d-c biasing for stable domain propagation;
(c) bias at the transit time frequency for resonant Gunn mode; (d) delayed or inhibited mode;
(e) quenched domain mode;
(f) LSA mode.

electron velocity vs. field for GaAs is repeated in Fig. 12-10a as a reference. When the field rises above the threshold value $\mathscr{E}_{th}$, negative conductance exists until a stable domain is formed with $v_d = v_s$ in both the high- and low-field regions. Domains drift along the sample until they reach the anode or until the low-field value drops below the *sustaining field* $\mathscr{E}_s$ required to maintain v_s (Prob. 12.4). We note that creation of a domain requires a larger field ($\mathscr{E}_{th}$) than that needed outside the domain to maintain it ($\mathscr{E}_s$). As the bias is reduced, the domain width decreases since less voltage can be maintained across it. Finally, the domain collapses (is *quenched*) when the field outside drops below $\mathscr{E}_s$, and a new domain cannot nucleate until the field is raised above $\mathscr{E}_{th}$ again. This distinction between the threshold and the sustaining fields is important in several resonant modes.

For a steady bias such that the average field is above the threshold value $\mathscr{E}_{th}$, stable (Gunn) domains form and drift toward the anode (Fig. 12-10b). This is the mode discussed in the previous section; the current waveform is composed of a series of spikes which appear with a frequency equal to the inverse transit time v_s/L. Obviously, this mode is not appropriate for efficient conversion of d-c to microwave power; a resonant circuit mode is more desirable.

An a-c *transit time* (Gunn) mode of operation can be achieved in a resonant circuit if the oscillation period is chosen to be essentially equal to the transit time. Assuming the field never drops below the sustaining field $\mathscr{E}_s$, a domain can form and propagate to the anode during each cycle (Fig. 12-10c). Again, the efficiency is low since the current is collected only when the domain reaches the anode. The best efficiency for this mode is attained when the $n_0 L$ product is chosen such that the domain width equals about half the sample length; for this case the current is collected during most of the negative half-cycle of the voltage. In this mode the d-c to microwave conversion efficiency can be about 10 percent under ideal conditions.

Another method for increasing the efficiency is to collect the domain early in the negative half-cycle of the voltage and delay the formation of a new domain. For example, if the transit time is chosen such that the domain is collected while $\mathscr{E} < \mathscr{E}_{th}$ in Fig. 12-10d, a new domain cannot form until the field rises above threshold again. During the portion of the negative half-cycle that no domain exists, there is an ohmic component of current higher than that flowing during the domain motion (Prob. 12.5). The efficiency of this *delayed domain mode* or *inhibited mode* is better than that of the simple transit time mode. If the bias drops below threshold early during the negative half-cycle and the domain is collected soon thereafter, the efficiency of the device can theoretically be about 20 percent. A further advantage of this mode is that the oscillation frequency can be determined within certain limits by the resonant circuit rather than by the transit time.

If the bias field drops below $\mathscr{E}_s$ during the negative half-cycle (Fig. 12-10e), the domain collapses before it reaches the anode. In this *quenched domain mode* the operating frequency must be higher than the transit time frequency so that the domain can be quenched before it is collected. In this mode also the frequency can be designated by the choice of resonant cavity, within the required limits.

One of the most efficient and useful modes of operation does not involve the formation of domains. In the *limited space charge accumulation (LSA) mode,* the frequency is so high that domains have insufficient time to form while the field is above threshold (Fig. 12-10f). As a result, most of the sample is maintained in the negative conductance state during a large fraction of the voltage cycle. We recall this is not possible in the other modes, since most of the sample drops back to the positive conductance state once a domain forms. The LSA mode is actually the simplest mode of operation; since domains do not form, electrons drift from cathode to anode through a sample which displays negative conductance and an essentially uniform field during much of the cycle. In this mode the frequency, determined by the resonant circuit, is much higher than the transit time frequency. This is a distinct advantage in achieving efficient conversion of power in the upper microwave range. The primary requirement for this mode is that the frequency be high enough that domains have insufficient time to form while the signal is above threshold. However, it is also necessary that any accumulation of electrons near the cathode have time to collapse while the signal is below threshold. Therefore, the oscillation period τ should be no more than several times larger than the magnitude of the dielectric relaxation time in the negative conductance regime, τ_d^*, but must be much larger than the positive resistance low-field value τ_d. Although this condition seems restrictive, LSA diodes can be operated over a very wide range of frequencies by proper choice of doping concentration. The efficiency of the LSA mode can be as high as 20 per cent, and LSA diodes can be operated at very high frequencies.

There are other modes of operation, including an *amplification mode* in which negative conductance is utilized without domain formation. In subcritically doped samples ($n_0 L \ll 10^{12}$), there are too few carriers for domain formation within the transit time. Therefore, amplification of signals near the transit time frequency can be accomplished.

To summarize the various modes of operation, we can construct an approximate mode diagram (Prob. 12.6) as shown in Fig. 12-11. In this diagram we plot the product of frequency and sample length fL vs. the product of doping and length $n_0 L$ for n-type GaAs. This type of figure helps in visualizing the rather complex interrelationships of resonant frequency, dielectric relaxation time, and transit time which give rise to the various modes. The first step in constructing such a diagram is to draw a vertical line at $n_0 L = 10^{12}$, where the transit time equals the negative conductivity dielectric relaxation time τ_d^* [Eq. (12-4)]. Domains are possible to the right of this line but not to the left. This is the first of two lines which separate regions of domain formation from regions with no domains. Next we can identify the fL product corresponding to the sustaining drift velocity ($v_s = 10^7$ cm/s). Along this horizontal line the frequency equals the inverse transit time ($f = v_s/L$). In the neighborhood of this line we can expect Gunn domains and the amplification mode.

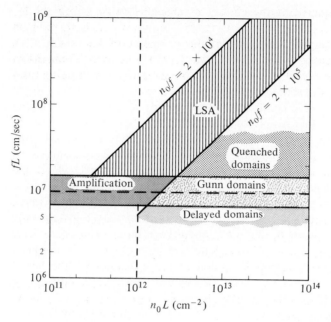

Figure 12-11
Mode diagram for
n-type GaAs.[†]

To identify the approximate LSA region, we recall that the oscillation period τ must be much greater than τ_d, but no greater than several τ_d^*. Using $\mu_n = 5000$ and $\mu^* = 100$ cm^2/V-s, we find[‡] (Prob. 12.6) that $\tau = \tau_d$ when $n_0/f = 1.4 \times 10^3$ and $\tau = \tau_d^*$ when $n_0/f = 7 \times 10^4$. In defining the LSA boundaries, we can choose the low-frequency limit such that the period τ is about $3\tau_d^*$. This gives the boundary $n_0/f = 2 \times 10^5$, indicating the limit for domain formation; this is the second line dividing the domain and no-domain regions. In the upper frequency limit for LSA of Fig. 12-11, we have taken $n_0/f = 2 \times 10^4$ as a somewhat arbitrary value to satisfy the requirement $\tau \gg \tau_d$. These two boundaries are the values quoted by Copeland as satisfying experimental evidence as well as falling within the general range predicted theoretically.

When $\tau \gg \tau_d^*$ (to the right of the $n_0/f = 2 \times 10^5$ line), domains have time to form while the field is above threshold. For frequencies near the inverse transit time we expect Gunn domains propagating to the anode during each cycle; for lower frequencies the domain can be collected during the negative half-cycle while $\mathcal{E} < \mathcal{E}_{\text{th}}$ (delayed domain); for higher frequencies domains can be quenched before they reach the anode.

[†]General features after J. A. Copeland, "LSA Oscillator Diode Theory," *Journal of Applied Physics,* 38, pp. 3096–3101 (July 1967).

[‡]As in Eq. (12-4), we have approximated the dielectric relaxation times and mobilities as discrete positive and negative resistance values. Actually, the electric field sweeps through the v_d-$\mathcal{E}$ diagram of Fig. 12-9a periodically so that these quantities are complex functions of time. A careful analysis would use a proper average of these quantities over the oscillation period.

We have omitted quite a few subtleties of mode behavior in Fig. 12-11, and the boundaries are more complicated than this; however, the basic division of the diagram should help to bring together the discussion of this section. It is clear from this diagram that the operation of devices related to the Gunn effect is complex. On the other hand, this complexity gives rise to a rich field of device applications.

12.2.4   Fabrication

Devices utilizing the Gunn effect and its variations can be made in a number of materials which have appropriate band structures. Although GaAs and InP are the most common materials, transferred electron effects have been observed in CdTe, ZnSe, GaAsP, and other materials. The band structure of some materials can be altered to exhibit properties appropriate for electron transfer. For example, the energy bands of InAs can be distorted by the application of pressure to the crystal, such that a set of satellite valleys becomes available for electron transfer, although these upper valleys are too far above the lower valley at normal pressures. We have discussed the device behavior in terms of GaAs, since this material can be prepared with good purity and is most widely used in microwave applications.

Gunn diodes and related devices are simple structures in principle, since they are basically homogeneous samples with ohmic contacts on each end. In practice, however, considerable care must be taken in fabricating and mounting workable devices. In addition to the obvious requirements on doping, carrier mobility, and sample length implied by Fig. 12-11, there are important problems with contacts, heat sinking, and parasitic reactances of the packaged device.

The samples must have high mobility, few lattice defects, and homogeneous doping in the range giving carrier concentrations $n_0 \simeq 10^{13} - 10^{16}$ cm^{-3}. Devices can be made from GaAs or InP bulk samples cut from an ingot, but it is more common to use ingot material as a substrate for an epitaxial layer, which serves as the active region of the device. The material properties of epitaxial layers are often superior to bulk samples, and the precise control of layer thickness is helpful in these devices, which require exact sample lengths. In a typical configuration, an n-type epitaxial layer about 10 μm thick is grown on an n$^+$ substrate wafer, which is perhaps 100 μm thick. The substrate serves as one of the contacts to the active region. A thin n^{++} (very heavily doped) layer is grown on top of the n region, so that an n$^+$-n-n^{++} sandwiched structure results. External contacts can be made by evaporating a thin layer of Au–Sn or Au–Ge on each surface, followed by a brief alloying step in a hydrogen atmosphere. The wafer is divided into individual devices by cutting or cleaving (giving a cube structure) or by selective etching (giving a mesa structure). Each device is mounted with the n^{++} side down on a copper stud or other heat sink, so that the active region can dissipate heat to the mount in one direction and to the substrate layer in the other direction. Then the substrate side can be contacted by a wire or pressure contact. In other configurations, planar fabrication techniques can be used to produce lateral devices in an n-type epitaxial layer grown on a high-resistivity substrate.

Removal of heat is a very serious problem in these devices. The power dissipation may be 10^7 W/cm^3 or greater (Prob. 12.7), giving rise to considerable heating of the sample. As the temperature increases, the device characteristics vary because of changes in n_0 and mobility. As a result of such heating effects, these devices seldom reach their theoretical maximum efficiency. Pulsed operation allows better control of heat dissipation than does continuous operation, and efficiencies near the theoretical limits can sometimes be achieved in the pulsed mode. The LSA mode is particularly suitable for microwave power generation because of its relatively high efficiency and high operating frequencies. If the application does not require continuous operation, peak powers of hundreds of watts can be achieved in pulses of microwave oscillation.

Many variations can be found on the basic principles presented here. For example, a bar supporting domains can be shaped so that its cross section varies in a predetermined way. The output waveform for such a device reflects the shape of the sample. This interesting *functional oscillator* effect can be used to produce a wide variety of waveforms. Additional flexibility of application can be found by adding biasing contacts along the length of the sample to further control space charge formation and propagation.

PROBLEMS

12.1.   (a) Calculate the ratio N_L/N_Γ of the effective density of states in the upper (L) valleys to the effective density of states in the lower (Γ) valley of the GaAs conduction band (Fig. 12-6).

   (b) Assuming a Boltzmann distribution $n_L/n_\Gamma = (N_L/N_\Gamma)\exp(-\Delta E/kT)$, calculate the ratio of the concentration of conduction band electrons in the upper valley to the concentration in the central valley in equilibrium at 300 K.

   (c) As a rough calculation, assume an electron at the bottom of the central valley has kinetic energy kT. After it is promoted to the satellite (L) valley, what is its approximate equivalent temperature?

12.2.   (a) Use Poisson's equation, the continuity equation, and the definition of current density in terms of the gradient of electrostatic potential to relate the time variation of space charge density ρ to the conductivity σ and the permittivity ϵ of a material, neglecting recombination.

   (b) Assuming a space charge density ρ_0 at $t = 0$, show that $\rho(t)$ decays exponentially with a time constant equal to the dielectric relaxation time τ_d.

   (c) Given a sample of thickness L and area A, calculate the inherent RC time constant if the conductivity is σ and the permittivity is ϵ.

12.3. Assuming that n_Γ electrons/cm^3 are in the lower (central) valley of the GaAs conduction band at time t and n_L are in the satellite (L) valleys, show that the criterion for negative differential conductivity ($dJ/d\mathscr{E} < 0$) is

$$\frac{\mathscr{E}(\mu_\Gamma - \mu_L)\dfrac{dn_\Gamma}{d\mathscr{E}} + \mathscr{E}\left(n_\Gamma \dfrac{d\mu_\Gamma}{d\mathscr{E}} + n_L \dfrac{d\mu_L}{d\mathscr{E}}\right)}{n_\Gamma\mu_\Gamma + n_L\mu_L} < -1$$

where μ_Γ and μ_L are the electron mobilities in the Γ and L valleys, respectively. *Note:* $n_0 = n_\Gamma + n_L$. Discuss the conditions for negative differential conductivity, assuming the mobilities are approximately proportional to $\mathscr{E}^{-1}$.

12.4. Explain why space charge in a domain is quenched when the electric field outside the domain drops below the sustaining field $\mathscr{E}_s$ (Fig. 12-10a).

12.5. Explain why the current increases during the period after a domain is collected (Fig. 12-10d) or quenched (Fig. 12-10e) and before a new domain is formed.

12.6. Draw a mode diagram for n-type GaAs (Fig. 12-11); show the calculations and explain each step in forming the diagram.

12.7. We wish to estimate the d-c power dissipated in a GaAs Gunn diode. Assume the diode is 5 μm long and operates in the stable domain mode.
   (a) What is the minimum electron concentration n_0? What is the time between current pulses?
   (b) Using data from Fig. 12-9a, calculate the power dissipated in the sample per unit volume when it is biased just below threshold, if n_0 is chosen from the calculation of part (a). In general, does operation at a higher frequency result in greater power dissipation?

READING LIST

Bate, R. T. "The Quantum Effect Device: Tomorrow's Transistor?" *Scientific American* 258, no. 3 (March 1988): 96–100.

Blakey, P. A., J. R. East, and **G. I. Haddad.** "Impact of Submicron Technology on Microwave and Millimeter-wave Devices." *VLSI Electronics: Microstructure Science,* vol. 2. New York: Academic Press, 1981. Ch. 4, pp. 105–147.

Bosch, B. G. and **R. W. N. Engelmann.** *Gunn Effect Electronics.* New York: John Wiley-Halstead Press, 1975.

Bulman, P. J., G. S. Hobson, and **B. C. Taylor.** *Transferred Electron Devices.* New York: Academic Press, 1972.

Capasso, F. and G. Margaritondo, eds. *Heterojunction Band Discontinuities: Device Physics and Applications.* New York: North Holland, 1987.

Ferry, D. K. ed. *Gallium Arsenide Technology.* Indianapolis: Howard W. Sams & Co., 1985.

Gibbons, G. *Avalanche-Diode Microwave Oscillators.* Oxford: Clarendon Press, 1973.

Hess, K. *Advanced Theory of Semiconductor Devices.* Englewood Cliffs, N.J.: Prentice Hall, 1988.

Hobson, G. S. *The Gunn Effect.* Oxford: Clarendon Press, 1974.

Liao, S. Y. *Microwave Devices and Circuits.* Englewood Cliffs, N.J.: Prentice Hall, 1980.

Special Issue on Microwave Semiconductor Devices, *IEEE Transactions on Electron Devices* ED-25, no. 6 (June 1978).

Special Issues of the *IEEE Transactions on Electron Devices* ED-13, no. 1 (January 1966) and ED-14, no. 9 (September 1967).

Shur, M. *GaAs Devices and Circuits.* New York: Plenum Press, 1987.

Sze, S. M. *Physics of Semiconductor Devices.* New York: John Wiley, 1981.

DEFINITIONS OF COMMONLY USED SYMBOLS[†]

a	Chapter 1: unit cell dimension (Å); Chapter 8: metallurgical channel half-width for an FET (cm)
$\mathbf{a}, \mathbf{b}, \mathbf{c}$	basis vectors
A	area (cm^2)
$\mathscr{B}$	magnetic flux density (Wb/cm^2)
B	base transport factor for a BJT
B, E, C	base, emitter, collector of a BJT
c	speed of light (cm/s)
C	capacitance/area in MOS (F/cm^2)
C_j	junction capacitance (F)
C_s	charge storage capacitance (F)
D_n, D_p	diffusion coefficient for electrons, holes (cm^2/s)
D, G, S	drain, gate, source of an FET
e	Napierian base
e^-	electron
$\mathscr{E}$	electric field strength (V/cm)

[†]This list does not include some symbols that are used only in the section where they are defined. Units are given in common semiconductor usage, involving cm where appropriate; it is important to note, however, that calculations should be made in the MKS system in some formulas.

E	energy $(\text{J}, \text{eV})^{\dagger}$; battery voltage (V)
E_a, E_d	acceptor, donor energy level (J, eV)
E_c, E_v	conduction band, valence band edge (J, eV)
E_F	equilibrium Fermi level (J, eV)
E_g	band gap energy (J, eV)
E_i	intrinsic level (J, eV)
E_r, E_t	recombination, trapping energy level (J, eV)
$f(E)$	Fermi–Dirac distribution function
F_n, F_p	quasi-Fermi level for electrons, holes (J, eV)
g, g_{op}	EHP generation rate, optical generation rate $(\text{cm}^{-3}\text{-s}^{-1})$
g_m	mutual transconductance (Ω^{-1}, S)
h	Planck's constant $(\text{J-s}, \text{eV-s})$; Chapter 8: FET channel half-width (cm)
$\hbar$	Planck's constant divided by 2π $(\text{J-s}, \text{eV-s})$
$h\nu$	photon energy (J, eV)
h, k, l	Miller indices
h^+	hole
$i, I^{\ddagger}$	current (A)
I (subscript)	inverted mode of a BJT
i_B, i_C, i_E	base, collector, emitter current in a BJT (A)
I_{CO}, I_{EO}	magnitude of the collector, emitter saturation current with the emitter, collector open (A)
I_{CS}, I_{ES}	magnitude of the collector, emitter saturation current with the emitter, collector shorted (A)
I_D	channel current in an FET, directed from drain to source (A)
I_0	reverse saturation current in a p-n junction (A)
j	$\sqrt{-1}$
J	current density (A/cm^2)
k	Boltzmann's constant $(J/\text{K}, \text{eV/K})$
$\mathbf{k}$	wave vector (cm^{-1})
k_d	distribution coefficient
K	$4\pi\epsilon_0$ (F/cm)
l, L	length (cm)
$\bar{l}$	mean free path for carriers in random motion (cm)
m, m^*	mass, effective mass (kg)
m_n^*, m_p^*	effective mass for electrons, holes (kg)

---

†In the Boltzmann factor $\exp(-\Delta E/kT)$, ΔE can be expressed in J or eV if k is expressed in J/K or eV/K, respectively.

‡See note at the end of this list.

m_0	rest mass of the electron (kg)
M	avalanche multiplication factor
m, n	integers; exponents
n	concentration of electrons in the conduction band (cm^{-3})
n	n-type semiconductor material
n^+	heavily doped n-type material
n_i	intrinsic concentration of electrons (cm^{-3})
n_n, n_p	equilibrium concentration of electrons in n-type, p-type material (cm^{-3})
n_0	equilibrium concentration of electrons (cm^{-3})
N (subscript)	normal mode of a BJT
N_a, N_d	concentration of acceptors, donors (cm^{-3})
N_a^-, N_d^+	concentration of ionized acceptors, donors (cm^{-3})
N_c, N_v	effective density of states at the edge of the conduction band, valence band (cm^{-3})
p	concentration of holes in the valence band (cm^{-3})
p	p-type semiconductor material
p^+	heavily doped p-type material
p	momentum (kg-m/s)
p_i	intrinsic hole concentration (cm^{-3}) $= n_i$
p_n, p_p	equilibrium concentration of holes in n-type, p-type material (cm^{-3})
p_0	equilibrium hole concentration (cm^{-3})
q	magnitude of the electronic charge (C)
Q_+, Q_-	total positive, negative charge (C)
Q_d	depletion region charge/area (C/cm^2)
Q_f	oxide fixed charge/area (C/cm^2)
Q_i	effective MOS interface charge/area (C/cm^2)
Q_{it}	interface trap charge/area (C/cm^2)
Q_m	mobile ionic charge/area (C/cm^2)
Q_n, Q_p	charge stored in an electron, hole distribution (C)
Q_n	mobile charge/area in FET channel (C/cm^2)
Q_{ot}	oxide trapped charge/area (C/cm^2)
r, R	resistance (Ω)
R_H	Hall coefficient (cm^3/C)
t	time (s)
t	sample thickness (cm)
$\bar{t}$	mean free time between scattering collisions (s)
t_{sd}	storage delay time (s)
T	temperature (K)

$v, V^\dagger$	voltage (V)
V	potential energy (J)
$\mathscr{V}$	electrostatic potential (V)
V_{CB}, V_{EB}	voltage from collector to base, emitter to base in a BJT (V)
V_D, V_G	voltage from drain to source, gate to source in an FET (V)
$\mathscr{V}_n, \mathscr{V}_p$	electrostatic potential in the neutral n, p material (V)
V_0	contact potential (V)
V_P	Chapter 8: pinch-off voltage for an FET; Chapter 11: forward breakover voltage for an SCR (V)
V_T	MOS threshold voltage (V)
$\mathbf{v}, \mathbf{v}_d$	velocity, drift velocity (cm/s)
w	sample width (cm)
W	depletion region width (cm)
W_b	base width in a BJT, measured between the edges of the emitter and collector junction depletion regions (cm)
x	distance (cm), alloy composition
x_n, x_p	distance in the neutral n region, p region of a junction, measured from the edge of the transition region (cm)
x_{n0}, x_{p0}	penetration of the transition region into the n region, p region, measured from the metallurgical junction (cm)
Z	atomic number; dimension in z-direction (cm)
α	emitter-to-collector current transfer ratio in a BJT
$\boldsymbol{\alpha}$	optical absorption coefficient (cm^{-1})
α_r	recombination coefficient (cm^3/s)
β	base-to-collector current amplification factor in a BJT
γ	emitter injection efficiency; in a p-n-p, the fraction of i_E due to the hole current i_{Ep}
δ, Δ	incremental change
$\delta n, \delta p$	excess electron, hole concentration (cm^{-3})
$\Delta n_p, \Delta p_n$	excess electron, hole concentration at the edge of the transition region on the p side, n side (cm^{-3})
$\Delta p_C, \Delta p_E$	excess hole concentration in the base of a BJT, evaluated at the edge of the transition region of the collector, emitter junction (cm^{-3})
$\epsilon, \epsilon_r, \epsilon_0$	permittivity, relative dielectric constant, permittivity of free space (F/cm); $\epsilon = \epsilon_r \epsilon_0$
λ	wavelength of light (μm, Å)
μ	mobility ($cm^2/V\text{-}s$)
ν	frequency of light (s^{-1})

---

†See note at the end of this list

ρ	resistivity (Ω-cm); charge density (C/cm^3)
σ	conductivity (Ω-cm)$^{-1}$
τ_d	dielectric relaxation time (s); in a BJT, delay time (s)
τ_n, τ_p	recombination lifetime for electrons, holes (s)
τ_t	transit time (s)
ϕ	flux density (cm^2-s)$^{-1}$; potential (V)
ϕ_F	$(E_i - E_F)/q$ (V)
ϕ_s	surface potential (V)
Φ	work function potential (V)
Φ_B	metal–semiconductor barrier height (V)
Φ_{ms}	metal–semiconductor work function potential difference (V)
ψ, Ψ	time-independent, time-dependent wave function
ω	angular frequency (s^{-1})
$\langle\ \rangle$	average of the enclosed quantity

Note: For d-c voltage and current, capital symbols with capital subscripts are used; lowercase symbols with lowercase subscripts represent a-c quantities; lowercase symbols with capital subscripts represent total (a-c + d-c) quantities. For voltage symbols with double subscripts, V is positive when the potential at the point referred to by the first subscript is higher than that of the second point. For example, V_{GD} is the potential difference $V_G - V_D$.

PHYSICAL CONSTANTS AND CONVERSION FACTORS[†]

Avogadro's number	$N_A = 6.02 \times 10^{23}$ molecules/mole
Boltzmann's constant	$k = 1.38 \times 10^{-23}$ J/K
	$= 8.62 \times 10^{-5}$ eV/K
Electronic charge (magnitude)	$q = 1.60 \times 10^{-19}$ C
Electronic rest mass	$m_0 = 9.11 \times 10^{-31}$ kg
Permittivity of free space	$\epsilon_0 = 8.85 \times 10^{-14}$ F/cm
	$= 8.85 \times 10^{-12}$ F/m
Planck's constant	$h = 6.63 \times 10^{-34}$ J-s
	$= 4.14 \times 10^{-15}$ eV-s
Room temperature value of kT	$kT = 0.0259$ eV
Speed of light	$c = 2.998 \times 10^{10}$ cm/s

Prefixes:

1 Å (angstrom) $= 10^{-8}$ cm	milli-,	m-	$= 10^{-3}$
1 μm (micron) $= 10^{-4}$ cm	micro-,	μ-	$= 10^{-6}$
1 mil $= 10^{-3}$ in.	nano-,	n-	$= 10^{-9}$
2.54 cm $= 1$ in.	pico-,	p-	$= 10^{-12}$
1 eV $= 1.6 \times 10^{-19}$ J	kilo-,	k-	$= 10^{3}$
	mega-,	M-	$= 10^{6}$
	giga-,	G-	$= 10^{9}$

A wavelength λ of 1 μm corresponds to a photon energy of 1.24 eV.

[†]Since cm is used as the unit of length for many semiconductor quantities, caution must be exercised to avoid unit errors in calculations. When using quantities involving length in formulas which contain quantities measured in MKS units, it is usually best to use all MKS quantities. Conversion to standard semiconductor usage involving cm can be accomplished as a last step. Similar caution is recommended in using J and eV as energy units.

PROPERTIES OF SEMICONDUCTOR MATERIALS

	E_g (eV)	μ_n (cm²/V-s)	μ_p (cm²/V-s)	ρ (Ω-cm)	Transition	Doping	Lattice	a (Å) ×10⁻⁸cm	ϵ_r	Density (g/cm³)	Melting point (°C)
Si	1.11	1350	480	2.5×10^5*	i	n, p	D	5.43	11.8	2.33	1415
Ge	0.67	3900	1900	43	i	n, p	D	5.66	16	5.32	936
SiC(α)	2.86	500		10^{10}	i	n, p	W	3.08	10.2	3.21	2830
AlP	2.45	80		10^{-5}	i	n, p	Z	5.46	9.8	2.40	2000
AlAs	2.16	180		0.1	i	n, p	Z	5.66	10.9	3.60	1740
AlSb	1.6	200	300	5	i	n, p	Z	6.14	11	4.26	1080
GaP	2.26	300	150	1	i	n, p	Z	5.45	11.1	4.13	1467
GaAs	1.43	8500	400	4×10^8*	d	n, p	Z	5.65	13.2	5.31	1238
GaSb	0.7	5000	1000	0.04	d	n, p	Z	6.09	15.7	5.61	712
InP	1.35	4000	100	8×10^{-3}	d	n, p	Z	5.87	12.4	4.79	1070
InAs	0.36	22600	200	0.03	d	n, p	Z	6.06	14.6	5.67	943
InSb	0.18	10^5	1700	0.06	d	n, p	Z	6.48	17.7	5.78	525
ZnS	3.6	110		10^{10}	d	n	Z, W	5.409	8.9	4.09	1650†
ZnSe	2.7	600		10^9	d	n	Z	5.671	9.2	5.65	1100†
ZnTe	2.25		100	100	d	p	Z	6.101	10.4	5.51	1238†
CdS	2.42	250	15		d	n	W, Z	4.137	8.9	4.82	1475
CdSe	1.73	650		10^5	d	n	W	4.30	10.2	5.81	1258
CdTe	1.58	1050	100	10^{10}	d	n, p	Z	6.482	10.2	6.20	1098
PbS	0.37	575	200	5×10^{-3}	i	n, p	H	5.936	161	7.6	1119
PbSe	0.27	1000	1000	10^{-3}	i	n, p	H	6.147	280	8.73	1081
PbTe	0.29	1600	700	10^{-2}	i	n, p	H	6.452	360	8.16	925

All values at 300 K.   *intrinsic resistivity.   †vaporizes.

Definitions of symbols: ρ is resistivity of high-purity material; i is indirect; d is direct; D is diamond; Z is zincblende; W is wurtzite; H is halite (NaCl). Values of mobility and resistivity are for material of available purity; these values are considered approximate (exception: Si and GaAs resistivities are extrapolated to intrinsic material). Most of the values in this table were taken from publications of the Electronic Properties Information Center (EPIC), Hughes Aircraft Co., Culver City, California; also, M. Neuberger, "III–V Semiconducting Compounds–Data Tables," published with permission from Plenum Publishing Corporation, copyright 1970.

Crystals in the wurtzite structure are not described completely by the single lattice constant given here, since the unit cell is not cubic. Several II–VI compounds can be grown in either the zincblende or wurtzite structures.

Many values quoted here are approximate or uncertain, particularly for the II–VI and IV–VI compounds.

DERIVATION OF THE DENSITY OF STATES IN THE CONDUCTION BAND

In this derivation we shall consider the conduction band electrons to be essentially free. Constraints of the particular lattice can be included in the effective mass of the electron at the end of the derivation. For a free electron, the three-dimensional Schrödinger wave equation becomes

$$-\frac{\hbar^2}{2m}\nabla^2\psi = E\psi \tag{IV-1}$$

where ψ is the wave function of the electron and E is its energy. The form of the solution to Eq. (IV-1) is

$$\psi = (\text{const.})e^{j\mathbf{k}\cdot\mathbf{r}} \tag{IV-2}$$

We must describe the electron in terms of a set of boundary conditions within the lattice. A common approach is to use periodic boundary conditions, in which we quantize the electron energies in a cube of material of side L. This can be accomplished by requiring that

$$\psi(x + L, y, z) = \psi(x, y, z) \tag{IV-3}$$

and similarly for the y- and z-directions. Thus our wave function can be written as

$$\psi_n = A \, \exp\left[j\frac{2\pi}{L}\,(\mathbf{n}_x x + \mathbf{n}_y y + \mathbf{n}_z z)\right] \tag{IV-4}$$

where the $2\pi\mathbf{n}/L$ factor in each direction guarantees the condition described by Eq. (IV-3), and A is a normalizing factor. Substituting ψ_n into the Schrödinger equation (IV-1), we obtain

$$-\frac{\hbar^2}{2m}A\nabla^2 \exp\left[j\frac{2\pi}{L}(\mathbf{n}_x x + \mathbf{n}_y y + \mathbf{n}_z z)\right] = EA \exp\left[j\frac{2\pi}{L}(\mathbf{n}_x x + \mathbf{n}_y y + \mathbf{n}_z z)\right]$$

$$(\text{IV-5})$$

$$E_n = \frac{\hbar^2}{2m}\left(\frac{2\pi}{L}\right)^2(\mathbf{n}_x^2 + \mathbf{n}_y^2 + \mathbf{n}_z^2) = \frac{h^2\mathbf{n}^2}{2mV^{2/3}} \qquad (\text{IV-6})$$

where V is the volume of the cube and $\mathbf{n}^2 = \mathbf{n}_x^2 + \mathbf{n}_y^2 + \mathbf{n}_z^2$ denotes the energy level.

Now let us find the number of allowed energy states per unit volume N as a function of energy. The number of electron states below the energy level $\mathbf{n}$ can be determined as shown in Fig. IV-1 by calculating the volume occupied in $\mathbf{n}$-space. Here we assume that the electron energy levels are filled in a spherical volume about the origin and the number of states within the volume is large. This allows us to consider an essentially continuous distribution of states in $\mathbf{n}$-space.

If we count the electron states per energy level in accordance with the Pauli principle, the number of states with energy level less than $\mathbf{n}$ is twice the volume in $\mathbf{n}$-space, $2(4\pi/3)\mathbf{n}^3$. Thus the total number of states in the volume V can be expressed as

$$NV = \frac{8\pi}{3}\mathbf{n}^3 \qquad (\text{IV-7})$$

From Eq. (IV-6), the energy of the $\mathbf{n}$th energy level is

$$E_n = \frac{h^2\mathbf{n}^2}{2mV^{2/3}} = \frac{h^2}{2mV^{2/3}}\left(\frac{NV}{8\pi/3}\right)^{2/3} = \frac{\hbar^2}{2m}(3\pi^2N)^{2/3} \qquad (\text{IV-8})$$

If $\mathbf{n}$ is some maximum value $\mathbf{n}_F$, such that all states below this level are filled and all states above it are empty, then the energy E_F as obtained from Eq. (IV-8) is the Fermi energy at 0 K.

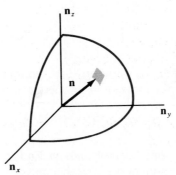

Figure IV-1.

Denoting the density of allowed states by $N(E)$, we use Eq. (IV-8) to write the total number of states per unit volume as

$$N = \int N(E)\, dE = \frac{1}{3\pi^2}\left(\frac{2mE}{\hbar^2}\right)^{3/2} \tag{IV-9}$$

Differentiating Eq. (IV-9), we obtain

$$N(E) = \frac{1}{3\pi^2}\left(\frac{3}{2}\right)\left(\frac{2m}{\hbar^2}\right)^{3/2}E^{1/2} = \frac{1}{2\pi^2}\left(\frac{2m}{\hbar^2}\right)^{3/2}E^{1/2} \tag{IV-10}$$

for the density of states as a function of energy.

To include the probability of occupation of any energy level E, we use the Fermi–Dirac distribution function:

$$f(E) = \frac{1}{e^{(E-E_F)/kT} + 1} \tag{IV-11}$$

The concentration of electrons in the range dE is given by the product of the density of allowed states in that range and the probability of occupation. Thus the density of occupied electron states N_e in dE is

$$N_e\, dE = N(E)f(E)\, dE \tag{IV-12}$$

We may calculate the concentration of electrons in the conduction band at a given temperature by integrating Eq. (IV-12) across the band:

$$n = \int_0^\infty N(E)f(E)\, dE = \frac{1}{2\pi^2}\left(\frac{2m}{\hbar^2}\right)^{3/2}e^{E_F/kT}\int_0^\infty E^{1/2}e^{-E/kT}\, dE \tag{IV-13}$$

In this integration we have referred the energies in the conduction band to the band edge (E_c taken as $E = 0$). Furthermore, we have taken the function $f(E)$ to be

$$f(E) = e^{(E_F-E)/kT} \tag{IV-14}$$

for energies such that $(E - E_F) \gg kT$.

The integral in Eq. (IV-13) is of the standard form:

$$\int_0^\infty x^{1/2}e^{-ax}\, dx = \frac{\sqrt{\pi}}{2a\sqrt{a}} \tag{IV-15}$$

Thus Eq. (IV-13) gives

$$n = 2\left(\frac{2\pi mkT}{h^2}\right)^{3/2}e^{E_F/kT} \tag{IV-16}$$

If we refer to the bottom of the conduction band as E_c instead of $E = 0$, the expression for the electron concentration is

$$n = 2\left(\frac{2\pi m_n^* kT}{h^2}\right)^{3/2}e^{(E_F-E_c)/kT} \tag{IV-17}$$

which corresponds to Eq. (3-15). We have included constraints of the lattice through the effective mass of the electron in the crystal, m_n^*.

appendix V

SOLID SOLUBILITIES OF IMPURITIES IN Si AND Ge[†]

[†]From F. A. Trumbore, "Solid Solubilities of Impurity Elements in Si and Ge," *Bell System Technical Journal* 39, no. 1, pp. 205–233, January 1960, copyright 1960, The American Telephone and Telegraph Co., reprinted by permission. Alterations have been made to include later data.

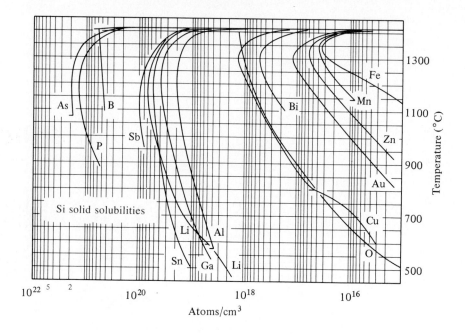

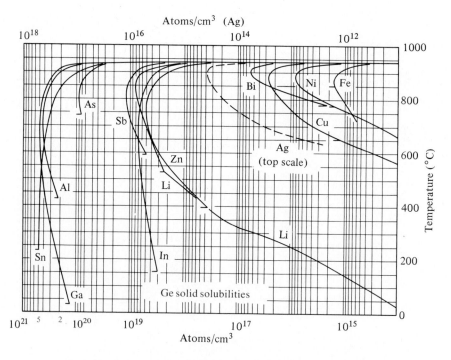

appendix VI

COMMON CIRCUIT SYMBOLS FOR SOLID STATE DEVICES[†]

[†]This table gives many of the commonly used symbols for devices discussed in this book. Standardization is incomplete, however, and other symbols are often used in the electronics literature.

Diodes

p-n diode, p-i-n, IMPATT,
Schottky barrier diode

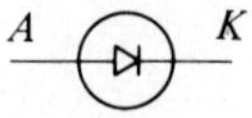

Breakdown ("Zener") diode

Bidirectional Zener diode

Varactor

Tunnel diode

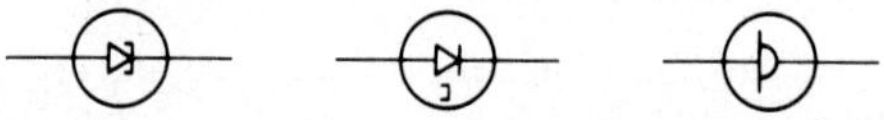

Photodiode, solar cell

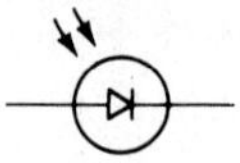

Light-emitting diode (LED)

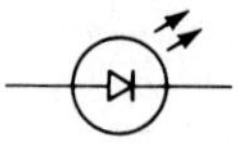

Transistors

Bipolar p-n-p

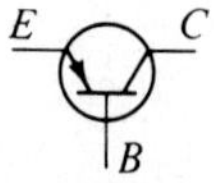

Bipolar n-p-n

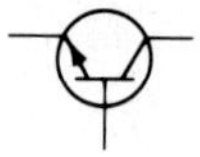

Transistors (Continued)

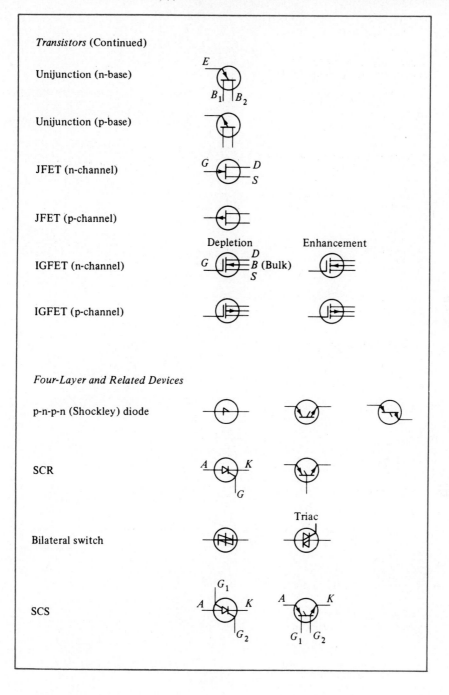

Unijunction (n-base)

Unijunction (p-base)

JFET (n-channel)

JFET (p-channel)

IGFET (n-channel)

IGFET (p-channel)

Four-Layer and Related Devices

p-n-p-n (Shockley) diode

SCR

Bilateral switch

SCS

INDEX